HANDBOOK OF

Dangerous Properties of Inorganic
and
Organic Substances
in
Industrial Wastes

HANDBOOK OF

Dangerous Properties of Inorganic and Organic Substances in Industrial Wastes

YA. M. GRUSHKO
Biology Research Institute
Irkutsk State University, Russia

English Edition Editor
A.P. KOTLOBYE

CRC Press is an imprint of the
Taylor & Francis Group, an **informa** business

This English edition of *Handbook of Dangerous Properties of Inorganic and Organic Substances in Industrial Wastes* was translated from the Russian edition by A. P. Kotlobye.

CRC Press
Taylor & Francis Group
6000 Broken Sound Parkway NW, Suite 300
Boca Raton, FL 33487-2742

First issued in paperback 2019

ISBN 13: 978-0-367-45031-1 (pbk)
ISBN 13: 978-0-8493-9300-6 (hbk)

Visit the Taylor & Francis Web site at
http://www.taylorandfrancis.com

and the CRC Press Web site at
http://www.crcpress.com

Library of Congress Cataloging-in-Publication Data

Catalog record is available from the Library of Congress.

CONTENTS

PREFACE

Owing to scientific and technological progress and rapid industrial growth, hundreds of thousands of chemicals devised and used by man enter the biosphere. Many of them have a harmful effect on humans and on plant and animal life. They also damage buildings and architectural and cultural monuments.

This handbook lists some of the physicochemical properties of inorganic and organic substances present in industrial wastes and describes their toxic effects on humans.

The author wishes to thank A. S. Chernyak of Irkutsk State University and M. F. Savchenkov and I. V. Khanygin of the Irkutsk Medical Institute for advice and help throughout the compiling of this handbook.

Ya. M. Grushko (Deceased)

Abbreviations and Designations

Classification of chemical substances according to their toxicity:

I - extremely toxic
II - very toxic
III - moderately toxic
IV - slightly toxic

CLE concentration limits of explosiveness, percent by volume

LD_{50} lethal dose of a chemical substance expected to cause the death of 50% of a test group of experimental animals

LC_{50}, LC_{100} lethal concentration of a chemical substance expected to cause the death of 50% and 100%, respectively, of a test group of experimental animals upon inhalation (mice, after two hours; rats, after four hours)

ASL approximate safe level of the effect of a chemical substance present in the air, calculated over a three-year period

MPC maximum permissible concentration of a chemical substance present in the air of populated localities that meets the sanitation standards for areas with highly favorable conditions for the dispersion of pollutants

MPSDC maximum permissible single-dose concentration of a chemical substance present in the air of populated localities; at this concentration there should be no reflex action in the human body after inhalation of the said substance for 30 minutes

MPADC maximum permissible average concentration per day of a chemical substance present in the air of populated localities; at this concentration there should be no direct or indirect harmful effect on the human body after prolonged exposure (for many years) to the said substance

TC threshold concentration

SC subthreshold concentration

SC_{odor}, TC_{odor} subthreshold and threshold concentration, respectively, of a chemical substance as detected by odor

$SSDC_{acute}$, $TSDC_{acute}$ subthreshold and threshold single-dose concentration, respectively, of a chemical substance having an acute negative effect on animals after inhalation in laboratory experiments

$SC_{irritation}$, $TC_{irritation}$ subthreshold and threshold concentration, respectively, of a chemical substance having an irritating effect on animals in laboratory experiments

$SC_{sensitivity\ to\ light}$, $TC_{sensitivity\ to\ light}$ subthreshold and threshold concentration, respectively, of a chemical substance adversely affecting the eye's sensitivity to light

$SC_{chronic}$, $TC_{chronic}$ subthreshold and threshold concentration, respectively, of a chemical substance having a chronic harmful effect on animals after inhalation in laboratory experiments

$SC_{central\ nervous\ system}$, $TC_{central\ nervous\ system}$ subthreshold and threshold concentration, respectively, of a chemical substance having a negative effect on the functioning of the central nervous system

f.p. flash point

s.-i.p. self-ignition point

d. decomposes

ρ relative density of a substance determined at room temperature

The number in parentheses given after the formula of a compound indicates its molecular weight.

Part I

Inorganic Substances

CHAPTER 1

PREVENTIVE MEASURES AGAINST ENVIRONMENTAL POLLUTION BY INDUSTRIAL WASTES

Many inorganic compounds that are present in industrial wastes emitted into the atmosphere have a harmful effect on humans and animals. Some of these compounds are poisonous; others are carcinogenic or mutagenic; still others are teratogenic or allergenic. Plants are particularly sensitive to the presence of lead, copper, zinc, cadmium, sulfur dioxide, chlorine, hydrogen chloride, hydrogen fluoride, ozone, fluorine, nitrogen oxides, and aerosols [106-108]. Some of these substances are absorbed by plants from the air and the soil, and they subsequently enter the body of man and animals as part of the food they consume. Some plants may absorb harmful substances without apparent damage to themselves, but can cause poisoning when consumed as food [119]. Some chemical substances can affect the climate, causing haze and reduced visibility [5,122].

Environmental pollution can be due to natural causes, such as volcanic eruptions, the escape of underground gas to the earth's surface, and so on. But the main source of environmental pollution is the impact of man's activity on nature.

For example, in the production of sulfuric acid by the nitrosation method, for every 1000 tons of sulfuric acid produced 20 tons of nitrogen oxides and 10 tons of sulfur dioxide are emitted. In the smelting of steel, for every 1000 tons of steel produced 40 tons of dust particles, 30 tons of sulfur dioxide, and 50 tons of carbon monoxide are discharged into the atmosphere. At thermal power stations fueled by coal, for every million kilowatt-hours of electricity generated 15 tons of sulfur dioxide, 10 tons of ashes, and 3 tons of nitrogen oxides are ejected into the atmosphere [10].

Eleven industrial nations in Europe annually discharge 3.8 million tons of sulfur into the atmosphere [120]. The United States annually discharges 142.2 million tons of carbon monoxide, 33.9 million tons of sulfur oxides, 22.7 million tons of nitrogen oxides, and 25.4 million tons of aerosols into the atmosphere [121], as well as poisonous metals (263,300 tons/year), chlorine (78,200 tons/year), and many other toxic substances [122,123]. The Clean Air Act and the Clean Air Amendments adopted in the United States in the years 1963-1977 set the limits of discharge for only five toxic inorganic substances: sulfur dioxide, nitrogen oxide, carbon monoxide, dust particles, and ozone. As has been pointed out by the Air Pollution Control Agency in the United States and by specialists, it is difficult as well as costly to observe these limits [124-126]. In the Federal Republic of Germany, according to estimates, the discharge of sulfur dioxide into the atmosphere would reach 3.24 million tons in 1990, and of nitrogen oxides - 2.32 million tons [127,128].

Each year the amount of sulfur dioxide carried by wind into the USSR across its western borders is 2 million tons, and of sulfates - 10 million tons. Of this total amount 4.1 million tons precipitate in the form of sulfuric acid in the western and central areas of the country [129].

In the Soviet Union the observance of air pollution control legislations is monitored by the Administration for Hydrometeorological Service and Environmental Protection and by sanitary epidemiologic stations. Thus far the level of discharge of wastes has been reduced at plants producing aluminum, sulfuric acid, and nitric acid and at some metallurgical plants [123,139,140].

Listed below are some of the harmful inorganic substances that are present in industrial wastes discharged into the atmosphere:

Industries	Substances present in wastes	References
Aluminum	F_2, fluorides, HF, SO_2, CO, dust particles, Si, Cr	80, 130-132
Ammonia	NH_3, CO	131
Asbestos	asbestos	80, 130
Asphalt concrete	nitrogen oxides, dust particles, smoke	130, 132
Beryllium	Be	80, 130
Brass	nitrogen oxides, dust particles	80
Bronze	nitrogen oxides, dust particles	80
Cellulose	S, sulfur oxides, H_2S, HCl, ClO_2 Cl_2, Hg, dust particles	120 133
Cement	SiO_2, Ca, Mg, Fe, As, Hg, F_2 fluorides, sulfur oxides, nitrogen oxides, CO, dust particles	80, 130, 131
Chlorine	Cl_2, Hg	80
Coal chemicals	sulfur oxides, H_2S, dust particles	130, 132
Fertilizers	phosphates, H_3PO_4, fluorides, HF, superphosphates, S, nitrogen oxides, CO, HNO_3, NH_3, NH_4Cl, H_2SO_4, P, dust particles	80, 131, 132

Galvanoplastics	Zn, other metals, Cl_2, dust particles, smoke	132
Ginning	dust particles	80
Hydrochloric acid	Cl_2, Hg	80
Iron	CO, nitrogen oxides, SO_2, F_2, dust particles	80, 130
Lead smelting	Pb, Zn, nitrogen oxides, SO_2, CO, dust particles	80, 130, 132
Lime	dust particles	80
Machine-building and metal-working	Cr, Ni, Hg	118
Mercury	Hg	130
Metallurgical	Cr, Ni, Zn, V, F_2, lead oxides, Mn, Hg, sulfur oxides, NH_3	118, 130-132
Nitric acid	nitrogen oxides, NH_3, CO_2	80, 130, 131
Nonferrous metals	SO_2, nitrogen oxides, F_2,Zn, Pb, Hg, Cd,CO	80, 131
Oxalic acid	nitrogen oxides	10
Paints and varnishes	phthalic anhydride, H_2S, dust particles	80, 132
Petrochemical	CO, H_2S, S, nitrogen oxides, dust particles	80
Soap manufacture	dust particles	80
Steel works	SO_2, nitrogen oxides, CO, F_2, dust particles	80, 130, 132
Sulfaminol	NH_3, H_2SO_4	127
Sulfuric acid	H_2SO_4, sulfur oxides, iron oxides	10, 80, 130
Synthetic fibers	CS_2	127
Synthetic rubber	SO_2 Cl_2, HCl, Cd, dust particles	131
Thermal power stations	SO_2, SiO_2, dust particles, As, V	80, 118, 131

The most effective ways to prevent atmospheric pollution by industrial wastes are the use of waste-free technological processes, the hermetic sealing of industrial equipment, and the utilization of the harmful waste materials.

Industrial wastes are often discharged through smokestacks [46]. This reduces somewhat atmospheric pollution in the immediate vicinity of the factory concerned. However, carried by the wind the wastes can cause atmospheric pollution over large areas. Thus, for example, it has been found that sulfates and metals ejected through chimneys are absorbed by soil and plants far away from the site of the discharge [17].

In the Soviet Union uniform standards have been established for the maximum permissible single-dose concentration (MPSDC) of harmful substances in the air of populated localities [16,17,34,35,117]. When several pollutants are present in the air simultaneously a synergistic effect of these pollutants is usually observed [5, 137]. In such cases the MPSDC is calculated according to a specially derived equation (Eq. 1) [35]. The sum of the concentrations of the pollutants, calculated with the help of Eq. 1, must not exceed unity:

$$(C^1/MPSDC^1) + (C^2/MPSDC) + \ldots (C^n/MPSDC^n); \quad (1)$$

where, C^1, C^2... C^n - the actual concentration of the pollutants in the atmosphere; $MPSDC^1$, $MPSDC^2$.. $MPSDC^n$ - maximum permissible concentration of the same substances.

When arsenous anhydride (As_2O_3) and germanium are present in the air at the same time they have a synergistic effect [117].

The MPSDC of each of the following substances remains unchanged under the conditions of their simultaneous presence in the atmosphere:

- carbon monoxide and sulfurous anhydrides;
- carbon monoxide, nitrogen dioxide, and sulfurous anhydride;
- phthalic anhydride, maleic anhydride, and 1-naphthoquinone.

It should be noted that the combined effect of harmful substances is usually investigated for mixtures consisting of two or three components, although in reality industrial wastes often contain many different pollutants [144-146].

In the Soviet Union the combined MPSDC for industrial wastes in resort regions and natural green areas is set at 0.8 [138,139].

The maximum permissible concentration of industrial wastes in the air of populated localities which meets USSR sanitation standards (MPC) is established by local hydrometeorological services and environmental protection bodies with account taken of the specific conditions - meteorological, geographical, etc - of the given region [14,15].

CHAPTER 2

AIR POLLUTANTS

ALUMINUM

Aluminum; Al (26.98)

Silver-white metal, m.p. 660.4°C, b.p. ≅2500°C, ρ 2.699.

Present in wastes from the production of aluminum, lacquers and paints, paper, textiles, synthetic rubber, ethanol, phenols, and some pharmaceuticals; and in wastes of machine-building, machine-tool, petrochemical, automobile, and aviation plants, and of shipyards and mining and construction works.

Toxicity: causes pneumosclerosis, aluminosis, liver damage, dermatitis, asthma, and changes in eye tissues, and has an irritating effect; $TC_{chronic}$ - 15 mg/m^3 [74, 102,103]. MPADC recommended in the USA - 0.1 mg/m^3 [76].

Removal: with the aid of cyclones, filters, and scrubbers [4,8,19,42,45,57,76].

Determination: colorimetric analysis (sensitivity 1 μg per sample) [47].

Aluminum ammonia sulfate (alum); $AlNH_4(SO_4)_2 \cdot 12H_2O$ (690.64)

Colorless crystals, m.p. 95°C, ρ 1.65; soluble in water.

Present in wastes from the production of dyes, printer's inks, pigments, baking yeast, artificial precious stones, cement for metallic articles and chinaware, glues, fire-retardants, and some pharmaceuticals; also in wastes of tanneries and metal-working plants.

Toxicity: causes vomiting and diarrhea, and has an irritating effect [74,81,113].

Aluminum arsenate (aluminum orthoarsenate); $AlAsO_4 \cdot 8H_2O$ (310.02)

Colorless crystals, ρ 3.011; insoluble in water.

Present in wastes of metallurgical and some chemical plants.

Toxicity: very toxic; causes allergy and damage to the nervous system, the lungs, liver, kidneys, and the gastrointestinal tract [74,81,113].

Aluminum arsenide; AlAs (101.90)

Greyish crystals, m.p. 1740°C, ρ 3.81; decomposes in water.

Present in wastes of metallurgical and radio engineering plants and plants making semiconductors.

Toxicity: very toxic; has a general irritating effect [74,81,113].

Aluminum bichromate; $Al(Cr_2O_7)_3$ (102.0)

Colorless crystals.

Present in wastes of some chemical and metallurgical plants.

Toxicity: very toxic; has a general irritating effect [74,81,113].

Aluminum boron hydride; $Al(BH_4)_3$ (71.53)

Colorless liquid, m.p. -64.5°C, b.p. 44.5°C, ρ 0.554; ignites in air in the presence of trace amounts of water.

Present in wastes from the production of rocket fuels, and in wastes of some chemical plants.

Toxicity: very toxic; has an irritating effect on the respiratory tracts, mucous membranes, and the eyes [74,81,113].

Aluminum bromide; $AlBr_3$ (266.72)

Colorless crystals, m.p. 98.0°C, b.p. 265°C, ρ 3.01; soluble in water.

Present in wastes of organic synthesis plants.

Toxicity: very toxic; has an irritating effect on the respiratory tracts and the digestive system [74,81,113].

Aluminum chlorate; $Al(ClO_3)_3 \cdot 6H_2O$ (385.43)

Colorless crystals; decomposes on heating; soluble in water.

Present in wastes from the production of chlorine dioxide, polyacrylic fibers, and some pharmaceuticals.

Toxicity: causes methemoglobinemia and damage to the heart and kidneys, and destroys erythrocytes [74,81,113]. MPCDC - 0.001 mg/m^3 [34].

Aluminum diboride; AlB_2 (48.05)

Colorless crystals, ρ 3.19; insoluble in water.

Present in wastes of some metallurgical plants.

Toxicity: has a general irritating effect, and is absorbed by undamaged skin [74,81,113].

Aluminum fluorosilicate; $Al_2(SiF_6)_3$ (480.23)

Colorless crystals, m.p. ≅1000°C(d.); soluble in water.

Present in wastes from the production of construction materials and glass.

Toxicity: very toxic; causes impairment of the respiratory tracts, cardiac depression, tremors, and cramps; and at low concentrations, over a prolonged period, causes destruction of bone tissues, mottled enamel, and osteosclerosis [74,81,113].

Aluminum hydride; AlH_3 (30.0)

Colorless powder, m.p. 105°C(d.); decomposes in water.

Present in wastes from the production of some polymerization catalysts, hydrides, and rocket fuels.

Toxicity: very toxic; has a general irritating effect; causes anemia and destruction of erythrocytes [74,81,113].

Aluminum hypophosphate; $Al(H_2PO_2)_3$ (221.96)

Colorless crystals, m.p. 220°C(d.); insoluble in water.

Present in wastes from the production of acrylonitrile.

Toxicity: has a general irritating effect [74,81,113].

Aluminum nitrate; $Al(NO_3)_3 \cdot 9H_2O$ (375.14)

Colorless crystals, m.p. 73.5°C; solubility in water: 43.14 g/100 g H_2O at 20°C.

Present in wastes of some petrochemical plants and from the production of textiles, catalysts, and leather.

Toxicity: causes dizziness, asthenia, headaches, cramps, collapse. LD_{50} for mice and rats - 204 and 210 mg/kg respectively [81,103].

Aluminum nitride; AlN (40.99)

Colorless crystals, m.p. 2000°C(d.), ρ 3.05; reacts with water.

Present in wastes from the production of refractory materials; and in wastes of some metallurgical and electronic plants.

Toxicity: causes irritation of the respiratory tracts, mucous membranes, and the skin.

Aluminum oxide (alumina); Al_2O_3 (101.96)

Colorless crystals, m.p. 2053°C, b.p. 3000!C; insoluble in water.

Present in wastes from the production of metallic aluminum, alumen, and other aluminum salts, refractory materials, electrical insolation materials, ceramics for radio equipment, electro-vacuum apparatus, some alloys, glass, lacquers and paints, synthetic precious stones, and tooth cement.

Toxicity: at high concentrations causes pneumosclerosis, emphysema, and irritation of the respiratory tracts [74,81,103]; ASL - 0.01 mg/m^3 [117].

Removal: with the aid of electrostatic precipitators (efficiency 96-98%), scrubbers, cyclones, cloth filters [8].

Aluminum phosphide; AlP (57.96)

Pale grey crystals, m.p. 1700°C, ρ 2.77; insoluble in water.

Present in wastes from the production of pesticides, semiconductors, and fumigants.

Toxicity: causes retrosternal pain, a burning sensation throughout the body, and loss of consciousness [74,103].

Aluminum selenide; Al_2Se_3 (290.84)

Light yellow crystals, ρ 3.43.

Present in wastes from the production of selenium hydride and semiconductors.

Toxicity: causes irritation of the eyes and mucous membranes [74,81,113].

Aluminum silicate (sillimanite, fibrolite); Al_2SiO_5 (162.05)

Colorless crystals, m.p. 1860°C, ρ 3.23; insoluble in water.

Present in wastes from the production of tooth cement, glass, enamels, ceramics, lacquers and paints, and detergents.

Toxicity: has a general irritating effect; causes pneumosclerosis [74,81,113].

Aluminum sulfate; $Al_2(SO_4)_3 \cdot 18H_2O$ (666.44)

Colorless crystals, m.p. 86.5°C(d.), ρ 1.69; soluble in water.

Present in wastes from the production of paper, textiles, leather, wood preservatives, lacquers and paints, fats and oils, pesticides, and aluminum salts.

Toxicity: LD_{50} for rats and mice - 980 and 370 mg/kg respectively [103].

Removal: with the aid of filters made of polyesters (including polyacrylates), polyethylene, teflon, nylon, polypropylene (efficiency - 99%) [57,76].

Aluminum thallium sulfate (rat poison); $AlTl(SO_4)_2 \cdot 12H_2O$ (639.66)

Colorless crystals, m.p. 91°C, ρ 2.32; solubility in water: 10 g/100 g H_2O at 20°C.

Present in wastes from the production of pesticides and in wastes of some metallurgical plants.

Toxicity: very toxic; causes polyneuritis, pains in limbs and joints; nephritis, vomiting, diarrhea, and falling out of hair [74,113].

Aluminum trichloride; $AlCl_3$ (133.34)

Colorless crystals, subl. p. 180°C, ρ 2.44; solubility in water: 45.1 g/100 g H_2O at 25°C.

Present in wastes from the production of lubricants, rubber, wool; some petrochemicals and catalysts.

Toxicity: at high concentrations causes vomiting; LD_{50} for rats and mice - 980 and 150 mg/kg respectively [74,81,103].

Aluminum trifluoride; AlF_3 (83.98)

Colorless crystals, m.p. 1279°C(subl.), ρ 3.07; solubility in water: 0.50 g/100 g

H_2O at 25°C.

Present in wastes of metallurgical plants, mining works, steel foundries, and wood-working plants; and in wastes from the production of aluminum, ceramics, catalysts, glass, and pesticides.

Toxicity: very toxic; causes weakness and debility, dermatitis, detrimentally affects the lungs, bones, joints, and ligaments; causes cachexia [74,81,103]. $SC_{irritation}$, $TC_{irritation}$, $TC_{c.n.s.}$ [165], MPSDC, and MPADC - 0.03, 0.3, 0.03-0.1, 0.2, and 0.03 mg/cu.m. respectively; toxicity classification II [34]. Recommended MPSDC and MPADC - 0.2 and 0.03 mg/m^3 respectively [33].

Aluminum triiodide; AlI_3 (407.69)

Colorless crystals, m.p. 188.3°C, b.p. 382.5°C, ρ 3.98; soluble in water.

Present in wastes from the production of some catalysts.

Toxicity: has an overall irritating effect; causes general depression, cachexia, anemia, and weakness and debility [74,81,113].

ANTIMONY

Antimonous oxychloride; SbOCl (173.22)

Colorless crystals, m.p. 170°C(d.); decomposes in water.

Present in wastes from the production of antimony derivatives.

Toxicity: causes irritation of mucous membranes and the skin [74,81].

Antimony; Sb (121.75)

Silvery white metal, m.p. 630.5°C, b.p. 1634°C, ρ 6.69; insoluble in water.

Present in wastes from the production of storage batteries, electro-vacuum equipment, explosives, thermoelectric piles, ceramics, dyes, articles made of rubber, type alloy, pyrotechnical compositions; also in wastes of some metallurgical and mining works.

Toxicity: has a general irritating effect; causes insomnia, headache, nausea, vomiting, irritation of the eyes and the respiratory tracts, conjunctivitis, inflammation of the cornea, ulceration, perforation of the nasal septum, general weakness, pneumoconiosis, hemolysis of erythrocytes; dysfunctioning of the liver, kidneys, and lungs [74,81,102,103,113]. MPSDC established in the USA - 0.05 mg/m^3 [102].

Determination: colorimetric analysis; spectroscopy [69].

Antimony bromide; $SbBr_3$ (361.51)

Colorless crystals or yellow powder, m.p. 96.6°C, b.p. 289°C, ρ 4.1; decomposes in water.

Present in wastes from the production of some analytical reagents.

Toxicity: causes irritation of mucous membranes and the skin, general depression and cachexia, mental disorders [74,81].

Antimony chloride (buttery antimony); $SbCl_3$ (228.11)

Colorless crystals, m.p. 73.2°C, b.p. 233°C, ρ 3.14; solubility in water: 98.81 g/100 g H_2O at 25°C.

Present in wastes from the production of antimony compounds, textiles, some pharmaceuticals, aromatic hydrocarbons, catalysts, and chemical reagents; microscopic instruments; also in wastes of metallurgical works.

Toxicity: causes irritation of mucous membranes, the eyes, and the skin; nausea, vomiting, conjunctivitis, inflammation of the cornea, perforation of the nasal septum, coughing, dyspnea, chemical burns of the skin [74,81,103].

Removal: with the aid of filters made of polyacrylates, poly(vinyl chloride), nylon, polyesters, polyethylene, teflon, polypropylene [57].

Antimony hydride (stibine); SbH_3 (124.77)

Colorless gas, m.p. -94.2°C, b.p. -18.4°C, ρ $5.6 \cdot 10^{-3}$; slightly soluble in water.

Present in wastes from the production of storage batteries, pesticides, pure antimony, and fumigants.

Toxicity: very toxic; has a general irritating effect on mucous membranes and the skin; causes chemical burns; impairment of internal organs, respiratory depression, headache, weakness. LC_{100} for rats - 5.8-6.6 mg/kg [81,103].

Antimony iodide; SbI_3 (502.45)

Red crystals, m.p. 170.5°C, b.p. 402°C, ρ 4.7; soluble in water.

Present in wastes from the production of semiconductors and some chemicals.

Toxicity: has a general irritating effect [74.81].

Antimony oxide (senarmontite); SbO_3 (291.50)

Colorless crystals, m.p. 656°C, b.p. 1456°C, ρ 5.1; insoluble in water.

Present in wastes from the production of glass, ceramics, lacquers and paints, textiles, articles made of rubber, luminescent lamps, some plastics and pharmaceuticals, enamels, pigments.

Toxicity: slightly toxic; has a general irritating effect [74,81,103].

Antimony pentachloride; $SbCl_5$ (299.02)

Light yellow liquid, m.p. 3°C, d.p.>140°C, ρ 2.35; soluble in water.

Present in wastes from the production of antimony compounds, textiles; some pharmaceuticals, aromatic hydrocarbons, catalysts, and chemical reagents; microscopic instruments; also in wastes of metallurgical works.

Toxicity: causes irritation of mucous membranes, the eyes, and the skin; also nausea, vomiting, conjunctivitis, inflammation of the cornea, perforation of the nasal septum, coughing, dyspnea, chemical burns on the skin [74,81,103].

Removal: with the aid of filters made of polyesters (including polyacrylates), poly(vinyl chloride), nylon, polyethylene, teflon, polypropylene [57].

Antimony pentafluoride; SbF_5 (216.74)

Colorless liquid, b.p. 8.3°C, m.p. 142.7°C, ρ 2.99; soluble in water.

Present in wastes from the synthesis of some organic compounds, fluorides, and fluorination catalysts; from the production of ceramics, porcelain, some pesticides, textiles, glass, pigments; also in wastes of steel foundries, metallurgical and woodworking plants.

Toxicity: causes irritation of the eyes, the respiratory tracts, mucous membranes and the skin; on prolonged exposure at low concentrations - fluorosis [74,81].

Antimony pentaiodide; SbI_5 (756.32)

Dark brown crystals, m.p. 79°C, b.p. 400.6°C; reacts with water.

Present in wastes from the production of semiconductors and some chemicals.

Toxicity: has a general irritating effect [74,81].

Antimony pentaoxide; SbO_5 (323.50)

Yellow crystals, d.p. 357°C, ρ 7.8; soluble in water.

Toxicity: causes irritation of the respiratory tracts, mucous membranes and the skin. LD_{50} for rats 4g/kg [74,81,103].

Antimony pentasulfide; Sb_2S_5 (403.82)

Orange-red powder, d.p. 120°C, ρ 4.12; insoluble in water.

Present in wastes from the production of articles made of rubber, matches, pyrotechnical compositions, ceramics, glass, and metallic antimony.

Toxicity: has a general irritating effect [74,81,103].

Antimony selenide; Sb_2Se_3 (480.40)

Gray crystals, m.p. 617°C, b.p. 1031°C; insoluble in water.

Present in wastes from the production of semiconductors, photoresistors, and electronic equipment.

Toxicity: is carcinogenic; causes irritation of the eyes, the respiratory tracts, and mucous membranes, pneumonia, damage to the liver, kidneys, and brain, coughing, pulmonary edema, cramps, death [74,81].

Antimony sulfate; $Sb_2(SO_4)_3$ (531.72)

Colorless crystals, ρ 3.6, decomposes on heating; decomposes in water.

Present in wastes from the production of pyrotechnical compositions.

Toxicity: causes irritation of the eyes, mucous membranes, the respiratory tracts and the skin [74,81]. LD_{50} for rats 1000 mg/kg [74].

Antimony trifluoride; SbF_3 (178.76)

Colorless crystals, m.p. 287°C, b.p. 319°C, ρ 4.3; soluble in water.

Present in wastes from the synthesis of fluorides, some organic compounds, and fluorination catalysts; from the production of ceramics, porcelain, pesticides, textiles, glass, and pigments; also in wastes of steel foundries, metallurgical and wood-working plants.

Toxicity: very posenous; has a general effect; causes dysfunctioning of the liver and kidneys, pneumosclerosis, fluorosis [74,81,103].

Antimony trisulfide (stibnite); Sb_2S_3 (339.70)

Gray crystals, m.p. 559.5°C, b.p. 1160°C, ρ 4.6; slightly soluble in water.

Present in wastes from the production of articles made of rubber, matches, pyrotechnical compositions, dyes, ceramics, glass, and metallic antimony.

Toxicity: has a general toxic and irritating effect [74,81,103].

ARSENIC

Arsenic; As (74.92)

Gray metal, m.p. 817°C, ρ 5.72; insoluble in water.

Present in wastes from the production of glass, some pharmaceuticals, semiconductors; also in wastes of metallurgical works.

Toxicity: slightly toxic in a very pure state; has a teratogenic, carcinogenic, and general irritating effect; at high concentrations causes weakness, and debility, cramps, loss of consciousness, paralysis of the respiratory tracts and vasomotor centers; on prolonged exposure at low concentrations causes emaciation, weakness, and debility, conjunctivitis, baldness, diseases of the skin and its pigmentation; polyneuritis, degenerative changes in the liver and kidneys [63,69,74,75,81,93,97, 103]. MPADC - 0.003 mg/m^3; toxicity classification - II (for arsenic and its inorganic derivatives, except arsine) [34]; MPADC established in GDR, Poland, Czechoslovakia, Yugoslavia - 0.0.003 mg/m^3; MPSDC and MPADC in Romania - 0.03 mg per m^3; and MPSDC in Poland - 0.01 mg/m^3 [88].

Determination: colorimetric analysis (sensitivity 0.5 mg/m^3) [69]; absorption spectroscopy (sensitivity 0.02 $\mu g/m^3$) [44,99].

Arsenic orthoacid; $H_3AsO_4 \cdot 0.5H_2O$ (150.93)

Colorless crystals, m.p. 35.5°C, b.p. ($-H_2O$) 160°C, ρ 2.0-2.5; soluble in water.

Present in wastes from the production of arsenates, organic arsenic derivatives, and wood preservatives.

Toxicity: has a general irritating effect. LD_{50} for rabbits - 8 mg/kg [74,81, 103].

Arsenic pentasulfide; As_2S_5 (310.14)

Yellow powder, m.p. 500°C(d); insoluble in water.

Present in wastes from the production of some dyes.

Toxicity: same as for arsenic.

Arsenic pentoxide; As_2O_5 (229.84)

Colorless powder, m.p. 315°C(d.), ρ 4.09; solubility in water: 65.8 g/100 g H_2O at 20°C.

Present in wastes from the production of arsenic derivatives, dyes, glass, furs, leather, pyrotechnical compositions, enamels, textiles; also in wastes of wood-working and metal-working plants.

Toxicity: same as for arsenic.

Arsenous bromide; $AsBr_3$ (314.7)

Colorless crystals, m.p. 31°C, b.p. 221°C, ρ 3.39; decomposes in water.

Present in wastes from some chemical works.

Toxicity: very toxic; has a general carcinogenic and allergenic effect; at high concentrations affects the stomach and the intestins, resulting in death; on prolonged exposure at low concentrations decreases appetite, and causes damage to the kidneys and liver [81].

Arsenous chloride (butter of arsenic) ; $AsCl_3$ (181.28)

Colorless liquid, m.p. -16°C, b.p. 130°C, ρ 2.17; soluble in water.

Present in wastes from the production of some pharmaceuticals, preparations for destroying the larve of malarial mosquitoes; ceramics, chlorine derivatives of arsenic; also in wastes of metallurgical works.

Toxicity: causes strong irritation of the eyes, mucous membranes of the upper respiratory tracts, pulmonary hemorrhage: and sometimes death from laryngospasms or laryngeal edema; fatty degeneration of the liver, kidneys, and heart [74,81,103].

Arsenous fluoride; AsF_3 (131.91)

Colorless liquid, m.p. -5.94°C, b.p. 57.8°C, ρ 3.01; soluble in water.

Present in wastes of some chemical plants.

Toxicity: causes nausea, vomiting, irritation of the eyes, the eyelids, the respiratory tracts, and the skin; on prolonged exposure at low concentrations causes fluorosis [74,81].

Arsenous hydride (arsine); AsH_3 (77.94)

Colorless gas, m.p. -116.92°C, b.p. -62.47°C, ρ 3.48; soluble in water.

Present in wastes from the production of hydrogen; and in wastes of metal-working and wood-working plants.

Toxicity: has a general irritating effect [74,81,103]; causes headache, dizziness, vomiting, dysfunctioning of the central nervous system; hemolysis of erythrocytes, methemoglobin formation, hepatitis, nephritis, necrosis of the ureters; drowsiness, respiratory depression; at high concentrations - cramps and death. $SC_{chronic}$ - 0.004-0.012 mg/l; $TC_{chronic}$ - 1:20000; LC_{100} for cats - 0.07 mg/l (after 9 days with one dose every 2 hrs); MPADC - 0.002 mg/m^3; toxicity classification - II [34].

Determination: see [92].

Arsenous iodide; AsI_3 (455.67)

Red crystals, m.p. 141°C, b.p. 371°C, ρ 4.68; soluble in water.

Present in wastes from the production of some types of films, semiconductors, and pharmaceuticals.

Toxicity: same as for arsenic.

Arsenous oxide; As_2O_3 (197.84)

Colorless crystals or amorphous powder, m.p. 314°C, b.p. 461°C, ρ 4.15; soluble in water.

Present in wastes from the production of arsenic derivatives, dyes, glass, furs, leather, pyrotechnical compositions, enamels, textiles; and in wastes of wood-working and metal-working plants.

Toxicity: very toxic. LD_{50} for rats - 19.1 mg/kg [103].

Removal: with the aid of electrostatic precipitators.

Arsenous sulfide; (orpiment); As_2S_3 (246.04)

Yellow crystals, m.p. 325°C, b.p. 707°C, ρ 3.46; slightly soluble in water.

Present in wastes from the production of glass, textiles, linoleum, semiconductors; some dyes, pyrotechnical compositions, and leather.

Toxicity: same as for arsenic.

BARIUM

Barium; Ba (137.34)

Silver-white metal, m.p. 727°C, b.p. ≅1860°C, ρ 3.63.

Present in wastes from the production of sulfuric acid; and in wastes of metallurgical plants and printing shops.

Toxicity: causes damage to the brain, heart, and blood vessels. MPSDC recommended in the USA - 0.01 mg/m^3 [99].

Determination: absorption spectroscopy (sensitivity - 0.02 $\mu g/m^3$) [99].

Barium bromate; $Ba(BrO_3)_2 \cdot 2H_2O$ (411.21)

Yellowish crystals, m.p. 260°C(d.), ρ 3.99; solubility in water: 0.44 g/100 g H_2O at 10°C.

Present in wastes from some chemical works.

Toxicity: has a general irritating effect [74,81].

Barium carbonate; $BaCO_3$ (197.35)

Colorless crystals, m.p. 1450°C(d.), ρ 4.43; slightly soluble in water.

Present in wastes from the production of some pesticides, ceramics, dyes, enamels, articles made of marble, electrical equipment, articles made of rubber, glass, barium salts, some chemical reagents.

Toxicity: causes damage to the kidneys, heart and blood vessels, paralysis; inhibits the action of enzymes; affects the gonads and embryos. $TC_{irritation}$, $TSDC_{acute}$ - 1 and 5 mg/m^3 respectively [74,81,103]; LD_{50} for rats - 800 mg/kg [74]; MPADC - 0.004 mg/m^3; toxicity classification - I [34].

Removal: with the aid of filters made of nylon, polyesters (including polyacrylates), polyethylene, teflon, or polypropylene [57].

Barium chlorate; $Ba(ClO_3)_2$ (322.29)

Colorless crystals, m.p. 414°C(d.), ρ 3.17; readily soluble in water.

Present in wastes from the production of explosives, matches, and some dyes.

Toxicity: has a general irritating effect [74,81].

Barium chromate (IV); $BaCrO_4$ (253.37)

Yellow crystals, m.p. 1380°C, ρ 4.50; slightly soluble in water.

Present in wastes from the production of glass, ceramics, and pyrotechnical compositions; and in wastes of metal-working plants.

Toxicity: has a general irritating effect [74,103].

Barium cyanide; $Ba(CN)_2$ (189.40)

Colorless crystals; soluble in water.

Present in wastes of some metallurgical and electrolytical, machine-building and metal-working plants.

Toxicity: very toxic; causes respiratory depression, asphyxia, and death [74, 81,109,113].

Barium dibromide (barium bromide); $BaBr_2$ (297.15)

Colorless crystals, m.p. 857°C, b.p. 1980°C, ρ 4.781; soluble in water (104 g per 100g).

Present in wastes from the production of bromides and phosphates.

Toxicity: causes damages to the brain and heart; has a general irritating effect [74,81].

Barium dichloride (barium chloride); $BaCl_2$ (208.25)

Colorless crystals, m.p. 961°C, b.p. ≅2050°C, ρ 3.86; soluble in water.

Present in wastes from the production of some pesticides, ceramics, textiles, mineral paints, leather, some pharmaceuticals, and aluminum.

Toxicity: causes general weakness, headache, insomnia, hemorrhage, pulmonary edema, degenerative changes in the liver, spleen sclerosis, and cardioplegia [74,81, 103].

Removal: by filtration [57].

Barium difluoride (barium fluoride); BaF_2 (175.36)

Colorless crystals, m.p. 1368°C, b.p. 2530°C, ρ 4.83; slightly soluble in water.

Present in wastes of some metallurgical, metal-working and wood-working plants, steel foundries; also in wastes from the production of glass, enamels, and some pesticides.

Toxicity: very toxic; causes weakness and debility, coughing, asphyxia, pneumosclerosis, calcification of the ligaments, inflammation of the kidneys; has a general irritating effect [74,81,102,103].

Barium diiodide (barium iodide); BaI_2 (391.15)

Colorless crystals, m.p. 740°C, ρ 4.92; solubility in water: 204.4 g/100 g H_2O at 20 °C.

Present in wastes from the production of iodides.

Toxicity: very toxic; causes respiratory depression, decubital paralysis, and irritation of the respiratory tracts, the eyes, mucous membranes, and the skin [74, 81].

Barium fluorosilicate; $Ba(SiF_6)$ (279.42)

Colorless crystals, m.p. 300°C (d.), ρ 4.29; slightly soluble in water.

Present in wastes from the production of silicon tetrachloride.

Toxicity: very toxic; causes abdominal pain, diarrhea, cramps and tremors, muscular paralysis; irritation of mucous membranes and the skin; on prolonged exposure at low concentrations causes fluorosis [74,81].

Barium hydroxide; $Ba(OH)_2$ (171.35)

Colorless crystals, m.p. 408°C, b.p. >1000°C(d.), ρ 4.5; solubility in water: 3.89 g/100 g H_2O at 20°C.

Present in wastes from the production of alkali, glass, articles made of rubber, some pesticides; sugar, fats, and oil.

Toxicity: causes dermatitis, irritation of mucous membranes and the skin; damage to the central nervous system and the heart [74,81,103].

Barium hypophosphate; $Ba(H_2PO_4)_2$ (267.34)

Colorless crystals, ρ 2.90; decomposes on heating; solubility in water: 28.6 g per 100 g H_2O at 17°C.

Present in wastes of electronic plants.

Toxicity: very toxic; causes respiratory depression, paralysis, cyanosis; irritation of the eyes, mucous membranes, the skin, and the respiratory tracts [74, 81].

Barium manganate; $BaMnO_4$ (256.29)

Gray-green crystals, ρ 4.85.

Present in wastes from the production of some pigments.

Toxicity: very toxic; causes damage to the nervous system, paralysis, cyanosis, death; has a general irritating effect [74,81,113].

Barium mercuric iodide; $Ba[HgI_4]\cdot 5H_2O$ (935.68)

Red crystals; readily soluble in water.

Present in wastes from the production of analytical reagents for analyzing alkaloids; also in wastes of some metallurgical works.

Toxicity: has a general irritating effect [74].

Barium nitrate; $Ba(NO_3)_2$ (261.38)

Colorless crystals, m.p. 594°C, ρ 3.24; soluble in water.

Present in wastes from the production of pyrotechnical compositions.

Toxicity: causes damages to the liver and spleen; has a general irritating effect [74,81,103].

Barium oxide; BaO (153.34)

Colorless crystals, m.p. ≅2020°C, ρ 5.8; solubility in water: 1.5 g/100 g H_2O at 0°C.

Present in wastes from the production of barium compounds.

Toxicity: causes ptyalism, cramps, changes in the blood pressure, and damage to the kidneys; has a general irritating effect [74,81,103].

Barium peroxide; BaO_2 (169.36)

Colorless crystals, m.p. ≅800°C(d.), ρ 4.96; slightly soluble in water.

Present in wastes from the production of some organic compounds, dyes, glass, Perhydrol, and bleaching agents.

Toxicity: has a general irritating effect [74,81,103].

Barium selenate; $BaSeO_4$ (280.30)

Colorless crystals, ρ 4.75; decomposes on heating; slightly soluble in water.

Present in wastes from the production of semiconductors.

Toxicity: very toxic; has a general irritating effect; causes damage to the liver, kidneys, and the nervous system, pneumonia, cramps; is carcinogenic [81,113].

Barium sulfate; $BaSO_4$ (233.43)

Colorless crystals, m.p. ≅1680°C, ρ 4.5; hardly soluble in water.

Present in wastes from the production of photographic paper, cellophane, articles made of rubber, linoleum, polymeric fibers, lithographic inks, mineral paints, textiles, ceramics, and some pharmaceuticals.

Toxicity: has a general irritating effect. $TSDC_{acute}$ - 250-300 mg/m^3 [74,103].

Barium sulfide; BaS (169.42)

Colorless crystals, m.p. 2200°C, ρ 4.36; slightly soluble in water.

Present in wastes from the production of lithopone (a mixture of zinc sulfide and barium sulfate), articles made of leather and rubber, some pesticides, and luminescent materials.

Toxicity: has a general irritating effect [74,81].

Barium zirconate; $BaZrO_3$ (276.56)

Colorless crystals; insoluble in water.

Present in wastes of electrical engineering and pyrotechnical plants; also in wastes from the production of special steels.

Toxicity: has a general toxic effect. $TC_{chronic}$ - 15 mg/m^3 [103].

BERYLLIUM

Beryllium; Be (9.012)

Silver-gray metal, m.p. 1287°C, b.p. 2507°C, ρ 1.844.

Present in wastes of metallurgical, machine-building, petrochemical, radio engineering, and electrical engineering plants; also in wastes from the production of beryllium, ceramics, and special instruments.

Toxicity: causes conjunctivitis and irritation of mucous membranes; damage to the central nervous system, the heart, liver, kidneys, and spleen; dermititis, ulceration of the cornea and perforation of the nasal septum; and has a carcinogenic, mutagenic, and teratogenic effect [63,69,93,97,102,103]. $TC_{chronic}$ - 0.01 $\mu g/m^3$ [33]; ASL - established for beryllium and its compounds (calculated for Be) - 10^{-5} mg/m^3 [117].

Determination: spectroscopic analysis [69,115].

Beryllium aluminate; $BeAl_2O_4$ (126.97)

Colorless crystals, ρ 3.76; insoluble in water.

Present in wastes of metallurgical works.

Toxicity: very toxic; has a general irritating effect [81,113].

Beryllium aluminum silicate, (silicate); $Be_2Al_2(SiO_4)_2(OH)_2$ (290.18)

Colorless crystals, ρ 3.1.

Present in wastes of metallurgical works.

Toxicity: has an irritating effect [81,113].

Beryllium boron hydride; $Be(BH_4)_2$ (38.70)

Colorless crystals, m.p. 91.3°C(subl.).

Present in wastes of some chemical plants.

Toxicity: very toxic; causes damage to the nervous system and lungs; destroys erythrocytes; and is absorbed by healthy skin [74,81,113].

Beryllium dichloride (beryllium chloride); $BeCl_2$ (79.93)

Colorless crystals, m.p. 415°C, b.p. 550°C, ρ 1.90; soluble in water.

Present in wastes from the production of beryllium oxide; also in wastes of metallurgical works.

Toxicity: has a general irritating effect. LD_{50} for mice and rats - 92 and 86 mg/kg respectively [74,81,103].

Beryllium difluoride (beryllium fluoride); BeF_2 (47.01)

Colorless crystals, m.p. 552-800°C, b.p. 1175°C, ρ 1.99; solubility in water: ≅846 g/l.

Present in wastes of metallurgical works; in wastes from the production of special types of glass.

Toxicity: has a general irritating effect. $TSDC_{acute}$ - 0.00002 g/m^3 [74,81,103].

Beryllium oxide; BeO (25.01)

Colorless crystals, m.p. 2580°C, b.p. 4260°C, ρ 2.86; hardly soluble in water.

Present in wastes from the production of refractory materials, certain organic compounds, and glass; also in wastes of metallurgical works.

Toxicity: very toxic; affects the respiratory organs. $TSDC_{acute}$ - 0.675 mg/m^3 [74,81,103].

Beryllium sulfate; $BeSO_4 \cdot 4H_2O$ (177.13)

Colorless crystals, m.p. 550-600°C(d.), ρ 1.71; solubility in water: 86.3 g per 100 g H_2= at 15°C.

Present in wastes of mining works.

Toxicity: has a general irritating effect. LD_{50} for mice and rats - 80 and 82 mg per kg respectively [74,81,103].

Beryllium sulfide; BeS (41.08)

Light gray crystals, ρ 2.36; solubility in water: 35.3 g/100 g H_2O at 0°C.

Present in wastes of metallurgical works.

Toxicity: very toxic; has a general irritating effect [81,103,113].

BISMUTH

Bismuth; Bi (208.980)

Silver-gray metal, m.p. 271.4°C, b.p. 1564°C, ρ 9.8; insoluble in water.

Present in wastes from the production of bismuth salts, some pharmaceuticals, fire retardants, perfumes, and electrical engineering goods; also in wastes of steel foundries.

Toxicity: affects the kidneys, gums; causes stomatitis and dermatitis [102]. MPSDC recommended in the USA - 0.005 mg/m^3 [76].

Bismuth hydroxide; $Bi(OH)_3$ (260.0)

White amorphous powder, ρ 4.36; slightly soluble in water.

Present in wastes from the production of biosynthetic substances; also in wastes of installations for the purification of plutonium.

Toxicity: causes stomatitis and damage to the kidneys; has a general irritating effect [74,81,113].

Bismuth oxychloride; BiOCl (260.48)

Colorless crystals, ρ 7.72; insoluble in water.

Present in wastes from the production of pigments, perfumes, and artificial pearls.

Toxicity: causes damage to the kidneys; has a general irritating effect [74,81, 113].

Bismuth pentafluoride; BiF_5 (266.0)

Colorless crystals, m.p. 550°C(subl.).

Present in wastes from the production of bismuth fluorides.

Toxicity: very toxic; has a general irritating effect [74].

Bismuth phosphate (bismuth orthophosphate); $BiPO_4$ (305.95)

Colorless crystals, ρ 6.32, decomposes on heating; insoluble in water.

Present in wastes from the production of optical-grade glass, cut glass; also in wastes of metallurgical works.

Toxicity: has a general irritating effect [74,81,113].

Bismuth selenide (bismuth triselenide; guanajuatite); Bi_2Se_3 (654.84)

Gray crystals, m.p. 706°C, b.p. 1007°C, ρ 6.82; insoluble in water.

Present in wastes from the production of semiconductors.

Toxicity: has a general irritating effect [74].

Bismuth subcarbonate; $(BiO)_2CO_3$ (509.97)

Colorless crystals, ρ 6.86, decomposes on heating: insoluble in water.

Present in wastes from the production of glazes, ceramics, and plastics.

Toxicity: has a general irritating effect [74,81,103,113].

Bismuth sulfate; $Bi_2(SO_4)_3$ (706.13)

Colorless crystals, ρ 5.08, decomposes on heating.

Present in wastes from the production of sulfate derivatives.

Toxicity: has a general irritating effect [74,81].

Bismuth telluride; Bi_2Te_3 (300.76)

Black crystals, m.p. 585°C, b.p. 1172°C, ρ 7.7.

Present in wastes from the production of semiconductors, refrigerators, and thermostats.

Toxicity: has a general irritating effect [74,113].

Bismuth tetraoxide; $Bi_2O_4 \cdot 2H_2O$ (517.99)

Brown powder, b.p. 305°C, ρ 5.6; insoluble in water.

Present in wastes from the production of lubricants.

Toxicity: has a general irritating effect [74,81,113].

Bismuth trichloride; $BiCl_3$ (315.34)

Colorless crystals, m.p. 233°C, b.p. 439°C; decomposes in water.

Present in wastes from the production of bismuth salts and catalysts for organic reactions.

Toxicity: affects the kidneys; has a general irritating effect [74,81].

Bismuth trifluoride; BiF_3 (265.98)

Gray crystals, m.p. 725-730°C, ρ 8.3; insoluble in water.

Present in wastes from the production of bismuth-containing compounds.

Toxicity: at high concentrations has a general irritating effect; at low concentrations causes fluorosis [74,81,113].

Bismuth trihydride (bismuth hydride); BiH_3 (212.0)

Colorless liquid, b.p. 22°C.

Present in wastes from the production of semiconductors.

Toxicity: causes anemia, pulmonary edema, death [74,81].

Bismuth trinitrate; $Bi(NO_3)_3 \cdot 5H_2O$ (485.10)

Colorless crystals, ρ 2.83; soluble in water.

Present in wastes from the production of bismuth salts, paints, medicinal preparations.

Toxicity: causes damage to the kidneys, dizziness, headache, and vomiting; has a general irritating effect [74,81,103].

Bismuth trioxide; Bi_2O_3 (466.0)

Yellow crystals, m.p. 825°C, b.p. 1890°C, ρ 8.9; insoluble in water.

Present in wastes from the production of disinfectants, glass, articles made of rubber, fire-resistant paper, and some polymers.

Toxicity: causes headache, vomiting; has a general irritating effect [74,81,103].

BORON

Boric acid (boracic acid; orthoboric acid); H_3BO_3 (61.89)

Colorless crystals, m.p. 171°C(d.), ρ 1.43; soluble in water.

Present in wastes from the production of glass, ceramics, foodstuffs, boron, cement, enamel, leather articles, soap, dyes, photographic chemicals; also in wastes of electrometallurgical and wood-working plants.

Toxicity: affects the central nervous system, the kidneys, and the liver; causes cramps, collapse; has a general irritating effect. $TC_{chronic}$ - 24.4-48.6 mg/m^3 [74,81,103]; MPADC - 0.02 mg/m^3; toxicity classification - III [35].

Removal: see [4,42].

Boron; B (10.81)

Dark gray crystals, m.p. 2075°C, b.p. 3700°C, ρ 2.37; insoluble in water.

Present in wastes of metallurgical works.

Toxicity: has a general toxic effect on embryos. ASL - 0.01 mg/m^3 [117].

Determination: spectrometric analysis [233].

Boron carbide; B_4C (55.25)

Black crystals, m.p. 2350°C, b.p. 3500°C, ρ 2.52; insoluble in water.

Present in wastes from the production of abrasive materials, base of electric bulbs, and accessories for furnaces.

Toxicity: has an irritating effect; is allergenic [81,103].

Boron nitride; BN (24.82)

Colorless crystals, m.p. 3200°C(D.), ρ 2.34; slightly soluble in water.

Present in wastes from the production of refractory materials, thermally stable fibers, dry lubricants, and abrasives.

Toxicity: affects the lungs; has a general irritating effect. $SC_{chronic}$ and $TC_{chronic}$ - 1-10 and 100-200 mg/m^3 respectively [103]; ASL - 0.02 mg per m^3 [117].

Boron oxide (boric anhydride); B_2O_3 (69.62)

Vitreous substance, m.p. 450°C, b.p. 1500°C, ρ 1.85; soluble in water.

Present in wastes from the production of special types of glass; also in wastes of metallurgical works.

Toxicity: causes retardation of growth, and irritation of mucous membranes, the skin, and the eyes. $SC_{chronic}$ and $TC_{chronic}$ - 0.077 and 0.47 mg/l respectively [103].

Boron tribromide (boron bromide); BBr_3 (250.57)

Colorless liquid, m.p. -45.84°C, b.p. 89.8°C, ρ 2.65.

Present in wastes from the production of diborane.

Toxicity: affects the central nervous system; has an irritating effect; has a cumilative effect [74,81,103].

Boron trichloride (boron chloride); BCl_3 (117.19)

Colorless liquid, m.p. -107°C, b.p. 12.5°C, ρ 1.43; reacts with water.

Present in wastes from the production of some catalysts and boron derivatives.

Toxicity: affect the central nervous system; has a general irritating effect; has a cumulative effect [74,81].

Boron trifluoride (boron fluoride); BF_3 (67.31)

Colorless gas, m.p. -128.36°C, b.p. -100°C, ρ 2.99; soluble in water.

Present in wastes of organic synthesis and metallurgical plants.

Toxicity: very toxic; causes damage to internal organs and bones; fluorosis; and irritation of mucous membranes, the respiratory tracts, and the eyes. $TC_{chronic}$ - 0.28 mg/m^3; LC_{50} for mice, rats, and guinea pigs - 3.45, 1.18, and 0.11 mg/l respectively [74,81,103].

Decaborane; $B_{10}H_{14}$ (122.2)

Solid substance, m.p. 99.5°C, b.p. 213°C, ρ 0.94.

Present in wastes from the production of some pharmaceuticals, perfumes, and articles made of rubber; also in wastes of metal-working and organic synthesis plants.

Toxicity: causes irritation of mucous membranes and the skin; general weakness, loss of appetite, vomiting, impairment of coordination of movements, cramps, pulmonary edema; damage to the kidneys, the liver and the central nervous system. TC_{odor} - 0.002 mg/l; $TSDC_{acute}$ - 0.0003 mg/l; LC_{50} for mice - 0.03 mg/l [81,103].

Diborane; B_2H_6 (27.67)

Colorless gas, m.p. -165°C, b.p. -92.5°C, ρ 0.44.

Present in wastes from the production of some pharmaceuticals, perfumes, and articles made of rubber; also in wastes of metal-working and organic synthesis plants.

Toxicity: has a general irritating effect. TC_{odor} - 0.004 mg/l; $TSDC_{acute}$ - 0.002 mg/l [84,103].

Pentaborane; B_5H_9 (63.13)

Liquid, m.p. -46.6°C, b.p. 60°C, ρ 0.63.

Present in wastes from the production of some pharmaceuticals, perfumes, and articles made of rubber; also in wastes of metal-working and organic synthesis plants.

Toxicity: causes weakness and debility; loss of appetite, headache, impairment of coordination of movements, cramps, pulmonary edema; and damage to the central nervous system, the liver, the kidneys, and the heart; has a general irritating effect; is absorbed by healthy skin. TC_{odor} and $TSDC_{acute}$ - 0.002 and 0.0025 mg/l respectively; LC_{50} for mice - 0.03 mg/l [103].

Tetrafluoroboric acid; $HBF_4 \cdot nH_2O$ (87.83)

Colorless liquid, b.p. 130°C(D.); soluble in water.

Present in wastes of organic synthesis plants, metal-working and galvanotechnical plants; also in wastes from the production of aluminum and some catalysts.

Toxicity: very toxic; causes skin burns, fluorosis, and irritation of mucous membranes and the respiratory tracts [74,81].

BROMINE

Bromine; Br_2 (159.81)

Red-brown liquid, m.p. -7.25°C, b.p. 59.82°C, ρ 3.102; soluble in water.

Present in wastes of organic synthesis plants; also in wastes from the production of textile fibers, dyes, some pharmaceuticals, gold, and platinum.

Toxicity: at low concentrations causes headache, dizziness, chest congestion, coughing, nose-bleed, ulceration of the cornea; at high concentrations - brown coloration of the mucous membrane inside the mouth, conjunctivitis, rhinitis, ptyalism, asphyxia, laryngospasm, bronchitis, pneumonia, bronchial asthma, and pulmonary edema. $TC_{irritation}$ and $TSDC_{acute}$ - 0.01 and 0.004 mg/l respectively [74,81,103]; MPADC - 0.04 mg/m^3; toxicity classification - II [34].

Removal: with the aid of filters made of polyesters or polypropylene [57]; by adsorption with activated carbon [76].

Determination: see [78].

Bromine pentafluoride; BrF_5 (174.92)

Colorless liquid, m.p. -61°C, b.p. 40.8°C, ρ 2.47; reacts with water.

Present in wastes from the production of some organic chemicals.

Toxicity: has a general irritating effect [74,81].

Bromine trifluoride; BrF_3 (136.90)

Light yellow liquid, m.p. 8.8°C, b.p. 126°C, ρ 2.84; reacts with water.

Present in wastes from the production of some organic chemicals.

Toxicity: has a general irritating effect [74,81].

Hydrobromic acid; HBr (80.92)

Colorless gas, m.p. -86.90°C, b.p. -66.77°C, ρ 2.71; solubility in water: 1200 g/l at 10°C.

Present in wastes from the production of bromine derivatives and some organic chemicals.

Toxicity: has a general irritating effect. $TSDC_{acute}$ and $TC_{c.n.s.}$ - 0.026 and 0.066 mg/l respectively [74,81,103,113].

Removal: with the aid of filters made of polyesters (including polyacrylates), polyethylene, teflon, polypropylene [57]; by scrubbing (efficiency 97-99%) [76].

CADMIUM

Cadmium; Cd (112.41)

Silver-white metal, m.p. 321.1°C, b.p. 766.5°C, ρ 8.65; insoluble in water.

Present in wastes of electrolytic metallurgical and metallurgical works; also in wastes from the production of storage batteries, radio engineering goods, dyes, enamels, plastics, articles made of rubber, ceramics; pesticides, some catalysts, and photographic chemicals.

Toxicity: very toxic; causes dizziness, headache, ptyalism, coughing, vomiting, nose-bleed, ulceration and perforation of the nasal septum, a metallic taste in the mouth, golden yellow coloration of the gums and dental neck, pulmonary emphysema, fibrosis of the lungs, and damage to bones; has a mutagenic, tetragenic, and carcinogenic effect [69,74,81,93,102,103,122,167]. $TC_{chronic}$ - 0.05-0.1 mg/m^3 [103]; MPSDC and MPADC established in Yugoslavia - 0.01 and 0.003 mg/m^3 respectively; MPADC established in the USA - 0.001 mg/m^3 [72,76,88].

Determination: colometric analysis [47,69]; absorption spectroscopy (sensitivity 0.0002 mg/m^3); emission spectroscopy (sensitivity 0.011 mg/m^3) [99]; fluorometric analysis (sensitivity 0.4 μg per 10 ml sample) [168]; extraction photometry [169].

Cadmium cyanide; $Cd(CN)_2$ (164.44)

Colorless crystals, m.p. 200°C(d.), ρ 2.226; solubility in water: 1.7 g/100 g at 15°C.

Present in wastes from galvanic processes.

Toxicity: very toxic; has a general irritating effect; at high concentrations causes death from asphyxia [74,81].

Cadmium dibromide; $CdBr_2$ (272.21)

Yellow crystals, m.p. 565°C, b.p. 863°C, ρ 5.192; soluble in water (98.4 g/100 g).

Present in wastes from the production of photographic materials, lithographic and engraving inks.

Toxicity: very toxic; has a general irritating effect [74,81].

Cadmium dichloride; $CdCl_2$ (182.31)

Colorless crystals, m.p. 568.5°C, b.p. 964°C, ρ 2.226; solubility in water: 53.3 g/100 g at 20°C.

Present in wastes from the production of some pharmaceuticals, dyes, yellow cadmium.

Toxicity: very toxic: has a carcinogenic effect [74,81,103].

Removal: with the aid of filters made of nylon, polyesters (including polyacrylates), polyethylene, teflon, polyvinyl chloride, or polypropylene [57].

Cadmium difluoride; CdF_2 (150.41)

Colorless crystals, m.p. 1072°C, b.p. 1753°C, ρ 6.33; soluble in water (43 g/l).

Present in wastes from the production of optical-grade glass, luminophors, materials for making lasers and solid cathodes.

Toxicity: very toxic; has a general irritating effect; found to cause in prolonged experiments on animals the formation of mottled enamel, sclerosis, osteomalacia, and calcification of the joints [74,81,103].

Cadmium diiodide; CdI (366.23)

Brown crystals, m.p. 388°C, b.p. 744°C, ρ 5.67; soluble in water (45.8%).

Present in wastes from the production of photographic materials, lithographic and engraving inks, some catalysts, cadmium, nematocides, phosphors, and some analytical reagents.

Toxicity: has a general irritating effect [74,81,103].

Cadmium hydroxide; $Cd(OH)_2$ (272.24)

Crystals or powder, m.p. 300°C(d.), ρ 4.79; insoluble in water.

Present in wastes from the production of cadmium salts and some types of electrotechnical equipment.

Toxicity: very toxic; has a general irritating effect [69,81,93,122].

Cadmium nitrate; $Cd(NO_3)_2$ (236.43)

Colorless crystals, m.p. 353°C, ρ 2.45; soluble in water.

Present in wastes from the production of glass, porcelain, photographic materials, and cadmium salts.

Toxicity: has a general irritating effect. LD_{50} for rats 100 mg/kg [74,81,103].

Cadmium oxide; CdO (128.39)

Brown crystals or amorphous mass, m.p. ≅900°C(subl.), ρ 8.15; insoluble in water.

Present in wastes from the production of phosphors, semiconductors, glass, silver alloys, radio engineering goods, ceramics, nematocides, and some organic reagents.

Toxicity: very toxic; causes ptyalism, dyspnea, pneumonia, emphysema, pathological changes in the liver and kidneys, and damage to the gonads and embryos; has a carcinogenic and mutagenic effect. $SSDC_{acute}$ and $TSDC_{acute}$ - 0.005 and 0.1 mg/m^3 respectively [170]; $TC_{chronic}$ - 15-20 mg/m^3; LC_{50} for rats 45 mg/m^3 [73, 74,76,81,103]; MPADC - 0.0001 mg/m^3; toxicity classification - II [34]; MPADC recommended in the USA - 0.0005 mg/m^3 [170], MPADC established in the USA - 0.001 mg/m^3 [78].

Cadmium selenide; CdSe (191.37)

Red or brown crystals, m.p. 1250°C, ρ 5.61; insoluble in water.

Present in wastes from the production of photographic materials, semiconductors, and phosphors.

Toxicity: has a general irritating effect; causes damages to the kidneys, the liver, and the central nervous system [74,81,103].

Cadmium sulfate; $CdSO_4$ (208.47)

Colorless crystals, m.p. 1135°C, ρ 4.69; solubility in water: 76.7 g/100 g H_2O at 20°C.

Present in wastes from the production of metallic cadmium and its salts, some catalysts, phosphors; also in wastes from galvanic processes.

Toxicity: very toxic; has a general mutagenic and teratogenic effect [97]. LD_{50} for rats 88 mg/kg [74,81,103].

Cadmium sulfide; CdS (144.47)

Orange crystals, m.p. ≅ 1480°C, ρ 4.8; insoluble in water.

Present in wastes from the production of some dyes, glass, porcelain, articles made of rubber, pyrotechnical chemicals, soap, textiles, ceramics, semiconductors, and photographic materials.

Toxicity: causes pulmonary emphysema and pneumonia [74,81,103].

Cadmium telluride; CdTl (240.02)

Dark brown crystals, m.p. 1090°C, ρ 6.2; insoluble in water.

Present in wastes from the production of semiconductors and phosphors.

Toxicity: has a general irritating effect [74,81].

CALCIUM

Calcium; Ca (40.08)

Silver-white metal, m.p. 850°C, b.p. 1480°C, ρ 1.54; reacts with water.

Present in wastes of metallurgical and metal-working plants; also in wastes from the production of organic solvents.

Toxicity: causes tracheobronchitis, pulmonary emphsema, sneezing and coughing [74,103].

Determination: absorption (sensitivity 0.0002 $\mu g/m^3$) [99].

Calcium aluminum hydride; $Ca(AlH_4)_2$ (102.10)

Gray powder; decomposes on heating; reacts with water.

Present in wastes from the production of aldehydes, ketones, acid chlorides, and agents for reducing esters into alcohols, nitriles into amines, and aromatic nitroderivatives into azoderivatives.

Toxicity: very toxic [74,81,113].

Calcium arsenate; $Ca_3(AsO_4)_2$ (398.08)

White amorphous powder, ρ 3.620; slightly soluble in water.

Present in wastes from the production of some pesticides.

Toxicity: has a general irritating, allergenic and carcinogenic effect; affects the circulatory system, the liver, and the kidneys; causes pigmentation of the skin [74,80,81,103]. MPSDC and MPADC - 0.009 and 0.004 mg/m^3 respectively [36].

Calcium borate (calcium tetraborate); CaB_4O_7 (195.36)

Colorless crystals, m.p. 986°C; slightly soluble in water.

Present in wastes from the production of porcelain and ethylene glycol; also in wastes of some metallurgical plants.

Toxicity: has a general irritating effect. MPADC - 0.02 mg/m^3; toxicity classification - III [117].

Calcium carbonate; $CaCO_3$ (100.09)

Colorless crystals, m.p. 1240°C(d.), ρ 2.71;slightly soluble in water.

Present in wastes from the production of paints, articles made of rubber, plastics, paper, ceramics, inks, pesticides, some foodstuffs, cosmetic preparations, and pharmaceuticals; matches, colored pencils, linoleum, some analytical reagents, construction materials, and cement; also in wastes of some metallurgical plants.

Toxicity: has a general irritating effect. $TC_{chronic}$ - 84 mg/m^3 [74,81,103].

Removal: with the aid of filters made of nylon, polyesters (including polyacrylates), polyethylene, poly(vinyl chloride), teflon, or polypropylene [57].

Calcium chlorate; $Ca(ClO_3)_2$ (206.99)

Colorless crystals, m.p. 340°C(d.), ρ 2.71; solubility in water: 63 g/100 g at 5°C.

Present in wastes from the production of some pesticides, defoliants, and pyrotechnical compositions.

Toxicity: causes hemolysis of erythrocytes and the formation of methemoglobin; has a poisenous effect on the heart muscle and the kidneys [74,81].

Calcium chromate; $CaCrO_4$ (156.09)

Yellow crystals, m.p. 1000°C(d.); solubility in water: 2.2 g/100 g at 18°C.

Present in wastes from the production of chromium, dyes; and in wastes of metallurgical works.

Toxicity: has a general irritating effect [74,81].

Calcium cyanamide; $Ca(CN)_2$ (80.1)

White crystals, m.p. 1340°C; soluble in water (2.5%); practically insoluble in hydrophobic organic solvents.

Present in wastes from the production of cyanides, melamine, defoliants, and some pesticides.

Toxicity: causes irritation of mucous membranes and the respiratory tracts, and dermititis [70]. MPSDC and MPADC - 0.02 mg/m^3 [24].

Calcium diborate; $Ca(BO_2)_2$ (125.70)

Colorless crystals, m.p. 1162°C, ρ 5.59; insoluble in water.

Present in wastes from the production of glass fibers, glaze, some plastics, articles made of rubber, and textiles.

Toxicity: causes pneumonia, dystrophy, and degeneration of internal organs; has a general irritating effect on the respiratory tracts. $TC_{chronic}$ - 120-150 mg/m^3 [103].

Calcium dichromate; $CaCr_2O_7$ (256.10)

Orange crystals, ρ 2.37; decomposes on heating; soluble in water.

Present in wastes from the production of some catalysts and chromium compounds.

Toxicity: has a general irritating effect [74,103].

Calcium fluoride; CaF_2 (78.88)

Colorless crystals, m.p. 1419°C, b.p. ≅2530°C, ρ 3.18; slightly soluble in water.

Present in wastes from the production of cement, glass, ceramics, articles made of rubber, some catalysts, pesticides, enamels, luminophors; also in wastes of metallurgical and mining works, and wood-working plants.

Toxicity: very toxic; causes asthemia, loss of weight, nausea, vomiting, asphyxia, cramps, and osteosclerosis; damage to the kidneys, and irritation of the respiratory tracts; on prolonged exposure at low concentrations causes fluorosis [74,81, 103]. $MPSDC_{acute}$ and TC_{acute} - 0.03 and 0.3 mg/m^3 respectively; MPSDC and MPADC - 0.2 and 0.03 mg/m^3 respectively; toxicity classification - II [34].

Calcium hydroxide; $Ca(OH)_2$ (74.08)

Colorless crystals, m.p. 580°c(d), ρ 2.24; slightly soluble in water.

Present in wastes from the production of cement, sugar, leather, some foodstuffs, and glass; also in wastes of some metallurgical plants.

Toxicity: causes irritation of the respiratory tracts, mucous memranes and the skin. LD_{50} for rats - 7.34 g/kg [79,103].

Removal: with the aid of filters made of polyesters (including polyacrylates), polyethylene, or poly(vinyl chloride) [57].

Calcium hypochlorite; $Ca(ClO)_2 \cdot 4H_2O$ (215.06)

Colorless crystals, m.p. 180°C(d.); soluble in water.

Present in wastes from the production of chloroform, chloropicrin, textiles, and paper.

Toxicity: irritates mucous membranes and the skin, and does damage to teeth.

Calcium nitrate; $Ca(NO_3)_2$ (164.09)

Colorless crystals, m.p. 560°C, ρ 3.36; solubility in water: 114.6 g/100 g H_2O at 20°C.

Toxicity: causes headache; impairment of vision; irritation of the eyes, mucous membranes, and the skin; and lowers blood pressure [74,81,113].

Calcium oxide (lime); CaO (56.08)

Colorless crystals, m.p. 2580°C, b.p. 2850°C, ρ 3.4; slightly soluble in water.

Present in wastes from the production of aluminum, magnesium, glass, paper, leather, sugar; in wastes of metallurgical works and steel foundries.

Toxicity: causes strong irritation and burns, ulceration of the mucous membrane inside the mouth, nose, and respiratory tracts, conjunctivitis; and damages the iris and the skin. $TSDC_{acute}$ - 8.5 mg/m^3 [74,81,102,103,113]; ASL - 0.3 mg/m^3 [117].

Calcium phosphide; Ca_3P_2 (182.20)

Red-brown crystals, m.p. 1250°C(d.), ρ 2.51; decomposes in water.

Present in wastes of metallurgical works and plants making pesticides.

Toxicity: very toxic; causes depression of central nervous system, irritation of the lungs, pulmonary edema; on prolonged exposure at low concentrations causes anemia, bronchitis, impairment of vision [74,81].

Calcium thiocyanate (calcium rhodonate); $Ca(CNS)_2$ (156.24)

Colorless crystals; decomposes on heating.

Present in wastes from the production of some polymers, acrylonitrile, textiles, pesticides, thiocyanates, some analytical reagents, pharmaceuticals, thiourea, and paints.

Toxicity: does damage to the thyroid gland; causes dyspnea, rales, cramps, cardiac depression [81,103,113].

Calcium selenate; $CaSeO_4$ (183.04)

Colorless crystals, ρ 2.93; soluble in water.

Present in wastes from the production of some pesticides, glass, dyes, plastics; also in wastes of steel foundries.

Toxicity: has a general irritating and carcinogenic effect; affects the brain, lungs, liver, and kidneys; causes dermatitis, pneumonia, and pulmonary edema [74, 81,113].

Calcium sulfate; $CaSO_4$ (136.6)

Colorless crystals, m.p. 1450°C(d.), ρ 2.32; slightly soluble in water.

Present in wastes from the production of cement, some dyes, paper, some fertilizers; also in wastes of mining works.

Toxicity: has a general toxic, irritating effect [102,103].

Calcium sulfide; CaS (72.14)

Colorless crystals, m.p. ≅2450°C, ρ 2.18.

Present in wastes from the production of phosphors, electronic equipment, lacquers and paints, some foodstuffs, and lubricants.

Toxicity: causes irritation of the eyes, the respiratory tracts, and the skin, pulmonary edema, and asphyxia [74,81].

Calcium tungstate; $CaWO_4$ (287,93)

Colorless crystals, m.p. 1580°C, ρ 6.06; slightly soluble in water: 0.2 g/100 g H_2O at 18°C.

Present in wastes from the production of photographic materials, luminescent materials, some medicinal preparations, and x-ray equipment, and cathodo-luminophors.

Toxicity: causes pneumosclerosis [103].

CARBON

Carbon dioxide; CO_2 (44.01)

Colorless gas, m.p. 56.6°C (in vacuo), subl.p. -78.50°C, ρ 1.5; soluble in water.

Present in wastes from the production of carbonates, sugar, beer, carbonated beverages, dry ice, soda, white lead, urea, refrigerants, some pharmaceuticals, fire-resistant fabrics, fire extinguishers; also in wastes from metallurgical works.

Toxicity: causes weakness and debility, dizziness, headache, high blood pressure; at high concentrations - has a narcotic and irritating effect; causes general depression and asphyxia [74,81,102,103]. $TC_{c.n.s.}$ - 0.1-0.5%; MPSDC (recommended - 0.05% [240].

Removal: by scrubbing (efficiency 97-99%) [76].

Determination: photocolorimetric analysis (sensitivity 0.025%) [24]; spectrophotometric analysis [69].

Carbon disulfide; CS_2 (76.14)

Colorless liquid, m.p. -11.9°C, b.p. 46.2°C, ρ 1.26; slightly soluble in water.

Present in wastes from the production of viscose, carbon tetrachloride, thiocyanates, xanthogenates, thiocarbonates, solvents, extracting agents, and electronic equipment.

Toxicity: has a general toxic, irritating effect on the gonados and embryos; causes impairment of central and peripheral nervous systems and internal organs, headache, dizziness, vomiting, tremors, cramps, hallucination, impairment of vision and hearing, unconsciousness, damage to the liver, kidneys, and heart. and irritation of mucous membranes and the skin; mental disorders. TC_{odor} - 0.04 mg/m^3 [74,81,103]; MPSDC and MPADC - 0.03 and 0.005 mg/m^3 respectively; toxicity classification - II [34]; MPSDC and MPADC established in Bulgaria, Romania, and Yugoslavia - 0.03 and 0.01 mg/m^3 respectively; and in Poland - 0.045 and 0.015 mg/m^3 respectively; MPADC established in GDR - 0.003 mg/m^3; MPC established in Czechoslovakia - 0.3 kg/hr [72].

Removal: see [43,48].

Determination: see [24,44,96,98].

Carbon monoxide; CO (28.04)

Colorless gas, m.p. -205.02°C, b.p. -191.5°C, ρ 0.97; slightly soluble in water.

Present in wastes from the production of cellulose sulfate, linoleum, tar paper, artificial parchment paper, plastic foam, aluminum, asphalt concrete, cement, ammonium nitrate, ammonia, methanol, synthetic benzene, and some organic compounds; also in wastes of by-product coke plants, metallurgical and metal-working plants, petroleum refineries, coal processing plants, and power stations run on coal.

Toxicity: very toxic, blood poison; causes headache, dizziness, vomiting, anxiety, dyspnea, slow breathing, cramps, death. $TSDC_{acute}$ - 0.01-0.05 mg/l [96]; MPSDC and MPADC - 5 and 3 mg/m^3 respectively; toxicity classification - IV [34, 117]. MPSDC and MPADC established in Bolgaria, GDR, Yugoslavia, and Poland - 3 and 1 mg/m^3 respectively; in Romania and Czechoslovakia - 6 and 0.5 mg/m^3 respectively [88,97].

Removal: see [43].

Determination: see [6,18,24,44,96,98].

Carbon oxychloride (carbonyl chloride; phosgene); $COCl_2$ (98.92)

Colorless gas, m.p. -118°C, b.p. -8.2°C, ρ 1.42; slightly soluble in water.

Present in wastes from the production of chloroform, carbon tetrachloride, trichloroethane, ketones, some plastics, synthetic subber, fire extinguishers, and some pharmaceuticals.

Toxicity: very toxic; cause pneumonia, pulmonary edema, death. TC_{odor}- 0.004-0.0050 mg/l; TC and $TSDC_{acute}$ - 0.005 and 0.0008 mg/l respectively [74,81, 103].

Carbon oxyfluoride (carbonyl fluoride); COF_2 (66.01)

Colorless gas, m.p. -114°C, b.p. -83°C.

Present in wastes from the production of fluoroorganic compounds.

Toxicity: has an irritating effect [74,81].

Carbon oxysulfide (carbonyl sulfide); COS (60.07)

Colorless gas, m.p. -138.82°C, b.p. -50.24°C, ρ 1.24 (at -87°C); soluble in water.

Present in wastes from the production of urea.

Toxicity: very toxic. LC_{100} - 3 mg/m^3 [81].

CESIUM

Cesium; Cs (132.905)

Silvery-white metal, m.p. 28.4°C, b.p. 667.5°C, ρ 1.90; reacts with water.

Present in wastes from the production of electronic equipment and radio engineering goods; ceramics, glass, some catalysts and hydrides; boron hydrides, and some pharmaceuticals.

Toxicity: has an allergenic effect [74,81,103].

Cesium bromide; CsBr (212.81)

Colorless crystals, m.p. 636°C, b.p. 1300°C, ρ 4.4; solubility in water: 55.29 g/100 g H_2O at 25°C.

Present in wastes from the production of fluorescent screens, spectrometric prisms, and absorption windows.

Toxicity: has a general toxic effect. LD_{50} for rats - 1.4 g/kg [74].

Cesium carbonate; Cs_2CO_3 (325.83)

Colorless crystals, m.p. 793°C; solubility in water: 75.5 g/100 g H_2O at 20°C.

Present in wastes from the production of catalysts used in polymerizing ethylene oxide; fuel elements; special types of glass; cesium compounds; also in wastes of steel foundries.

Toxicity: has a general toxic effect. LD_{50} for mice - 2.3 g/kg [74,103].

Cesium chloride; CsCl (168.36)

Colorless crystals, m.p. 645°C, b.p. 1302°C, ρ 3.97; solubility in water: 65.7 g/100 g H_2O at 25°C.

Present in wastes from the production of vacuum tubes for radio and TV, fluorescent screens, cesium, radiographic contrasting media, fuel elements, and phosphors; also in wastes of some metallurgical plants.

Toxicity: has a general toxic effect. $TC_{chronic}$ - 30 mg/m^3; LD_{50} for mice - 2000 mg/kg [74,81,103].

Cesium hydroxide; CsOH (149.91)

Colorless crystals, m.p. 272°C, ρ 3.68; solubility in water: 385.6 g/100 g H_2O

at 15°C.

Present in wastes from the production of some electrolytes and catalysts, and cyclic siloxanes.

Toxicity: very toxic; has a general irritating and allergenic effect [74,100,103, 113].

Cesium iodide; CsI (259.82)

Colorless crystals, m.p. 632°C, b.p. 1280°C, ρ 4.5; solubility in water: 46.1 g/100 g H_2O at 25°C.

Present in wastes from the production of prisms for infrared spectrometers and fluorescent screens; scintillation counters, cathode-ray equipment; and luminophors.

Toxicity: has a general toxic and allergenic effect. LD_{50} for rats - 1400 mg per kg [74,81,100].

Cesium nitrate; $CsNO_3$ (194.91)

Colorless crystals, m.p. 414°C, b.p. 490°C(d.), ρ 3.68; solubility in water: 22.4 g/100 g H_2O at 25°C.

Present in wastes from the production of cesium salts, luminophors, and magnetohydrodynamic generators.

Toxicity: at high concentrations causes vomiting, cramps, collapse, and death; at low concentrations has a general toxic and allergenic effect. LD_{50} for rats - 12-00 mg/kg [74,81,100,113].

Cesium nitrite; $CsNO_2$ (178.91)

Yellow crystals; decomposes on heating; soluble in water.

Present in wastes from the production of some organic dyes.

Toxicity: has a general toxic and irritating effect; at high concentrations causes vomiting, cyanosis, the formation of methemoglobin in the blood, collapse, and coma; at low concentrations - a lowering of blood pressure, rapid pulse, headache, and impairment of vision [81,113].

CHLORINE

Chlorine; Cl_2 (70.91)

Yellowish green gas, m.p. -101°C, b.p. -34.1°C, ρ 3.21; soluble in water.

Present in wastes from the synthesis of some organic chemicals, and from the production of pigments, paper, some pesticides and desinfectants, chlorinated hydrocarbons, calcium hypochlorite, carbontetrachloride, chlorine, some plastics, magnesium, potassium chlorate, hydrochloric acid, and textiles; also in wastes of metallurgical works.

Toxicity: very toxic; causes strong irritation of the respiratory tracts, the eyes, and mucous membranes, dyspnea, pulmonary edema, collapse, and death; at high concentrations causes instant death owing to reflex inhibition of the respiratory center; at low concentrations - coughing, asphyxia, impaired breathing, chemical burns of the eyes, and dermatitis. TC_{odor}, TC_{acute}, and $TC_{chronic}$ - 0.05mg/m^3, 0.01 mg/l, and 0.1 mg/m^3 respectively [74, 81, 87,102,103]; established MPSDC and MPADC - 0.1 and 0.03 mg/m^3 respectively [34]. In Bulgaria, Hungary, Poland, GDR, Czechoslovakia, and Yugoslavia established MPSDC and MPADC - 0.1 and 0.03 mg/m^3 and in Romania - 0.3 and 0.1 mg/m^3 respectively [88]. MPS established in Czechoslovakia - 1 mg/hr [88].

Removal: see [7,76].

Determination: see [24,98].

Chlorine dioxide; ClO_2 (67.45)

Greenish yellow gas, m.p. -59°C, b.p. 9.7°C, ρ 1.6; reacts with water.

Present in wastes from the production of bleaching agents, some foodstuffs, cellulose, paper, leather, fats and oils, textiles, waxes, bactericides, antiseptics, and chlorinating agents; also in wastes of flour mills.

Toxicity: causes irritation of mucous membranes and the skin, pulmonary edema. $TC_{chronic}$ - 0.003-0.005 mg/l [74,81,102,103].

Chlorine heptaoxide; Cl_2O_7 (182.90)

Colorless liquid, m.p. -90°C, b.p. 80°C, ρ 1.8; reacts with water.

Present in wastes from the production of bleaching agents, some foodstuffs, cellulose, paper, leather, fats and oils, textiles, waxes, bactericides, antiseptics, and chlorinating agents; also in wastes of flour mills.

Toxicity: causes irritation of mucous membranes and the skin; pulmonary edema [74,81,102,103].

Chlorine monoxide; Cl_2O (86.91)

Yellowish brown gas, m.p. -116°C, b.p. 2°C, ρ 3.89 g/l; reacts with water.

Present in wastes from the production of bleaching agents, some foodstuffs, cellulose, paper, leather, fats and oils, textiles, waxes, bactericides, antiseptics, and chlorinating agents; also in wastes of flour mills.

Toxicity: causes irritation of mucous membranes and the skin; pulmonary edema [74,81,102,103].

Hydrochloric acid (hydrogen chloride); HCl (36.46)

Colorless gas, m.p. -114.2°C, b.p. -85.01°C, ρ 1.639 g/l; solubility in water: 45.15 g/100 g H_2O at 0°C.

Present in wastes from the production of metal chlorides, some pesticides, chloroprene, dyes, ethanol, glucose, some plastics, sugar, gelatin, glues, leather, activated carbon, textiles, some reagents and organic compounds, and articles made of rubber; also in wastes of metal-working and metallurgical plants, and from galvanic processes and the saponification of fats.

Toxicity: very toxic; has an irritating effect; at high concentrations causes necrosis of mucous membranes, nebula, pharyngismus, and pulmonary edema; at low concentrations - irritation of the nasal mucous membrane, perforation of the nasal septum, and ptyalism. $SC_{chronic}$ and $TC_{chronic}$ - 0.015 and 0.15-0.21 mg/l respectively [74,81,103]; MPSDC and MPADC - 0.05 and 0.015 mg/m^3 respectively; toxicity classification - II [34].

Removal: with the aid of srubbers (efficiency 97-99%) [76], and of filters made of polyesters (including polyacrylates), polyethylene, poly(vinyl chloride), teflon, or polypropylene [57].

Perchloric acid; $HClO_4$ (100.47)

Colorless liquid, m.p. -102°C, b.p. 110°C, ρ 1.76; soluble in water.

Present in wastes from the production of perchlorates, some analytical reagents, and catalysts used in hydrolysis and esterification; also in wastes from galvanic processes.

Toxicity: very toxic; has a general irritating effect; causes corrosion of mucous membranes and the skin [74,81].

CHROMIUM

Chromic ammonium sulfate; $Cr_2(SO_4)_3(NH_4)_2SO_4 \cdot 24H_2O$ (956.66)

Violet crystals, m.p. 94°C, ρ 1.72; soluble in water: 15.7 g/100 g H_2O at 40°C.

Present in wastes from the production of tanning agents, textiles, motion-picture films, and electrolytic chromium.

Toxicity: has a general toxic effect; causes irritation of the eyes, the respiratory tracts, mucous membranes, and the skin. $TC_{chronic}$ - 0.2 mg/m^3 [74,81,103].

Chromic boride; CrB (62.83)

Silvery crystals, m.p. 2760°C, ρ 6.17; insoluble in water.

Present in wastes from the production of some chemicals.

Toxicity: causes irritation of mucous membranes, the respiratory tracts, and the skin; and degeneration of the liver, the kidneys, and the heart [84,103].

Chromic bromide; $CrBr_3$ (291.71)

Dark green crystals, m.p. 842°C, ρ 4.25; soluble in water.

Present in wastes from the production of catalysts used in the polymerization of olefins.

Toxicity: causes irritation of the respiratory tracts, mucous membranes, and the skin; the perforation of the nasal septum, skin ulceration; and general depression and mental disorders [74,81].

Chromic chloride; $CrCl_3$ (158.36)

Violet-pink crystals, m.p. 1152°C (under pressure), ρ 3.03; slightly soluble in water.

Present in wastes from the production of chromium, catalysts used in the polymerization of olefins, textiles, and tanning agents.

Toxicity: has a carcinogenic and allergenic effect; causes irritation of the respiratory tracts, perforation of the nasal septum, allergic eczema, and branchogenic carcinoma. $TC_{chronic}$ - 0.1 mg/m^3; LC_{100} - 31 mg/m^3 [74,81,103].

Removal: with the aid of filters made of polyesters (including polyacrylates), polyethylene, poly(vinyl chloride), teflon, or polypropylene [57].

Chromic diboride; CrB_2 (73.62)

Gray crystals, m.p. 2200°C, ρ 2.2; insoluble in water.

Present in wastes from the production of cermets; also in wastes of metallurgical and metal-working plants.

Toxicity: has a general toxic effect; causes dysfunctioning of the respiratory tracts, fibrosis, pulmonary emphysema; dystrophic changes in the liver, kidneys, and heart muscle; muscle weakness and incoordination, muscular twitching, and paralysis of limbs. $TC_{chronic}$ - 300-500 mg/m^3 [103].

Chromic fluoride; CrF_3 (108.99)

Green crystals, m.p. 1400°C, ρ 3.7; solubility in water: 4 g/100 g H_2O at 20°C.

Present in wastes from the production of some pigments and catalysts; also in wastes of metal-working and wool-processing plants.

Toxicity: has a carcinogenic and allergenic effect; causes irritation of the respiratory tracts, mucous membranes, and the skin, perforation of the nasal septum, allergenic eczema, lung cancer; on prolonged exposure at low concentrations - fluorosis [74,81].

Chromic hydroxide; $Cr(OH)_3$ (103.03)

Grayish green powder, d.p. >100°C; insoluble in water.

Present in wastes from the production of dyes, tanning agents, and some catalysts used in organic synthesis.

Toxicity: has an irritating effect [74].

Chromic nitrate; $Cr(NO_3)_3$ (238.03)

Greenish powder, decomposes on heating; soluble in water.

Present in wastes from the production of chromium catalysts and textiles; also in wastes of metal-working plants.

Toxicity: has a general toxic and irritating effect; causes asthenia, general depression, and headache [74,81,103].

Chromic oxide; Cr_2O_3 (152.02)

Dark green crystals, m.p. 2340°C, b.p. 3000°C, ρ 5.2; insoluble in water.

Present in wastes from the production of chromium by the aluminothermic method, chromium carbide, polishing pastes, paints, glass, ceramics, chromium catalysts, and electric semiconductors.

Toxicity: has an irritating and carcinogenic effect [103,113].

Chromic potassium sulfate; $Cr_2(SO_4)_3 \cdot K_2SO_4 \cdot 24H_2O$ (998.78)

Violet or green crystals, m.p. 89°C, ρ 1.83; soluble in water.

Present in wastes from the production of tanning agents, leather, textiles, motion picture films, pigments, inks, chromium salts, and photographic materials.

Toxicity: has a general toxic effect; causes irritation of the eyes, the respiratory tracts, mucous membranes, and the skin [74,81,103].

Chromic sulfate; $Cr_2(SO_4)_3$ (392.17)

Violet or red powder, ρ 3.012; soluble in water.

Present in wastes from the production of gelatin, some catalysts, textiles, leather, lacquers and paints, inks, and glass; also in wastes of metal-working and metallurgical plants.

Toxicity: has a general toxic and irritating effect [74,81].

Chromium; Cr (51.996)

Silvery gray metal, m.p. 1890°C, b.p. 2680°C, ρ 7.19; insoluble in water.

Present in wastes of steel foundries and metal-working and metallurgical plants; also in wastes from the production of ceramics.

Toxicity: slightly toxic [103].

Chromium hexacarbonyl; $Cr(CO)_6$ (214.06)

Colorless crystals, m.p. 151-152°C, ρ 1.77; insoluble in water.

Present in wastes in the production of petrochemicals, some catalysts used in polymerizing olefins, and chromium oxide.

Toxicity: has a general toxic effect. LD_{50} for mice - 100 mg/kg [74,81].

Chromium tetrafluoride; CrF_4 (128.01)

Dark green crystals, m.p. 200°C, b.p. 400°C, ρ 2.89; decomposes in water.

Present in wastes from the production of some pigments and catalysts; also in wastes of metal-working and wool-processing plants.

Toxicity: has a carcinogenic and allergenic effect; causes irritation of the respiratory tracts, mucous membranes, and the skin, perforation of the nasal septum, allergic eczema, and lung cancer; on prolonged exposure at low concentrations - fluorosis [74,81].

Chromium trioxide (chromic anhydride); CrO_3 (99.99)

Dark red crystals, m.p. 198°C, ρ 2.8; solubility in water: 62.8 g/100 g H_2O at 20°C.

Present in wastes from the production of photographic materials, some pharmaceuticals, and pigments; also in wastes of steel foundries and metal-working and metallurgical plants.

Toxicity: has a toxic, carcinogenic, and teratogenic effect [63]; causes allergic eczema, irritation of the respiratory tracts, perforation of the nasal septum, bronchogenic carcinoma, ulceration of the skin, and pathological changes in the kidneys. $SC_{irritation}$ and $TC_{irritation}$ - 0.0015 and 0.0025 mg/m^3 respectively [69,74, 81,89,97]; MPSDC and MPADC - 0.0015 mg/m^3; toxicity classification - I [34]; MPSDC established in Yugoslavia, Romania, and GDR - 0.0015 mg/m^3, MPADC established in these countries - 0.0015, 0.0015, and 0.001 mg/m^3 respectively [12,88].

Removal: by filtration (efficiency 99%) [57,76].

Determination: see [44,47,69,99,115].

Chromyl chloride; CrO_2Cl_2 (154.90)

Dark red liquid, m.p. -96.5°C, b.p. 116°C, ρ 1.91; reacts with water.

Present in wastes from the production of catalysts used in polymerizing olefins and of hydrocarbon oxidants.

Toxicity: very toxic; causes chemical burns of the skin. $TC_{chronic}$ - 0.1 mg/m^3; LC_{50} - 31 mg/m^3 [74,81,103]; MPSDC and MPADC - 0.0015 mg/m^3; toxicity classification - I [34].

Removal: with the aid of filters made of polyesters (including polyacrylates), polyethylene, poly(vinyl chloride), teflon, or polypropylene [57].

Determination: see [44,47,69,99,115].

COBALT

Cobalt; Co (58.93)

Pale yellow metal, m.p. 1494°C, b.p. 2960°C, ρ 8.9; insoluble in water.

Present in wastes from the production of cobalt salts, nitric acid, and oil and fats; and in wastes of some metallurgical plants.

Toxicity: has a general irritating, allergenic, carcinogenic, and mutagenic effect [63,69,74,81,93,103]. MPADC - 0.001 mg/m^3; toxicity classification - I [34].

Determination: spectroscopic analysis [44,69], chromatographic analysis [74].

Cobalt arsenate; $Co(AsO_4)_2$ (218.77)

Purplish red crystals; decomposes on heating; ρ 2.948.

Present in wastes from the production of glass and porcelain.

Toxicity; has a general irritating and allergenic effect; causes nausea, vomiting, loss of appetite, and degeneration of the liver and kidneys [74,81].

Cobalt bromide (cobalt dibromide); $CoBr_2$ (218.77)

Reddish purple crystals, m.p.678°C, b.p. 927°C, ρ 4.90; soluble in water.

Present in wastes from the production of some catalysts and hydrometers.

Toxicity: has a general irritating and allergenic effect [74,81].

Cobalt carbonate; $CoCO_3$ (118.95)

Red crystals, m.p. 400°C(d.), ρ 4.13; insoluble in water.

Present in wastes from the production of ceramics, cobalt compounds, cobalt dyes, and pigments.

Toxicity: has a general irritating and allergenic effect; does damage to the liver and kidneys [74,81,103].

Cobalt chromate; $CoCrO_4$ (174.95)

Brown or dark gray crystals; decomposes on heating.

Present in wastes from the production of ceramics.

Toxicity: causes allergy and irritation of the respiratory tracts, mucous membranes, and the skin [74,81].

Cobalt dichloride; $CoCl_2$ (129.84)

Blue crystals, m.p. 740°C, b.p. 1050°C, ρ 3.56; soluble in water.

Present in wastes from the production of hydrometers, invisible inks, some catalysts, glass, porcelain, fertilizers, foaming agents in beer-brewing, and vitamins; also in wastes from electrolytic metal-plating processes.

Toxicity: has a general toxic effect [74,81,103].

Cobalt difluoride; CoF_2 (96.94)

Pink crystals, m.p. 1127°C, b.p. ≅1740°C, ρ 4.43; slightly soluble in water.

Present in wastes from the production of some catalysts.

Toxicity: at high concentrations has a general irritating effect; under prolonged exposure at low concentrations causes fluorosis [74,81].

Cobalt disulfide (cobalt (II) sulfide); CoS_2 (91.0)

Pinkish gray crystals, m.p. 953°C, ρ 5.45; insoluble in water.

Present in wastes from the production of some catalysts.

Toxicity: has a general irritating and allergenic effect; causes dermatitis, irritation of the skin; and damage to the liver and kidneys [74,81,113].

Cobalt hydroxide; $Co(OH)_2$ (92.96)

Pink or blue powder, m.p. >150°C(d.), ρ 3.59; insoluble in water.

Present in wastes from the production of cobalt compounds, pigments, and lithographic and colored inks.

Toxicity: causes nausea, vomiting, irritation of the lungs and skin [74,81].

Cobalt iodide (cobalt diiodide); CoI_2 (312.74)

Dark yellow crystals, m.p. 516°C, b.p. 760°C, ρ 5.54; soluble in water.

Present in wastes from the production of some catalysts and humidity indicators, also in wastes during the determination of water in solvents.

Toxicity: has a general irritating and allergenic effect; causes dizziness, lacrimation, skin irritation, dermatitis, damage to the liver and kidneys, and irritation of mucous membranes [74,81,103,113].

Cobalt nitrate; $Co(NO_3)_2 \cdot 6H_2O$ (291.03)

Present in wastes from the production of cobalt dyes and pigments, invisible inks, and some vitamins.

Toxicity: has a general irritating effect; at high concentrations causes dizziness, vomiting, cramps, and collapse; at low concentrations - weakness, general depression, and headache [74,81,103].

Cobalt (II,III) oxide; Co_3O_4 (240.82)

Black crystals, m.p. 900°C(d.), ρ 6.07; reacts with water [63,69,74,81,93,103].

Present in wastes from the production of enamels, semiconductors, dyes, ceramics, glass, and pigments.

Toxicity: same as that of Co.

Cobalt sulfate; $CoSO_4$ (154.99)

Pink crystals, m.p. 989°C(d.), ρ 3.71; solubility in water: 35.5 g/100 g H_2O at 20°C.

Present in wastes from the production of storage batteries, lithographic inks, lacquers and paints, ceramics, and enamels; also in wastes from galvanotechnological processes.

Toxicity: causes an irritation of mucous membranes and the respiratory tracts.

LD_{50} for mice - 54 mg/kg [74,81,103,113]; MPSDC (recalculated for Co) - 0.001 mg/m^3; MPADC - 0.0004 mg/m^3; toxicity classification - II [35].

Cobaltic oxide (cobalt(III) oxide); CoO_3 (165.88)

Black powder, m.p. 600°C, ρ 5.18 [74,81,103].

Present in wastes from the production of enamels, semiconductors, dyes, ceramics, glass, and pigments.

Toxicity: same as that of Co.

Cobaltous cyanide (cobalt(II) cyanide); $Co(CN)_2 \cdot 2H_2O$ (147.0)

Bluish purple powder, m.p. 280°C(d.), ρ 1.87; insoluble in water.

Present in wastes from the production of cobalt catalysts.

Toxicity: causes damage to the liver, the kidneys, and the skin; and has a general irritating effect [74,81].

Cobaltous oxide (cobalt(II) oxide); CoO (74.93)

Brown or olive-green crystals, m.p. 1810°C, ρ 5.7; insoluble in water [74,103].

Present in wastes from the production of enamels, semiconductors, dyes, ceramics, glass, and pigments.

Toxicity: same as that of Co.

COLUMBIUM

Columbium (Niobium); Cb (92.91)

Light gray metal, m.p. 2500°C, b.p. 4927°C, ρ 8.57; insoluble in water.

Present in wastes from the production of jet engines, rockets, gas turbines, and some chemical equipment; also in wastes of some metallurgical and radio engineering plants.

Toxicity: suppresses the action of ferments [113].

Columbium chloride; $CbCl_5$ (270.20)

Light yellow crystals, m.p. 205°C, b.p. 247.5°C, ρ 2.75; reacts with water.

Present in wastes of electrolytic plating and some metallurgical plants.

Toxicity: causes irritation of mucous membranes, the eyes, and the respiratory tracts; and damage to the kidneys and liver. LD_{50} for mice and rats - 829.6 and 1400 mg/kg respectively [74,81,103].

Columbium fluoride; CbF_5 (187.91)

Colorless crystals, m.p. 80°C, b.p. 234.9°C, ρ 3.3; reacts with water.

Present in wastes from electrolytic processes and of metallurgical plants.

Toxicity: very toxic; has a general irritating effect; on prolonged exposure at low concentrations - mottled enamel, osteosclerosis, and calcification of the joints [74,81].

Columbium nitride; CbN (106.92)

Light gray crystals, m.p. 2300°C(d.), ρ 8.4; insoluble in water.

Present in wastes from the production of television components and radiators.

Toxicity: $SC_{chronic}$ and $TC_{chronic}$ - 10 and 40 mg/m^3 respectively [103].

Columbium pentaoxide; Cb_2O_5 (265.81)

Colorless crystals, m.p. 1490°C, ρ 4.95; insoluble in water.

Present in wastes from the production of columbium and its compounds, refractory materials, some catalysts, and ceramics.

Toxicity: suppresses the activity of ferments [103].

COPPER

Copper; Cu (63.55)

Reddish metal, m.p. 1084.5°C, b.p. 2540°C, ρ 8.26; insoluble in water.

Present in wastes of metallurgical, machine-building, metal-working, and some chemical plants.

Toxicity: very toxic; has a mutagenic effect; causes headache, dizziness, asthenia, and muscular pain; dysfunctioning of the liver, the kidneys, and the central nervous system; irritation of mucous membranes, the skin and the eyes; ulceration and perforation of nasal septum; ulceration of the cornea, a sweet taste in the mouth, and rise in temperature to 38-39°C. $TSDC_{acute}$ - 0.22 mg/m^3 [74,81,91,103].

Determination: colorimetric analysis (sensitivity 2 μg); chromatographic analysis (sensitivity 0.2 μg); paper chromatography [44]; absorption spectroscopy (sensitivity 0.001 μg/m^3); and emission spectroscopy (sensitivity 0.002 μg/m^3) [99,115].

Copper iodide; CuI (190.46)

Colorless crystals, m.p. 588-606°C, b.p. ≅1290°C, ρ 5.63; insoluble in water.

Present in wastes from the production of some catalysts.

Toxicity: causes irritation of mucous membranes and the respiratory tracts; hemolysis of erythrocytes; and damage to the liver and pancreas [74,81].

Cupric arsenite (Paris green); $CuHAsO_3$ (187.46)

Yellowish green crystals; insoluble in water.

Present in wastes from the production of some dyes and pesticides; also in wastes of wood-working plants.

Toxicity: affects the kidneys, liver, and the nervous system; has an irritating and allergenic effect [81,113].

Cupric bromide; $CuBr_2$ (223.37)

Black crystals, m.p. 498°C, b.p. 900°C, ρ 4.71; soluble in water.

Present in wastes from the production of photographic materials, some organic compounds, and electric batteries; also in wastes of wood-working plants.

Toxicity: has a general irritating effect; causes depression, mental disorders; rash, and furunculosis [74,81].

Cupric carbonate (basic malachite); $Cu_3(OH)_2CO_3$ (221.11)

Dark green crystals, m.p. 200°C(d.); insoluble in water.

Present in wastes from the production of pesticides, pyrotechnical compositions, lacquers and paints, petrochemicals, and copper salts.

Toxicity: has an irritating and allergenic effect; causes hemolysis of erythrocytes; cirrhosis of liver and pancreas; and irritation of the eyes, mucous membranes, and the respiratory tracts [81]; LD_{50} for rats - 159 mg/kg [103].

Cupric chloride (eriochalcite); $CuCl_2$ (134.45)

Brownish yellow crystals, m.p. 596°C, ρ 3.39; solubility in water: 74.5 g/100 g H_2O at 20°C.

Present in wastes from the production of some catalysts, petrochemicals, textiles, dyes, photographic materials, pyrotechnical compositions, acrylonitrile, melamine, glass, ceramics, and desinfectants; also in wastes of wood-working plants.

Toxicity: causes irritation of mucous membranes and the skin [74,81,103]. LD_{50} for rats - 140 mg/kg [103]; MPADC - 0.002 mg/m^3; toxicity classification - II [34].

Removal: with the aid of filters made of polyesters (including polyacrylates), polyethylene, poly(vinyl chloride), teflon, or polypropylene [57].

Cupric hexahydrosulfate; $CuSO_4 \cdot 3Cu(OH)_2$ (452.39)

Bluish particles, m.p. 200°C, ρ 2.28; insoluble in water.

Present in wastes from the production of pesticides, pigments, and synthetic fibers; also in wastes of concentrating mills.

Toxicity: causes irritation of mucous membranes and the skin. LD_{50} for rats - 300 mg/kg [74,103].

Cupric nitrite; $Cu(NO_3)_2$ (187.56)

Bluish green crystals, m.p. 256°C; soluble in water.

Present in wastes from the production of pigments, ceramics, textiles, aluminum, pesticides, motor fuels, organosilicons, and some catalysts; also in wastes of wood-working plants.

Toxicity: has a general toxic and irritating effect [74,81,103].

Cupric oxide; CuO (79.5)

Black crystals, m.p. 1026°C(d.), ρ 6.45; insoluble in water.

Present in wastes from the production of glass, enamel, synthetic silk, copper compounds, ceramics, synthetic precious stones, pyrotechnical compositions, some catalysts, dyes, and pesticides; also in wastes of metal-working and galvanotechnical plants.

Toxicity: has a general toxic effect; causes irritation of mucous membranes and the skin [74,81,103]. $SC_{chronic}$ and $TC_{chronic}$ - 0.002 and 0.1 mg/m^3 respectively [175]; MPADC - 0.002 mg/cu.m; toxicity classification - II [134].

Cupric phosphate; $Cu_3(PO_4)_2 \cdot 3H_2O$ (434.61)

Bluish green crystals; insoluble in water.

Present in wastes from the production of pesticides, some catalysts, and fertilizers.

Toxicity: has a general toxic and allergenic effect; causes irritation of mucous membranes and the skin; hemolysis of erythrocytes, and cirrhosis of liver and pancreas [81].

Cupric selenate; $CuSeO_4 \cdot 5H_2O$ (296.58)

Light blue crystals; decomposes on heating, ρ 2.56; solubility in water: 26.3 g/100 g H_2O at 20°C.

Present in wastes from the production of metal coatings.

Cupric sulfate; $CuSO_4 \cdot 5H_2O$ (249.68)

Blue crystals, ρ 2.28; soluble in water.

Present in wastes from the production of pesticides, bactericides, fertilizers, copper salts, dyes, leather, inks, petrochemicals, lacquers and paints, photographic materials, and pyrotechnical compositions; also in wastes of electrotechnical, metallurgical and wood-working plants.

Toxicity: causes irritation of mucous membranes and the skin. LD_{50} for rats - 300 mg/kg [74,81,103]; MPSDC and MPADC - 0.003 and 0.001 mg/m^3 respectively; toxicity classification - II [35].

Cupric sulfide (convellite); CuS (95.61)

Black crystals, m.p.>450°C(d.), ρ 4.68; insoluble in water.

Present in wastes from the production of pigments, textiles, and some catalysts.

Toxicity: causes irritation of mucous membranes, the respiratory tracts, and the skin [74,81].

Cuprous bromide; CuBr (143.46)

Colorless crystals, m.p. 488°C, b.p. 1350°C, ρ 4.77; insoluble in water.

Present in wastes from the production of some catalysts.

Toxicity: causes mental disorders and skin irritation [74,81].

Cuprous chloride; CuCl (99.00)

Colorless crystals, m.p. 450°C, b.p. 1212°C, ρ 3.7; slightly soluble in water.

Present in wastes from the production of some catalysts, petrochemicals, soap, cellulose, and fats.

Toxicity: has a general toxic and irritating effect [74,81,103]. MPSDC (recalculated for Cu) - 0.003 mg/m^3; MPADC - 0.001 mg/m^3; toxicity classification - II [35].

Cuprous cyanide; CuCN (89.56)

Colorless crystals, m.p. 473°C; insoluble in water.

Present in wastes from the production of nitriles and pesticides; also in wastes from electrolytic plating of metals.

Toxicity: very toxic; at high concentrations causes respiratory standstill, cramps, paralysis; at low concentrations - headache, nausea, vomiting; in experiments with animals subjected to prolonged exposure - fatigue and asthenia [74,81].

Cuprous oxide; Cu_2O (143.08)

Red crystals, m.p. 1240°C, ρ 6.0; insoluble in water.

Present in wastes in the production of glass, enamel, synthetic silk, copper compounds, ceramics, synthetic precious stones, pyrotechnical compositions, some catalysts, dyes, and pesticides; also in wastes of metal-working and galvanotechnical plants.

Toxicity: has a general toxic effect; causes irritation of mucous membranes and the skin [74,81,103].

Cuprous selenide; Cu_2Se (206.04)

Black crystals, m.p. 1112°C, ρ 6.84.

Present in wastes from the production of semiconductors, glass, photoelectric cells, and solar cells.

Toxicity: causes paleness, dysfunctioning of the nervous system, general depression; irritation of mucous membranes and the skin; and damage to the liver [74, 81].

Cuprous sulfide; Cu_2S (159.15)

Black crystals, m.p. 1130°C, ρ 5.7; insoluble in water.

Present in wastes from the production of luminescent compositions, thermoelectric equipment, copper salts, solid lubricants, and some catalysts.

Toxicity: causes irritation of the eyes, mucous membranes, and the skin; on prolonged exposure - damage to the lungs [74,81]. MPSDC (recalculated for Cu) - 0.003 mg/m^3, MPADC - 0.001 mg/m^3; classification according to toxicity - II [35].

Cuprous sulfite; $Cu_2SO_3 \cdot H_2O$ (225.16)

Colorless crystals; decomposes on heating, ρ 3.83; slightly soluble in water.

Present in wastes from the production of some pesticides, grape wines, and dyes for polyacrylic fibers.

Toxicity: has a general toxic and irritating effect [74,81].

FLUORINE

Fluorine; F_2 (38.00)

Light yellow gas, m.p. -219.699°C, b.p. -188.200°C, ρ 1.70 g/l; reacts with water.

Present in wastes from the production of glass, cement, ceramics, textiles, lacquers and paints, uranium fluorides, sulfur and some hydrocarbons; also in wastes of mining and metallurgical plants, refineries, and chemical and wood-working plants.

Toxicity: very toxic; causes chemical burns, strong irritation of the eyes, mucous membranes, the respiratory tracts and the skin, hemorrhage and pulmonary edema; on prolonged exposure at low concentrations - fluorosis. $TC_{chronic}$ and TC_{odor} - 0.0008 and 0.015 mg/l respectively [74,81,102,103,113]; at concentrations 0.2-0.5 μg/m^3 affects the biosynthesis of plants [103]; MPSDC and MPADC - 0.02 and 0.005 mg/m^3 respectively; toxicity classification - II [34,88]. MPSDC and MPADC established in Bolgaria, GDR, and Romania - 0.02 and 0.005 mg/m^3 respectively; in Hungary and Czechoslovakia - 0.03 and 0.01 mg/m^3 respectively [72,88]; MPC recommended ≦3 t/year [104].

Removal: with the aid of scrubbers [105].

Determination: see [47,222].

Fluorine dioxide; F_2O_2 (70.00)

Brown gas, m.p. -163.5°C, b.p. -57°C(d.), ρ 1.4.

Present in wastes from the production of rocket fuels.

Toxicity: has a general toxic and irritating effect; causes fluorosis.

Fluorine oxide; F_2O (54.00)

Colorless gas, m.p. -224°C, b.p. -145.05°C, ρ 1.5; soluble in water.

Present in wastes from the production of rocket fuels.

Toxicity: has a general toxic and irritating effect; causes fluorosis.

Hydrofluoric acid (hydrogen fluoride); HF (20.1)

Colorless gas or liquid, m.p. -83.36°C, b.p. 19.52°C, ρ 0.98; soluble in water.

Present in wastes from the synthesis of organic compounds, superphosphates, fluorides, aluminum, uranium, beryllium, manganese; some alcohols, aldehydes, esters, pharmaceuticals, and plastics; dyes, glass; and nuclear equipment; also in wastes of refineries and metal-working plants.

Toxicity: very toxic; causes strong irritation and chemical burns of the eyes, mucous membranes, the respiratory tracts, and the skin; hemorrhage and ulceration of the respiratory tracts, pulmonary edema; ptyalism, lacrimation, nasal hemorrhage, purulent bronchitis; dysfunctioning of the heart, pharyngospasm, asphyxia, and cramps [74,81,102,103]. $SC_{chronic}$ and $TC_{chronic}$ - 0.01 and 0.1-3.0 mg/m^3 respectively [223]; MPSDC and MPADC - 0.02 and 0.005 mg/m^3 respectively; toxicity classification - II [34].

Removal: with the aid of filters made of polyethylene, polypropylene, or teflon (efficiency 97-99%) [86]; and scrubbers [57,87,105].

Determination: see [47,224].

GALLIUM

Gallium; Ga (69.72)

Silvery gray metal, m.p. 29.78°C, b.p. 2205°C, ρ 5.9; insoluble in water.

Present in wastes of metallurgical and instrument-making plants; also in wastes from the production of glass and electrotechnical equipment.

Toxicity: causes damage to the liver, kidneys, and bone marrow; and hemorrhagic nephritis [81,102,103].

Gallium arsenide; GaAs (144.53)

Dark gray crystals tinged with purple, m.p. 1238°C, ρ 5.31; insoluble in water.

Present in wastes from the production of photoelectric equipment, semiconductors, lasers, and radio equipment.

Toxicity: causes damage to the nervous system, blood vessels, the heart, the liver, and the kidneys [103].

Gallium oxide; Ga_2O_3 (167.44)

Colorless crystals, m.p. 1725°C, ρ 5.88; insoluble in water.

Present in wastes from the production of optical-grade glass, explosives, and radio equipment.

Toxicity: has a general toxic and irritating effect [74,103].

Gallium phosphide; GaP (100.69)

Yellowish orange crystals, m.p. 1465°C; insoluble in water.

Present in wastes from the production of semiconductors.

Toxicity: has a general toxic and irritating effect [74,81,113].

GERMANIUM

Germanium; Ge (72.59)

Silvery white metal, m.p. 958.5°C, b.p. ≅2850°C, ρ 5.35; insoluble in water.

Present in wastes from the production of semiconductors, instruments, film resistors, radar detectors, and some special alloys.

Toxicity: slightly toxic.

Removal: see [8].

Germanium dibromide (germanium bromide); $GeBr_2$ (232.41)

Colorless crystals, m.p. 122°C; insoluble in water.

Present in wastes from the production of some organic compounds.

Toxicity: has a general toxic and irritating effect [8].

Germanium dioxide (germanium(IV) oxide); GeO_2 (104.59)

Colorless crystals, m.p. 1116°C, ρ 6.23; slightly soluble in water.

Present in wastes from the production of ultra-pure gellium, optical-grade glass, enamels, ceramics, and luminophors.

Toxicity: causes damage to the liver and spleen; irritation of mucous membranes, the respiratory tracts, and the eyes. MPADC - 0.04 mg/m^3; toxicity classification - III [35].

Germanium tetrachloride (germanium chloride); $GeCl_4$ (214.43)

Colorless liquid, m.p. -49.5°C, b.p. 83.1°C, ρ 1.874; decomposes in water.

Present in wastes from the production of glass fibers and semiconductors.

Toxicity: has a general toxic and irritating effect [74,81,103].

Germanium tetrahydride (germanium(IV) hydride); GeH_4 (76.60)

Colorless gas, m.p. -165°C, b.p. -88.5°C, ρ 3.42.

Present in wastes from the production of ultra-pure germanium and semiconductors.

Toxicity: has a general toxic effect. $TSDC_{acute}$ - 0.013-0.07 mg/m^3 [74].

HYDROGEN

Hydrazoic acid (azoimide; hydrogen azide); HN_3 (43.03)

Colorless liquid with a strong odor, m.p. -80°C, b.p. 36°C, ρ 1.13; soluble in water.

Present in wastes from the production of some pharmaceuticals.

Toxicity: has a general toxic effect. $TC_{chronic}$ - 0.00053-0.0068 ml/l [103].

Hydrocyanic acid; HCN (27.03)

Colorless liquid with the odor of the almond kernel, m.p. -13.3°C, b.p. 25.7°C, ρ 0.68; soluble in water.

Present in wastes from the production of hydrocyanic acid, benzene, toluene, xylene, illuminating gas, cyanides, thiocyanates, oxalic acid, rubber, synthetic fibers, plastics, Plexiglas, lactic acid, and pesticides; also in wastes of plants of the by-product coke industry, and from galvanic and gold-, silver- and copper-plating processes.

Toxicity: very toxic; at high concentrations causes palpitation, impairment of respiration, and paralysis; at low concentrations - dizziness, general weakness, impairment of the nervous system, tremors, cramps, and irritation of the respiratory tracts, mucous membranes, and the skin [74,81,103,113]. MPADC - 0.01 mg/m^3; toxicity classification - II [34].

Hydrogen; H_2 (2.014)

Colorless gas, m.p. -259.19°C, b.p. -252.7°C, ρ 0.089 g/l; solubility in water: 1.82 g/100 g H_2O at 20°C.

Hydrogen peroxide; H_2O_2 (34.01)

Colorless liquid, m.p. 0.43°C, b.p. 150.2°C(d.), ρ 1.45; infinitely soluble in water.

Present in wastes from the production of some pharmaceuticals, vat dyes, bleaching agents, surface-active substances, and rocket fuels.

Toxicity: has a toxic effect on the gonads and a mutagenic effect [62]; causes lacrimation and irritation of the respiratory tracts and the skin. LC_{100} for mice - 13.2-19.0 g/m^3; $TC_{chronic}$ - 0.01 mg/l [74, 81,103,113].

Thiocyanic acid (thiocyanate); HSCN (59.06)

Colorless gas, m.p. -110°C; soluble in water.

Present in wastes from the production of analytical reagents, pesticides, textiles, photographic chemicals.

Toxicity: has a general toxic and irritating effect; on prolonged exposure at low concentrations causes dizziness, cramps, damage to the nervous system, and irritation of mucous membranes and the skin [74,81,103,113].

INDIUM

Indium; In (114.82)

Silver-gray metal, m.p. 156.4°C, b.p. 2000°C, ρ 7.3; insoluble in water.

Present in wastes from the production of semiconductors and of radio and electrotechnical equipment.

Toxicity: causes pain in the joints and bones, tooth decay, general weakness, pain in the heart, nervous disorders, changes in the composition of blood; damage to the heart, liver and kidneys [74,81,103].

Indium arsenide; InAs (189.73)

Dark gray crystals with a metallic sheen, m.p. 943°C; insoluble in water.

Present in wastes from the production of semiconductors, research instruments, and radio and electrotechnical equipment.

Toxicity: causes nausea, vomiting, skin rash and pigmentation; degeneration of the liver and kidneys, and impairment of the central nervous system [74,81,103].

Indium bromide; InBr (194.73)

Reddish brown crystals, m.p. 220°C, b.p. 658°C, ρ 4.96; decomposes in water.

Present in wastes from the production of semiconductors and of radio and electrotechnical equipment.

Toxicity: affects the heart, liver, and kidneys; causes irritation of mucous membranes and the respiratory tracts [81].

Indium dibromide; $InBr_2$ (274.64)

Yellowish crystals, m.p. 235°C, b.p. 630°C(d.), ρ 4.22; decomposes in water.

Present in wastes from the production of semiconductors and of radio and electrotechnical equipment.

Toxicity: affects the composition of blood and the heart; causes impairment of the central nervous system, skin eruption, furunculosis, and damage to the liver and kidneys [8].

Indium dichloride; $InCl_2$ (185.73)

Colorless crystals, m.p. 235°C, b.p. 488°C, ρ 3.65; decomposes in water.

Present in wastes from the production of semiconductors and of radio and electrotechnical equipment.

Toxicity: affects the heart, liver, and kidneys; causes changes in the composition of the blood [81].

Indium sesquioxide; In_2O_3 (277.64)

Light yellow crystals, m.p. 1910°C, b.p. ≅3300°C, ρ 7.179; insoluble in water.

Present in wastes from the production of indium, glass, semiconductors, and radio and electrotechnical equipment.

Toxicity: has a general toxic effect [74,103].

Indium sulfide; In_2S_3 (325.83)

Reddish brown crystals, m.p. 1050°C, ρ 4.90; insoluble in water.

Present in wastes from the production of semiconductors.

Toxicity: affects the nervous system and protein metabolism; has an irritating effect [103].

IODINE

Hydrogen iodide (hydroiodic acid); HI (127.91)

Colorless gas, m.p. -50.9°C, b.p. 35.4°C, ρ 5.7; soluble in water.

Present in wastes from the production of iodine compounds, some reagents, and some products of organic synthesis.

Toxicity: very toxic; at high concentrations causes necrosis of the nasal mucous membrane, nebula, spasm and pulmonary edema; at low concentrations - irritation of the nasal mucous membrane, ptyalism, and lacrimation [74,81,103,113].

Iodic acid; HIO_3 (175.91)

Colorless crystals, m.p. 110°C, b.p. 195°C(d.), ρ 4.62; soluble in water.

Present in wastes from the production of some pharmaceuticals.

Toxicity: very toxic; has a general irritating effect [74,81,103,113].

Iodine; I_2 (253.81)

Purplish black crystals with a metallic sheen, m.p. 113.6°C, b.p. 184.35°C, ρ 4.94; slightly soluble in water.

Present in wastes from the synthesis of some organic and inorganic compounds, and from the production of dyes and some pharmaceuticals.

Toxicity: causes caughing, rhinitis, lacrimation, conjunctivitis, headache, murmur in the ears, and dizziness; has a general irritating effect [74,81]; MPADC - 0.03 mg/m^3; toxicity classification - II [34].

Iodine azide; IN_3 (168.93)

Yellow crystals; explodes when heated.

Present in wastes from the production of explosives.

Toxicity: very toxic; has a general toxic and irritating effect [81].

Iodine dioxide; IO_2 (158.0)

White powder; decomposes at >75°C, ρ 4.2.

Present in wastes from the production of some pharmaceuticals.

Toxicity: has an irritating effect [74,81].

Iodine heptafluoride; IF_7 (259.91)

Colorless liquid, m.p. 6.4°C, ρ 2.8.

Present in wastes from the production of explosives and inorganic fluorides.

Toxicity: very toxic; has a general toxic and irritating effect [74,81].

Iodine monochloride; ICl (162.36)

Dark red crystals, m.p. 27.2°C, b.p. 98°C(d.), ρ 3.18.

Present in wastes from the synthesis of some organic compounds.

Toxicity: very toxic; has a general toxic and irritating effect [8,74,113].

Iodine pentabromide; IBr_5 (526.5)

Colorless crystals; decomposes on heating.

Present in wastes from the synthesis of some organic compounds.

Toxicity: very toxic; has a general toxic and irritating effect [81,113].

Iodine pentafluoride; IF_5 (221.90)

Colorless liquid, m.p. 9.421°C, b.p. 104.48°C, ρ 3.21; decomposes in water.

Present in wastes from the production of fluorides.

Toxicity: has a general toxic and irritating effect [74,81,113].

Iodine pentaoxide; I_2O_5 (333.84)

Colorless crystals, m.p. 300°C(d.), ρ 4.79; soluble in water.

Present in wastes of some chemical plants.

Toxicity: very toxic; has a general toxic and irritating effect [74,81,113].

Iodine trichloride; ICl_3 (233.26)

Yellow crystals, m.p. 101°C (at 16 kPa), b.p. 64°C(d.), ρ 3.12.

Present in wastes from the production of some organic chemicals.

Toxicity: very toxic; has a general toxic and irritating effect [74,81,113].

IRON

Ferric arsenate; $Fe_3(AsO_4)_2 \cdot 6H_2O$ (553.47)

Greenish brown powder; decomposes on heating; insoluble in water.

Present in wastes from the production of some pharmaceuticals.

Toxicity: very toxic; affects the nervous system and the liver; at high concentrations causes collapse and death; has an allergenic and carcinogenic effect [74, 81,103,113].

Ferric chloride (molysite); $FeCl_3$ (162.22)

Reddish brown crystals, m.p. 309°C, b.p. 316°C, ρ 2.90; soluble in water.

Present in wastes from the production of photographic materials, iron salts, dyes, inks, catalysts, some organic chemicals, and textiles.

Toxicity: has a general toxic and irritating effect [74,81,102]. MPADC - 0.004 mg/m^3; toxicity classification - II [34].

Removal: by filtering [57].

Ferric fluoride; FeF_3 (112.85)

Greenish crystals, m.p. 1027°C, b. p. 1327°C, ρ 3.87; slightly soluble in water.

Present in wastes from the production of catalysts, ceramics, and cement.

Toxicity: causes changes in bone tissues, fluorosis; has an irritating effect [74,81].

Ferric nitrate; $Fe(NO_3)_3 \cdot 6H_2O$ (349.95)

Colorless crystals, m.p. 47.2°C, b.p. 125°C, ρ 1.68; soluble in water.

Present in wastes from the production of dyes, silk, leather, and analytical reagents.

Toxicity: has a general toxic and irritating effect [74,81].

Ferrous iodide; FeI_2 (309.66)

Reddish brown crystals, m.p. 594°C, b.p. 935°C, ρ 5.315; very slightly soluble in water.

Present in wastes from the production of some organic chemicals.

Toxicity: has an irritating effect [74,81,113].

Ferrous oxide; FeO (71.85)

Black crystals, m.p. ≅1360°C, ρ 5.7; insoluble in water.

Present in wastes from the production of glass, enamels, and catalysts; also in wastes of metallurgical plants.

Toxicity: has a general toxic effect [74,113]. MPADC - 0.04 mg/m^3; classification - III [34].

Iron; Fe (55.85)

Silvery gray metal, m.p. 1539°C, b.p. 2870°C, ρ 7.87; insoluble in water.

Present in wastes of metallurgical plants.

Toxicity: causes retardation of growth and irritation of mucous membranes and the eyes; affects the lungs; has a carcinogenic effect [69,93,103,239].

Removal: by scrubbing [76].

Determination: see [99].

KRYPTON

Krypton; Kr (83.80)

Colorless gas, m.p. -157.37°C, b.p. -153.22°C, ρ 3.708 g/l; solubility in water: 6 ml/100 ml H_2O at 25°C.

Present in wastes from the production of electrotechnical equipment.

Toxicity: causes flaccidity and asphyxia [74,81,103,113].

LANTHANUM

Lanthanum; La (138.91)

Silvery gray metal, m.p. 920°C, b.p. 3470°C, ρ 6.16.

Present in wastes from the production of glass, ceramics, some chemicals, and textiles; also in wastes of metallurgical plants.

Toxicity: causes headache and irritation of the eyes and respiratory tracts [81,102].

LEAD

Lead; Pb (207.2)

Bluish gray metal, m.p. 237.4°C, b.p. 1745°C, ρ 11.34; insoluble in water.

Present in wastes from the manufacture of corrosion-resistant pipes, storage batteries, cables, and chemical equipment; and in wastes from the production of purified petroleum, lead compounds, tetramethyl lead, alloys, lead paints, and ceramics; also in wastes of some metallurgical plants.

Toxicity: very toxic; has a carcinogenic, mutagenic, and teratogenic effect [51, 103,196,197], also a cumulative effect [198]; causes dysfunctioning of the central and peripheral nervous systems (polyneuritis, paralysis), damage to the composition of the blood and vessels; a metallic taste in the mouth, ptyalism, vomiting, headache, dizziness, and "optical neurosis" [102]; affects the gonads; increases intracranial pressure [74]. $SC_{chronic}$ and $TC_{irritation}$ - 0.001-0.003 and 0.0039 mg/m^3 respectively [5]; $TC_{c.n.s.}$ and $TC_{chronic}$ - 0.001-0.04 and -0.010-0.011 mg/m^3 respectively; LC_{100} - 271-795 mg/m^3 [103]; MPADC - 0.0003 mg/m^3; toxicity classification - I [34]; MPADC established in Bulgaria, GDR, Czechoslovakia, and Yugoslavia - 0.0007 mg/m^3, in Hungary, Poland, and Romania - 0.001 mg/m^3 [72,88]. In the presence of other pollutants the MPADC recommended - $\leqq$ 0.0001 mg/m^3 [104]. MPC recommended - $\leqq$ 0.6 t/year [104].

Removal: by filtration [1,8,45,199].

Determination: colorimetry (sensitivity 2 μg); polarography (sensitivity under alkaline and acidic conditions - 5 and 2 μg respectively) [47]; absorption spectroscopy (sensitivity 0.01 $\mu g/m^3$) [18,44,99,200].

Lead antimonate; $Pb_2Sb_2O_7$ (769.87)

Orange-yellow crystals, m.p. >600°C(d.), ρ 6.72; insoluble in water.

Present in wastes from the production of pigments, dyes, glass, porcelain, and ceramics.

Toxicity: very toxic; causes irritation of the respiratory tracts, mucous membranes, and the skin [74,81].

Lead arsenate; $PbHAsO_4$ (347.13)

Colorless crystals, ρ 5.79; insoluble in water.

Present in wastes from the production of special paints (for ships) and pesticides.

Toxicity: very toxic; has a general irritating effect; causes polyneuritis, degeneration of the liver and kidneys; and dysfunctioning of the central nervous system [31,74].

Lead arsenite; $Pb(AsO_2)_2$ (421.03)

Colorless crystals, ρ 5.85; insoluble in water. Present in wastes from the production of pesticides.

Toxicity: very toxic; has an overall toxic effect [74,81].

Lead azide (azoimide); $Pb(N_3)_2$ (291.26)

Colorless crystals, ρ 4.71-4.93; slightly soluble in water.

Present in wastes from the production of explosives.

Toxicity: very toxic; supresses the action of ferments [74,81].

Lead carbonate (cerusite); $PbCO_3$ (267.20)

Colorless crystals, m.p. >300°C(d.), ρ 6.56; slightly soluble in water.

Present in wastes from the production of white lead.

Toxicity: very toxic [103].

Lead chloride (cotunnite); $PbCl_2$ (278.10)

Colorless crystals, m.p. 501°C, b.p. 953°C, ρ 5.85; solubility in water: 0.98 g/100 g H_2O at 20°C.

Present in wastes from the production of lead pigments and luminophors; also in wastes of some metallurgical plants.

Toxicity: very toxic [74,81,104].

Lead chromate (crocoite); $PbCrO_4$ (323.18)

Yellow crystals, m.p. 844°C, ρ 6.12; insoluble in water.

Present in wastes from the production of dyes, oxidizers, porcelain, and reagents used in organic synthesis.

Toxicity: very toxic; causes irritation of the respiratory tracts, mucous membranes, and the skin. LD_{50} for guinea pigs - 400 mg/kg [74,81,103].

Lead cyanide; $Pb(CN)_2$ (259.25)

Colorless crystals, soluble in water.

Present in wastes from the production of insecticides; also in wastes from galvanic processes.

Toxicity: very toxic; has an irritating effect; on prolonged exposure at low concentrations causes loss of appetite, headache, weakness and debility, vomiting, dizziness, and irritation of the upper respiratory tracts and the eyes [81].

Lead dichromate; $PbCr_2O_7$ (423.18)

Red crystals; reacts with water.

Present in wastes of some chemical plants.

Toxicity: very toxic; causes irritation of the respiratory tracts, mucous membranes, and the skin; ulceration of the skin and eczema; and perforation of the nasal septum [81,102,103,113].

Lead dioxide (plattnerite); PbO_2 (239.21)

Dark brown crystals, d.p. 220-280°C, ρ 9.33-9.67; insoluble in water.

Present in wastes from the production of some plastics, pharmaceuticals, and reagents; lead compounds, lacquers and paints, glass, cement, articles made of rubber, storage batteries, electrodes, enamels, polygraphic compositions, and matches; also in wastes of some metallurgical plants.

Toxicity: very toxic. $SC_{chronic}$ and $TC_{chronic}$ - 0.00027 and 0.010 mg/m^3 respectively [201]; MPADC - 0.0003 mg/m^3; toxicity classification - I [34].

Lead fluoride; PbF_2 (245.19)

Colorless crystals, m.p. 822°C, b.p. ≅1290°C, ρ 8.24; slightly soluble in water.

Present in wastes from the production of pesticides, some catalysts, cathodes, ceramics, and enamels; also in wastes of metallurgical plants.

Toxicity: very toxic; has an overall toxic and irritating effect; on prolonged exposure at low concentrations causes fluorosis [74,81,103].

Lead iodide; PbI_2 (461.0)

Golden yellow crystals, m.p. 412°C, b.p. 954°C, ρ 6.16; slightly soluble in water.

Present in wastes from the production of some reagents, pigments, photographic materials; also in wastes of metal-working plants.

Toxicity: very toxic; has an overall toxic and irritating effect [74,81,103].

Lead metaborate; $Pb(BO_2)_2 \cdot H_2O$ (310.82)

Colorless crystals, m.p. 600°c(d.), ρ 5.59; insoluble in water.

Present in wastes from the production of paints, glass, ceramics, and porcelain.

Toxicity: very toxic; has an overall toxic effect [74,81,103].

Lead monoxide; PbO (223.19)

Litharge-red crystals, m.p. 886°C, b.p. 1535°C, ρ 9.51; insoluble in water.

Massicot-yellow crystals, m.p. 600°C, ρ 8.70; insoluble in water.

Present in wastes from the production of some plastics, pharmaceuticals, and reagents, lead compounds, lacquers and paints, glass, cement, articles made of rubber, sorage batteries, electrodes, enamels, polygraphic compositions, and matches; also in wastes of metallurgical plants.

Toxicity: very toxic [74,103].

Lead nitrate; $Pb(NO_3)_2$ (331.20)

Colorless crystals, m.p. >200°C(d.), ρ 4.53; solubility in water: 52.2 g/100 g H_2O at 20°C.

Present in wastes from the production of matches, explosives, pigments, textiles, and photographic materials.

Toxicity: has a general toxic and irritating effect [74,81,103].

Lead oxide (minum); Pb_3O_4 (685.6)

Red crystals, m.p. 830°C, ρ 8.79; insoluble in water.

Present in wastes from the production of some plastics, pharmaceuticals, and reagents, lead compounds, lacquers and paints, glass, cement, articles made of rubber, storage batteries, electrodes, enamels, polygraphic compositions, and matches; also in wastes of some metallurgical plants.

Toxicity: very toxic [103].

Lead selenide (clausthalite); PbSe (286.17)

Gray crystals, m.p. 1065°C, ρ 8.1.

Present in wastes from the production of semiconductors, infrared radiators, photoresistors, and lasers.

Toxicity: has a general toxic and irritating effect; causes pneumonia, fatty degeneration of the liver, and degeneration of the kidneys [81].

Lead silicate; $PbSiO_3$ (283.27)

Colorless crystals, m.p. 1770°C(d.), ρ 6.35; very slightly soluble in water.

Present in wastes from the production of sulfuric acid, lacquers and paints, metallic lead, pigments, galvanic dry cells, and lithographic materials.

Toxicity: has a general toxic and irritating effect. LD_{50} for guinea pigs - 300 mg/kg [74,81,103].

Lead sulfide (galena); PbS (239.25)

Grayish black crystals, m.p. 1114°C, b.p. 1281°C(subl.), ρ 7.59; insoluble in water.

Present in wastes of metallurgical plants and from the production of ceramics.

Toxicity: has a general toxic and irritating effect. $SC_{chronic}$ and $TC_{chronic}$ - 0.0135 and 0.0483 mg/m^3 respepectively [202]; MPADC - 0.0007 mg/m^3; toxicity classification - I [34]; MPADC established in GDR and Yugoslavia - 0.0017 mg/m^3, and in Bulgaria - 0.0007 mg/m^3 [72,88].

Lead telluride; PbTl (334.82)

Dark red crystals, m.p. 860°C, ρ 8.16.

Present in wastes from the production of photographic materials, semiconductors, and infrared optical equipment.

Toxicity: has a general toxic and irritating effect; is absorbed by healthy skin; causes loss of appetite, hidrosis, and ptyalism [74,81].

Lead tetrafluoride; PbF_4 (283.21)

Colorless crystals, m.p. 600°C, ρ 6.7.

Present in wastes from processes involving fluorination of hydrocarbons.

Toxicity: has a general toxic and irritating effect; on prolonged exposure at low concentrations causes fluorosis [74].

Lead thiocyanate; $Pb(CNS)_2$ (231.25)

Colorless crystals, m.p. 190°C(d.), ρ 3.82; slightly soluble in water.

Present in wastes from the production of dyes, matches, and explosives.

Toxicity: has a general toxic and irritating effect; causes impairment of the nervous system; on prolonged exposure causes damage to skin, nasal mucous membrane, dizziness, and cramps [74,81,103].

Lead titanate; $PbTiO_3$ (303.11)

Yellow crystals, m.p. ≅1290°C, ρ 7.52; insoluble in water.

Present in wastes from the production of pigments, dielectrics, and piezoelectric devices.

Toxicity: very toxic [81].

LITHIUM

Lithium; Li (6.94)

Silvery white metal, m.p. 180.5°C, b.p. 1336.6°C, ρ 0.53; reacts with water.

Present in wastes from the production of some metals and of pharmaceutical plants.

Toxicity: at high concentrations causes general weakness and debility, loss of appetite, thirst, xerostomia, and opacification of the cornea; on prolonged exposure at low concentrations - has a general irritating effect; affects the central nervous system and the kidneys [74,81,103].

Lithium bromide; LiBr (86.85)

Cololess crystals, m.p. 552°C, b.p. 1310°C, ρ 3.46; solubility in water: 160 g/100 g H_2O at 20°C.

Present in wastes from the production of air conditioners.

Toxicity: at high concentrations suppresses the activity of the central nervous system; on prolonged exposure at low concentrations has a general irritating effect [74,81].

Lithium carbonate; Li_2CO_3 (73.89)

Colorless crystals, m.p. 735°C(d.), ρ 2.11; slightly soluble in water.

Present in wastes from the production of lithium compounds, ceramics, glass, some catalysts and pharmaceuticals.

Toxicity: has a general irritating effect; is absorbed by healthy skin; causes damage to the kidneys and the central nervous system. LD_{50} for mice - 53 g/kg [74,81,103].

Lithium chloride; LiCl (42.39)

Colorless crystals, m.p. 610°C, b.p. 1380°C, ρ 2.07; solubility in water: 83.2 g/100 g H_2O at 20°C.

Present in wastes from the production of metallic lithium, air conditioners, photographic chemicals, dry batteries, fusing agents, mineral waters, explosives, aluminium, and refrigerators.

Toxicity: causes impairment of the central nervous system, damage to the kidneys, and irritation of the skin. LD_{50} for mice - 1.17 g/kg [74,81,103].

Removal: by filtration [57].

Lithlum chromate; $Li_2CrO_4 \cdot 2H_2O$ (165.90)

Orange-yellow crystals, m.p. 150°C(d.); solubility in water: 141 g/100 g H_2O at 18°C.

Present in wastes from the production of corrosion inhibitors for atomic reactors.

Toxicity: causes irritation of the respiratory tracts, the eyes, and mucous membranes, and perforation of the nasal septum [74,81].

Lithium fluoride; LiF (25.94)

Colorless crystals, m.p. 848.1°C, b.p. 1676°C, ρ 2.63; slightly soluble in water.

Present in wastes from the production of ceramics, aluminum, glass, and enamels; also in wastes of metallurgical plants.

Toxicity: has a general toxic and irritating effect; on prolonged exposure at low concentrations causes fluorosis [74,81].

Lithium hydride; LiH (7.95)

Colorless crystals, m.p. 688°C, b.p. 850°C(d.), ρ 0.82; reacts with water.

Present in wastes from the production of boron hydrides and some organic compounds.

Toxicity: has a general toxic and irritating effect. $TSDC_{acute}$ - 0.2-0.5 mg/m^3 [74,81,103].

Lithium hydroxide; LiOH (23.95)

Colorless crystals, m.p. 471°C, b.p. 925°C(d.), ρ 1.46; soluble in water.

Present in wastes from the production of storage batteries and lubricants.

Toxicity: has a general toxic and irritating effect. $TC_{irritation}$ and TSDC (acute) - 0.01 mg/l and 0.49 mg/m^3 respectively [74,81,103].

Lithium iodide; $LiI \cdot 3H_2O$ (187.89)

Colorless crystals, m.p. 73°C, b.p. 300°C(d.), ρ 3.48; soluble in water.

Present in wastes from the production of photographic materials.

Toxicity: has an irritating effect [74,81,103,113].

Lithium metasilicate; Li_2SiO_3 (89.96)

Colorless crystals, m.p. 1202°C, ρ 2.52; slightly soluble in water.

Present in wastes from the production of calibration thermoelements and ceramics.

Toxicity: causes pneumosclerosis and irritation of the respiratory tracts [74,81].

Lithium oxide; Li_2O (29.88)

Colorless crystals, m.p. 1570°C, b.p. ≅2600°C, ρ 2.01.

Present in wastes of metallurgical plants; also in wastes from the production of plastics and enamels.

Toxicity: causes dizziness, prostration, and damage to the kidneys [74,103,113].

Lithium perchlorate; $LiClO_4$ (106.39)

Colorless crystals, m.p. 236°C, b.p. 400°C(d.); solubility in water: 37.5 g/100 g H_2O at 25°C.

Present in wastes from the production of oxidizers.

Toxicity: causes irritation of the skin and mucous membranes [74,81,113].

Lithium tetraborate; $Li_2B_4O_7 \cdot 5H_2O$ (259.19)

Colorless crystals, m.p. 200°C(d.); soluble in water.

Present in wastes from the production of enamel.

Toxicity: has a general toxic and irritating effect, and also a cumulative effect [74,81].

MAGNESIUM

Magnesium; Mg (24.31)

Silvery white metal, m.p. 650°C, b.p. 1095°C, ρ 1.74; insoluble in water.

Present in wastes from the production of optical-grade glass, pyrotechnical compositions, and instruments; also in wastes of some metallurgical plants.

Toxicity: on prolonged exposure causes chronic atrophic inflammation of the mucous membrane of the nose and throat, nosebleed, and rhinitis; shedding of hair, hidrosis, and cyanosis, tremor of the hands, tongue and eyelids; and skin irritation. $TC_{chronic}$ - 6.7 mg/m^3 [74,81,103].

Magnesium chlorate; $Mg(ClO_3)_2 \cdot 6H_2O$ (299.30)

Colorless crystals, m.p. 35°C(d.), ρ 1.80; solubility in water: 86.7 g/100 g H_2O at 20°C.

Present in wastes from the production of defoliants and some pesticides.

Toxicity: has a general toxic and irritating effect [74,81]. MPADC - 0.3 mg/m^3; toxicity classification - IV [34].

Magnesium dibromide; $MgBr_2 \cdot 6H_2O$ (292.20)

Colorless crystals, m.p. 172.4°C; soluble in water.

Present in wastes from the production of some organic compounds and pharmaceuticals.

Toxicity: on prolonged exposure causes mental disorders and skin irritation [74, 81].

Magnesium dichloride; $MgCl_2$ (95.23)

Colorless crystals, m.p. 707°C, b.p. 1412°C, ρ 2.32; solubility in water: 54.8

g/100 g H_2O at 20°C.

Present in wastes from the production of wood preservatives, desinfectants, magnesium cement, parchmentized paper, artificial leather, casein glue, and some analytical reagents.

Toxicity: same as that of Mg.

Removal: with the aid of filters made of nylon, polyesters, polyethylene, poly(vinyl chloride), teflon, or polypropylene [57].

Magnesium difluoride; MgF_2 (62.32)

Colorless crystals, m.p. 1263°C, b.p. ≅2250°C, ρ 3.13; slightly soluble in water.

Present in wastes from the production of ceramics, glass, and leather.

Toxicity: causes severe burns of the eyes, mucous membranes, and the skin; at low concentrations causes fluorosis [74,81]. LD_{50} for guinea pigs - 1 g/kg [74].

Magnesium hexafluosilicate; $Mg[SiF_6]\cdot 6H_2O$ (274.51)

Colorless crystals, ρ 1.78; soluble in water.

Present in wastes from the production of insulation materials, magnesium cement, wood preservatives, ceramics, and textiles.

Toxicity: has a general toxic and irritating effect; on prolonged exposure at low concentrations causes mottled enamel, bone sclerosis, and calcification of ligaments [74].

Magnesium metaborate; $Mg(BO_2)_2\cdot 8H_2O$ (254.09)

Colorless crystals, m.p. 988°C(d.), ρ 2.9; slightly soluble in water.

Present in wastes from the production of desinfectants, pesticides, micro-fertilizers, lacquers and paints.

Toxicity: has a general toxic and irritating effect [74,81,103].

Magnesium nitrate; $Mg(NO_3)_2\cdot 6H_2O$ (256.43)

Colorless powder, m.p. 89.9°C, ρ 1.46; soluble in water.

Present in wastes from the production of pyrotechnical compositions.

Toxicity: has a general toxic and irritating effect; at high concentrations causes dizziness, cramps, collapse; at low concentrations - asthenia, general depression, mental disorders [74,81].

Magnesium oxide (magnesia); MgO (40.31)

Colorless crystals, m.p. ≅2825°C, b.p. 3600°C, ρ 3.58; slightly soluble in water: 0.0086 g/100 g H_2O at 30°C.

Present in wastes from the production of refractory materials, magnesium cements, and casein glue.

Toxicity: causes glandular fever and pneumosclerosis; and irritation of mucous membranes and the respiratory tracts [74,81,103]. MPSDC and MPADC - 0.4 and 0.05 mg/m^3 respectively; toxicity classification - III [34]; MPADC recommended - 0.05 mg/m^3 [174].

Magnesium sulfate; $MgSO_4$ (120.38)

Colorless crystals, m.p.≅1140°C(d.), ρ 2.66; soluble in water.

Present in wastes from the production of leather, fire-resistant fabrics, paper, some pharmaceuticals, fertilizers, explosives, matches, and dyes.

Toxicity: causes zinc-fume fever, chronic inflammatory atrophy of mucous membranes, and nosebleed [74,81,103].

Manganese

Manganese; Mn (54.94)

Silvery white metal, m.p. 1245°C, b.p. ≅2080°C, ρ 7.44; insoluble in water.

Present in wastes of metallurgical and machine-building works.

Toxicity: has a general toxic, irritating, carcinogenic, and mutagenic effect; causes asthenia, sleepiness, mental disorders, changes in the gait, paralysis, and symptoms of Parkinson's disease [74,81,103]. $TC_{chronic}$ - 0.033 mg/m^3 [172]; MPADC - 0.01 mg/m^3; toxicity classification - II [34]; MPADC established in GDR, Bulgaria, Czechoslovakia, Yugoslavia, and Romania - 0.01 mg/m^3 [72]; MPC established in Czechoslovakia - 0.1 kg/hr [72,87].

Determination: colorimetric analysis [47], absorption spectroscopy (sensitivity 0.001 μg/m^3); and emission spectroscopy [99,115].

Manganese borate $MnB_4O_7 \cdot 8H_2O$ (354.34)

Powder; insoluble in water.

Present in wastes from the production of lacquers and paints, oils, and leather.

Toxicity: affects the central nervous system; is absorbed by healthy skin [81, 113].

Manganese bromide; $MnBr_2 \cdot 4H_2O$ (380.9)

Pink crystals; decomposes on heating, ρ 4.38; soluble in water.

Present in wastes from the production of some organic compounds.

Toxicity: very toxic; affects the central nervous system; causes symptoms of Parkinson's disease [74,81].

Manganese carbonate (rhodochrosite); $MnCO_3$ (114.95)

Pink crystals, m.p. >300°C(d.), ρ 3.12; slightly soluble in water.

Present in wastes from the production of pigments and desiccants.

Toxicity: has a general toxic effect [74,81].

Manganese chloride (scacchite); $MnCl_2$ (125.84)

Pink crystals, m.p. 650°C. b.p. 1238°C, ρ 2.97; solubility in water: 72.3 g/100 g H_2O at 25°C.

Present in wastes from the production of manganese salts, dyes, desinfectants, and electric batteries.

Toxicity: same as for Mn.

Manganese difluoride; MnF_2 (92.93)

Pink crystals, m.p. 930°C, ρ 3.98; slightly soluble in water.

Present in wastes from the production of fluorides.

Toxicity: very toxic; causes Parkinson's disease and fluorosis; affects the central nervous system [81].

Manganese dioxide (pyrolusite; polianite); MnO_2 (86.94)

Black or brownish black crystals, m.p. >535°C(d.), ρ 5.02; insoluble in water [74,81,103,113].

Present in wastes from the production of manganese, manganese steel, galvanic batteries, pigments, glass, and some catalysts; also in wastes of metallurgical works.

Toxicity: very toxic; causes Parkinson's disease; affects the central nervous system. MPSDC recommended and MPADC recommended - 0.001 and 0.0005 mg/m^3 respectively [174].

Manganese heptaoxide; Mn_2O_7 (221.87)

Dark red liquid, m.p. 5.9°C, ρ 2.4; soluble in water [81].

Present in wastes from the production of manganese, manganese steel, galvanic batteries, pigments, glass, and some catalysts; also in wastes of metallurgical works.

Toxicity: very toxic; causes Parkinson's disease; affects the central nervous system.

Manganese hydroxide (manganous hydroxide; pyrochroite); $Mn(OH)_2$ (88.95)

Light pink crystals;decomposes on heating, ρ 3.25;slightly soluble in water.

Present in wastes from the production of manganese oxide.

Toxicity: very toxic; affects the central nervous system [81].

Manganese oxide (hausmannite); Mn_3O_4 (228.81)

Brownish black crystals, m.p. ≅1560°C, ρ 4.72; insoluble in water [74,81,103].

Present in wastes from the production of manganese, manganese steel, galvanic batteries, pigments, glass, and some catalysts; also in wastes of metallurgical works.

Toxicity: very toxic; causes Parkinson's disease; affects the central nervous system.

Manganese silicate (hermannite; rhodonite); $MnSiO_3$ (131.2)

Red crystals, m.p. 1291°C, ρ 3.72; insoluble in water.

Present in wastes from the production of dyes, special-grade glass, and earthenware.

Toxicity: affects the central nervous system; causes pneumosclerosis [81,113].

Manganic nitrate; $Mn(NO_3)_3 \cdot 4H_2O$ (251.01)

Pinkish crystals, m.p. 25.8°C, ρ 1.82; soluble in water.

Present in wastes from the production of manganese dioxide, some catalysts, and micro-fertilizers.

Toxicity: has a general toxic effect [74,81].

Manganic oxide; Mn_2O_3 (157.87)

Brownish black crystals, m.p. >750°C(d.), ρ 4.5; insoluble in water [74,81,103].

Present in wastes from the production of manganese, manganese steel, galvanic batteries, pigments, glass, and some catalysts; also in wastes of metallurgical works.

Toxicity: very toxic; causes Parkinson's disease; impairs the nervous system.

Manganous nitrate; $Mn(NO_3)_2 \cdot 6H_2O$ (287.04)

Pink crystals, m.p. 25.3°C, b.p. 129.4°C, ρ 1.82; solubility in water: 132.3 g/100 g H_2O at 20°C.

Present in wastes from the production of manganese dioxide, some catalysts, and micro-fertilizers.

Toxicity: very toxic; at high concentrations has a general toxic effect; on prolonged exposure at low concentrations causes impairment of speech and gait, a monotonous voice, general depression, headache, and dizziness [81,113].

Manganous oxide (manganosite); MnO (70.94)

Grayish green crystals, m.p. 1842°C, ρ 5.18; insoluble in water [81,103,113].

Present in wastes from the production of manganese, manganese steel, galvanic batteries, pigments, glass, and some catalysts; and in wastes of metallurgical works.

Toxicity: very toxic; causes Parkinson's disease; affects the central nervous system.

Manganous sulfate; $MnSO_4$ (151.00)

Colorless crystals, m.p. 700°C, ρ 3.25; solubility in water: 62.9 g/100 g H_2O

at 20°C.

Present in wastes from the production of textiles, porcelain, lacquers and paints, and some fertilizers.

Toxicity: same as for Mn.

Manganous sulfide; MnS (87.00)

Green crystals, m.p. 1615°C, ρ 3.99; insoluble in water.

Present in wastes from the production of some chemicals.

Toxicity: very toxic; affects the central nervous system; causes Parkinson's disease [74,81].

MERCURY

Mercuric arsenate; $Hg_3(AsO_4)_2$ (879.61)

Yellow crystals, m.p. >300°C(d.); insoluble in water.

Present in wastes from the production of special paints.

Toxicity: very toxic; has a general toxic effect [74,81,102].

Mercuric bromide; $HgBr_2$ (360.44)

Colorless crystals, m.p. 238.1°C, b.p. 319°C, ρ 6.11; soluble in water.

Present in wastes from the production of some organic compounds; also in wastes from the manufacture of cathodes for converters.

Toxicity: very toxic [74,81].

Mercuric chloride; $HgCl_2$ (271.50)

Colorless crystals, m.p. 280°C, b.p. 302°C, ρ 5.44; soluble in water.

Present in wastes from the production of explosives.

Toxicity: causes irritation of mucous membranes, conjunctivitis, rhinitis, and catarrh of the upper respiratory tracts; dermititis and itching; edema around the eyes, at the back of the head, around the ears, and of the forearms; and small skin ulcers [103].

Mercuric cyanide; $Hg(CN)_2$ (252.63)

Colorless crystals, m.p. 320°C(d.), ρ 3.99; soluble in water.

Present in wastes from the production of dicyanogen, photographic materials, antiseptic soap, and some pesticides.

Toxicity: very toxic [74,81,103].

Mercuric fluoride; HgF_2 (238.59)

Colorless crystals, m.p. 645°C(d.), ρ 8.9; reacts with water.

Present in wastes from the production of mercury compounds and electrodes.

Toxicity: very toxic; has a general irritating effect; on prolonged exposure at low concentrations causes fluorosis [74,81].

Mercuric fulminate; $Hg(CNO)_2$ (284.62)

Colorless crystals; soluble in water.

Present in wastes from the production of explosives.

Toxicity: causes irritation of mucous membranes, conjunctivitis, rhinitis, and catarrh of the upper respiratory tracts; dermititis, and itching; edema around the eyes, at the back of the head, around the ears, and of the forearms; small skin ulcers [103].

Mercuric iodide; HgI_2 (454.40)

Red or yellow crystals, m.p. 256°C, b.p. 354°C, ρ 6.36; slightly soluble in water.

Present in wastes from the production of some chemical reagents and pharmaceuticals, and cathodes for converters.

Toxicity: very toxic [81].

Mercuric nitrate; $Hg(NO_3)_2 \cdot 0.5H_2O$ (331.61)

Colorless crystals, m.p. 145°C, ρ 4.3; soluble in water [81,103].

Present in wastes from the production of some organic compounds and pharmaceuticals; porcelain, glazes, and pyrotechnical compositions; also in wastes of metal-working plants.

Toxicity: very toxic; has a general toxic and irritating effect; on prolonged exposure at low concentrations causes weakness, general depression, headache, and mental disorders [74,81,103]. $TSDC_{acute}$ - 0.25 mg/m^3 [192].

Mercuric oxide (montroydite); HgO (216.61)

Red or yellow crystals, m.p. >400°C(d.), ρ 11.3-11.8; slightly soluble in water.

Present in wastes from the production of special paints (for ships), paints for chainaware, galvanic dry cells, some analytical reagents and organic compounds, and disinfectants.

Toxicity: very toxic [74,81].

Mercuric oxycyanide; $Hg(CN)_2 \cdot HgO$ (469.26)

Colorless crystals, m.p. >320°C(d.), ρ 4.44; slightly soluble in water.

Present in wastes from the production of disinfectants and sergical instruments.

Toxicity: very toxic; has a general toxic and irritating effect [74,81].

Mercuric selenide; HgSe (279.57)

Black crystals, m.p. 799°C, ρ 7.1-8.9; insoluble in water.

Present in wastes from the production of semiconductors, photoresistors, and transducers for measuring magnetic fields.

Toxicity: very toxic [81].

Mercuric sulfate; $HgSO_4$ (296.65)

Colorless crystals, m.p. 550°C(d.), ρ 6.47; reacts with water.

Present in wastes from the production of wines (for achieving the desired coloration); some organic catalysts and analytical reagents; and galvanic dry cells; also in wastes of some metallurgical works (during the extraction of noble metals).

Toxicity: very toxic [74,81].

Mercuric sulfide; HgS (232.68)

Red crystals, m.p. 825°C, ρ 8.10; insoluble in water.

Present in wastes from the production of pigments for plastics, wax, dyes, articles made of rubber; phosphors, some catalysts, and mercury compounds.

Toxicity: very toxic [74,81,103].

Mercuric thiocyanate; $Hg(SCN)_2$ (316.75)

Colorless crystals, m.p. >165°C(d.); slightly soluble in water.

Present in wastes from the production of photographic materials and some analytical reagents.

Toxicity: very toxic [102,103].

Mercurous chloride (calomel); Hg_2Cl_2 (472.09)

Colorless crystals, m.p. 525°C, ρ 7.15; slightly soluble in water.

Present in wastes from the production of electrodes, some pharmaceuticals, storage batteries, special paints (for ships), leather, photographic materials, lithographic materials, pyrotechnical compositions, pesticides, some reagents, galvanic dry cells, and porcelain; also in wastes of wood- and metal-working and metallurgical plants, and from electrolytic plating of metals.

Toxicity: very toxic; has a general toxic and irritating effect [74,81,103].

Mercurous fluoride; Hg_2F_2 (439.22)

Yellow crystals, m.p. 570°C, ρ 8.7; decomposes in water.

Present in wastes from the production of mercury compounds and electrodes.

Toxicity: very toxic; has a general toxic and irritating effect; on prolonged exposure at low concentrations causes fluorosis [74,81].

Mercurous nitrate; $Hg_2(NO_3)_2 \cdot 2H_2O$ (561.22)

Colorless crystals, m.p. 70°C(d.), ρ 4.78; reacts with water [81].

Present in wastes from the production of some organic compounds and pharmaceuticals; porcelain, glazes, and pyrotechnical compositions; also in wastes of metal-working plants.

Toxicity: very toxic; has a general toxic and irritating effect; on prolonged exposure at low concentrations causes weakness, general depression, headache, and mental disorders [74,81,103]. $TSDC_{acute}$ - 0.34 mg/m^3 [192].

Mercurous oxide; HgO (417.22)

Black crystals. m.p. 100°C(d.), ρ 9.8; insoluble in water.

Present in wastes from the production of special paints (for ships), paints for chinaware, galvanic dry cells, some analytical reagents and organic compounds, and disinfectants.

Toxicity: very toxic [74,81].

Mercurous sulfate; Hg_2SO_4 (497.24)

Colorless crystals, ρ 7.56; decomposes on heating; slightly soluble in water.

Present in wastes from the production of wines (to achieve the desired coloration); some organic catalysts and analytical reagents; and galvanic dry cells; also in wastes of metallurgical works (during the extraction of noble metals).

Toxicity: very toxic [74,81].

Mercurous sulfide; Hg_2S (433.24)

Black powder; insoluble in water.

Present in wastes from the production of pigments for plastics, wax, dyes, articles made of rubber; phosphors, some catalysts; and mercury compounds.

Toxicity: very toxic [74,81,103].

Mercury; Hg (200.59)

Silvery white liquid metal, m.p. -38.86°C, b.p. 356.66°C, ρ 13.59; insoluble in water.

Present in wastes from the manufacture of physical research equipment, mercury UV lamps, fluorescent lamps, and mirrors; also in wastes from the production of some chemicals and pharmaceuticals, electrotechnical lacquers and paints, chlorine and sodium hydroxide, acetic acid (from acetylene), organomercuric and other mercury compounds, some chemical reagents, catalysts for the enrichment of noble metals, amalgams; and dental prosthesis.

Toxicity: very toxic; absorbed by healthy skin; has an allergenic and carcinogenic effect; causes extreme fatigue, drowsiness, loss of appetite, headache, dizziness, tremors of the hands, eyelids, and tongue, lacrimation, ptyalism, vomiting, pains when swallowing, bleeding of the gums, formation of a dark border on the gums; nervous disorder, skin irritation, and damage to the liver and kidneys [69, 74,81,93,102,103,113]. $TC_{chronic}$ - 0.01 mg/m^3 [103]; $TC_{c.n.s.}$ - 0.002-0.005 mg/m^3 [188]; MPADC - 0.0003 mg/m^3; toxicity classification - I [34]; MPADC established in Bulgaria, Hungary, GDR, and Yugoslavia - 0.0003 mg/m^3 [72].

Removal: see [189,193].

Determination: colorimetric analysis (sensitivity 0.1 μg) [47]; spectrophotometric analysis (sensitivity 0.1 mg/m^3) [18,44,98,191].

MOLYBDENUM

Molybdenum; Mo (95.94)

Light gray metal, m.p. 2620°C, b.p. 4600°C, ρ 10.2; insoluble in water.

Present in wastes of metallurgical and electrotechnical and radio engineering plants [87].

Toxicity: has a carcinogenic and mutagenic effect. $TC_{chronic}$ - 120-160 g/m^3 [74,81,93,103].

Molybdenum boride; Mo_2B_5 (245.93)

Gray crystals, m.p. 2200°C(d.), ρ 8.01; insoluble in water.

Present in wastes from the production of some catalysts and semiconductors; also in wastes of metallurgical works.

Toxicity: causes severe and chronic poisoning; disturbance of coordination of movements, muscular twitching, paralysis of arms and legs; purulent bronchitis; and dystrophic changes in the liver, kidneys, and heart. $TC_{chronic}$ - 60-70 mg/m^3 [103].

Molybdenum disulfide; MoS_2 (160.08)

Gray crystals, m.p. >1300°C(d.), ρ 4.8; insoluble in water.

Present in wastes from the production of lubricants.

Toxicity: causes damage to the liver and kidneys; has a carcinogenic and mutagenic effect [74,81,93,103].

Molybdenum pentachloride; $MoCl_5$ (273.20)

Purplish black crystals, m.p. 194°C, b.p. 268°C, ρ 2.92.

Present in wastes from the production of some catalysts; also in wastes of metallurgical works.

Toxicity: has a general toxic and irritating effect [103].

Molybdenum trioxide (molybdite); MoO_3 (143.94)

Colorless crystals, m.p. 801°C, b.p. 1155°C, ρ 4.69; slightly soluble in water.

Present in wastes from the production of some catalysts; also in wastes of metallurgical works.

Toxicity: has a general toxic effect. $TSDC_{acute}$ - 30 mg/m^3 [103].

NICKEL

Nickel; Ni (58.70)

Silvery white metal, m.p. 1455°C, b.p. 2900°C, ρ 8.9; insoluble in water.

Present in wastes from the production of alkali storage batteries and from electrolytic plating of metals, silver-alloying processes, and the hydrogenation of fats; also in wastes of electrotechnical, chemical, and machine-building plants and steel foundries.

Toxicity: has a general toxic and irritating effect, and a carcinogenic, mutagenic, and teratogenic effect. $SC_{chronic}$ and $TC_{chronic}$ - 0.0002 and 0.001 mg/m^3 respectively [5,63,69,79,93,100,103]; MPSDC and MPADC for metallic Ni - 0.001 mg/m^3; MPSDC and MPADC for water-soluble nickel salts - 0.0002 mg/m^3; toxicity classification - I [34].

Determination: paper chromatography (sensitivity 0.1 μg); absorption spectroscopy (sensitivity 0.006 μg/m^3) [44,99,115].

Nickel antimonide; NiSb (180.45)

Red crystals, m.p. 1153°C, ρ 7.54; insoluble in water.

Present in wastes of metallurgical works.

Toxicity: very toxic; causes irritation of the nasal mucous membrane, dizziness, fatigue, drowsiness, eczema, and fatty degeneration of the liver and heart; and impairs the nervous system [81,113].

Nickel arsenate (xanthiosite); $Ni_3(AsO_4)_2$ (453.91)

Yellow powder, ρ 4.98; insoluble in water.

Present in wastes of some chemical plants.

Toxicity: very toxic; has a general toxic, allergenic and irritating effect [74, 81,113].

Nickel arsenide (nicollite); NiAs (133.60)

Yellow or red crystals, m.p. 964°C, ρ 7.5; insoluble in water.

Present in wastes from the production of infrared instruments; also in wastes of some chemical plants.

Toxicity: has a general toxic, irritating, and allergenic effect [81,113].

Nickel bromate; $Ni(BrO_3)_2 \cdot 6H_2O$ (422.62)

Green crystals, ρ 2.57; decomposes on heating; soluble in water.

Present in wastes of some chemical plants.

Toxicity: has a general toxic and a carcinogenic effect [81].

Nickel bromide; $NiBr_2$ (218.51)

Yellowish brown crystals, m.p. 963°C(under pressure); ρ 4.64; solubility in water: 57 g/100 g H_2O at 20°C.

Present in wastes of some chemical plants.

Toxicity: causes general depression, irritation of mucous membranes and the eyes, and mental disorders [74,81,113].

Nickel bromoplatinate; $Ni[PtBr_6] \cdot 6H_2O$ (841.51)

Colorless crystals, ρ 3.71.

Present in wastes of some metallurgical and chemical plants.

Toxicity: has a general toxic and irritating effect; causes asthma, coughing, and mental disorders [81,113].

Nickel chloride; $NiCl_2$ (129.62)

Yellow or brown crystals, m.p. 1009°C(under pressure), ρ 3.35; solubility in water: 59.5 g/100 g H_2O at 10°C.

Present in wastes from the production of invisible inks, some catalysts, and ceramics; also in wastes of metallurgical plants and from electrolytic plating of metals.

Toxicity: same as that of Ni.

Nickel monoxide (bunsenite); NiO (74.70)

Grayish green crystals, m.p. 1955°C, ρ 7.45; insoluble in water.

Present in wastes from the production of pigments, ceramics, glass, and some catalysts.

Toxicity: same as that of Ni. MPSDC and MPADC - 0.001 mg/m^3; toxicity classification - II [34].

Nickel nitrate; $Ni(NO_3)_2 \cdot 6H_2O$ (290.82)

Green crystals, m.p. 56.7°C, b.p. 140°C(d.), ρ 2.05; soluble in water.

Present in wastes from the production of nickel salts, nickel catalysts, and ceramic pigments; also in wastes from electrolytic plating of metals.

Toxicity: has a general toxic and irritating effect, also an allergenic and carcinogenic effect [74,81,102,103].

Nickel sesquioxide; Ni_2O_3 (165.42)

Grayish black crystals, m.p. 600°C(d.), ρ 4.8; insoluble in water.

Present in wastes from the production of pigments, ceramics, glass, and some catalysts.

Toxicity: same as that of Ni. MPSDC and MPADC - 0.001 mg/m^3; toxicity classification - II [34].

Nickelic hydroxide; $Ni(OH)_3$ (109.72)

Black powder; insoluble in water.

Present in wastes from the production of alkali storage batteries.

Toxicity: has a general toxic and a carcinogenic effect [74,81,102,103].

Nickelous hydroxide; $Ni(OH)_2$ (92.73)

Light green powder, m.p. 230°C(d.); insoluble in water.

Present in wastes from the production of alkali storage batteries.

Toxicity: has a general toxic and a carcinogenic effect [74,81,102,103].

NITROGEN

Ammonia; NH_3 (17.03)

Colorless gas with a pungent odor, m.p. -80°C, b.p. -36°C; CLE - 15-28%; ρ 0.771 g/l; solubility in water: 87.5 g/100 g H_2O at °C and 52.6 g/100 g H_2O at 20°C; also soluble in ethanol and diethyl ether.

Present in wastes from petrochemical and metallurgical works, organic synthesis plants, and plants of the by-product coke industry; also in wastes from the production of fertilizers, lacquers and paints, photographic materials, nitric acid, ammonium nitrate, ammonium sulfate, explosives, synthetic fibers, and some plastics [74,103].

Toxicity: causes asphyxia, pharyngospasm, pulmonary edema, lacrimation, coughing, conjunctivitis, and damage to the cornea. TC_{odor} and $TC_{irritation}$ - 0.037 and 0.01 mg/l respectively [103]; has a toxic effect on plants (TC - 1 mg/m^3); MPSDC and MPADC - 0.2 and 0.04 mg/m^3 respectively; toxicity classification - IV [34]; MPSDC and MPADC established in Bulgaria, Hungary, and Yugoslavia - 0.2 mg/m^3, and in GDR, Romania, and Czechoslovakia - 0.3 and 0.1 mg/m^3 respectively [72]; MPC established in Czechoslovakia - 3 kg/hr [88]; MPADC recommended - 0.1 mg/m^3 [157].

Removal: by adsorption [158], burning [76,95,105,112,158].

Determination: colorimetric analysis [35], photocolorimetric analysis (in the range of 0-20 mg/m^3) [98], and automatic means and devices [18,92,160-162].

Ammonium arsenate (ammonium orthoarsenate); $(NH_4)_2AsO_4 \cdot 3H_2O$ (247.08)

Colorless crystals; decomposes on heating with the evolution of ammonia; soluble in water.

Present in wastes of metallurgical plants.

Toxicity: causes damage to blood vessels, the heart, liver, kidneys, and respiratory tracts [74,81,113].

Ammonium arsenite (ammonium metaarsenite); NH_4AsO_2 (124.96)

Colorless crystals, soluble in water.

Present in wastes of metallurgical plants.

Toxicity: very toxic; causes damage to blood vessels, the heart, liver, kidneys, the nervous system, and respiratory tracts [74,81,113].

Ammonium azide; NH_3N_3 (60.06)

Colorless crystals, m.p. 160°C (on heating above this temperature undergoes sublimation accompanied by explosion), ρ 1.346; solubility in water: 27.07 g/100 g H_2O at 40°C; also soluble in ethanol.

Present in wastes from the production of explosives.

Toxicity: very toxic; has a general toxic and irritating effect [81].

Ammonium bicarbonate (ammonium hydrocarbonate); NH_4HCO_3 (79.06)

Colorless crystals, m.p. 70°C(d.), ρ 1.586; solubility in water: 16.1 g/100 g H_2O at 10°C.

Present in wastes from the production of plastics, fertilizers, textiles, and leather.

Toxicity: has a general toxic and irritating effect. LD_{50} for mice - 245 mg/kg [113].

Removal: with the aid of filters made of nylon, polyesters, or polyethylene [57].

Ammonium bichromate (ammonium dichromate); $(NH_4)_2Cr_2O_7$ (252.10)

Orange-red crystals, m.p. ≅180°C(d.), ρ 2.155; solubility in water: 29.18 g/100 g H_2O at 25°C.

Present in wastes of chemical and metal-working plants; also in wastes from the production of pharmaceuticals, textiles, leather, lacquers and paints, ceramics, matches, photographic materials, pyrotechnical materials, and some catalysts.

Toxicity: causes dermatitis, ulcer formation, and perforation of the nasal bone; irritation of mucous membranes and the skin. $TC_{chronic}$ - 1 mg/m^3 [[74,81,103].

Ammonium bifluoride (ammonium hydrofluoride); NH_4HF_2 (57.05)

Colorless crystals; sublimes on heating; ρ 1.21; solubility in water: 39.76 g/100 g H_2O at 0°C.

Present in wastes from the production of plastics, some foodstuffs, glass, porcelain, enamel, and articles made of rubber; also in wastes of metallurgical works.

Toxicity: has a general toxic and irritating effect; at large concentrations causes ptyalism, vomiting, general depression, death; at low concentrations - destruction of the dental enamel, and sclerotic changes in the bones and ligaments [74,81,113].

Ammonium bisulfate (ammonium hydrosulfate); NH_4HSO_4 (115.11)

Colorless crystals, m.p. 146.9°C, b.p. 490°C, ρ 1.787; soluble in water.

Present in wastes from the production of some catalysts used in organic synthesis and from the manufacture of cosmetics.

Toxicity: has a general toxic and irritating effect [74,81].

Ammonium bisulfide (ammonium hydrosulfide); NH_4HS (51.11)

Colorless crystals, m.p. 120°C(subl.), ρ 1.17; soluble in water.

Present in wastes from the production of lubricants.

Toxicity: has an irritating effect [74].

Ammonium bisulfite (ammonium hydrosulfite); NH_4HSO_3 (99.11)

Colorless crystals, m.p. ≅150°C(d.), ρ 2.03; soluble in water.

Present in wastes from the production of paper.

Toxicity: very toxic; has a general toxic and irritating effect. LD_{50} for mice, rats, and guinea pigs - 70, 60, and 50 mg/kg respectively [74,81].

Ammonium bromate; NH_4BrO_3 (145.96)

Colorless crystals which decompose on heating; soluble in water.

Present in wastes of chemical plants.

Toxicity: causes paralysis and the formation of methemoglobin; and affects the nervous system [81,113].

Ammonium bromide; NH_4Br (97.94)

Colorless crystals, m.p. 394°C(d.), ρ 2.40; solubility in water: 75.5 g/100 g H_2O.

Present in wastes from the production of photographic materials and paper, and from lithographic processes.

Toxicity: has a general toxic effect; causes cachexia, mental disorders, and presenility [74,81].

Ammonium bromoplatinate; $(NH_4)_2[PtBr_6]$ (710.8)

Reddish brownish crystals; ρ 4.265.

Present in wastes of some metallurgical works.

Toxicity: has a general irritating effect; causes asthma, respiratory impairment, cachexia, skin eruption, cyanosis, and dermatitis [81,113].

Ammonium bromoselenate; $(NH_4)_2SeBr_6$ (594.5)

Red crystals; ρ 3.326.

Present in wastes from the production of glass, dyes, plastics, photoelectric equipment; also in wastes of some metallurgical works.

Toxicity: causes pneumonia, dermatitis, damage to the liver and kidneys, and irritation of mucous membranes and the respiratory tracts [81,113].

Ammonium bromostanate; $(NH_4)_2[SnBr_6]$ (634.3)

Colorless crystals; decomposes on heating; ρ 3.50.

Present in wastes of some metallurgical works.

Toxicity: has a general irritating effect; causes psychosis and also furunculosis [81,133].

Ammonium cadmium chloride; $4NH_4Cl \cdot CdCl_2$ (397.27)

Colorless crystals, ρ 2.01; soluble in water.

Present in wastes of metallurgical works.

Toxicity: causes damage to the liver and kidneys, pulmonary edema, death; xerostomia and respiratory depression [81].

Ammonium chloride (salammonial); NH_4Cl (53.49)

Colorless crystals, m.p. 338°C(subl.), ρ 1.526; solubility in water: 39.3 g/100 g H_2O at 25°C.

Present in wastes from the production of dyes, explosives, leather, and cement; also in wastes of metallurgical and electrotechnical plants.

Toxicity: has a general toxic and irritating effect [74,81].

Removal: with the aid of scrubbers [105] and filters made of nylon, polyesters, polyethylene, teflon, or polypropylene [57].

Ammonium cobalt phosphate; $NH_4CoPO_4 \cdot H_2O$ (190.0)

Purplish crystals; insoluble in water.

Present in wastes from the production of ceramic pigments, ceramics, enamel, textiles, fertilizers, and some analytical reagents.

Toxicity: not very toxic; causes damage to the liver and kidneys, and dermatitis; also has an allergenic effect [74,81,113].

Ammonium fluoride; NH_4F (37.04)

Colorless crystals, m.p. 168°C(d.), ρ 0.015; soluble in water (452.5 g/l).

Present in wastes of breweries; also in wastes from the production of wood preservarives, dyes, textiles, and some pharmaceuticals.

Toxicity: causes vomiting, ptyalism, cramps, and death; upon prolonged exposure at low concentrations - mottled tooth enamel and changes in the bones and ligaments [74,81,103].

Ammonium hexachloroplatinate(IV), (ammonium chloroplatinate); $(NH_4)_2[PtCl_6]$ (443.91)

Yellow crystals, m.p. 360°C(d.), ρ 3.06; slightly soluble in water.

Present in wastes from the production of platinum and from electrolytic platinizing.

Toxicity: has a general toxic and irritating effect. $TC_{irritation}$ - 5 mg/m^3 [74,81,103].

Removal: see [4,42,45].

Ammonium hexafluoroaluminate, (aluminum cryolite); $(NH_4)_3[AlF_6]$ (195.10)

Colorless crystals, m.p. 305°C(d.), ρ 1.78; soluble in water (768.5 g/l).

Present in wastes from the production of ammonium fluoride; also in wastes of some other chemical plants.

Toxicity: causes acute poisoning, respiratory depression, ptyalism, vomiting, diarrhea, asthenia, death; also causes chronic poisoning: toothe decay, sclerotic changes in the bones, and calcification of the ligaments [74].

Ammonium hexafluorogallate; $(NH_4)_3[GaF_6]$ (237.84)

Colorless crystals; soluble in water.

Present in wastes from the production of gallium fluoride.

Toxicity: causes respiratory impairment, general depression, death; also causes chronic poisoning resulting in fluorosis, skin eruption, and impairment of bone growth [74].

Ammonium hexafluorophosphate, (ammonium phosphofluoride) $NH_4[PF_6]$ (163.02)

Colorless crystals, ρ 2.180; decomposes on heating; solubility in water: 74.8 g/100 g H_2O at 20°C.

Present in wastes of some chemical plants.

Toxicity: at high concentrations causes ptyalism, vomiting, general depression, death; at low concentrations - fluorosis [74,81,113].

Ammonium hexafluosilicate, (ammonium silicofluoride); $(NH_4)_2[SiF_6]$ (178.11)

Colorless crystals, m.p. ≅319°C(d.), ρ 2.01; soluble in water (187.5 g/l).

Present in wastes from the production of some pesticides and superphosphate fertilizers; also in wastes of metallurgical works.

Toxicity: very toxic; has a general toxic and irritating effect; on prolonged exposure at low concentrations causes fluorosis. LD_{50} for guinea pigs - 150 mg/kg [74,81,113].

Ammonium hydroarsenate, (ammonium hydroorthoarsenate); $(NH_4)_2HAsO_4$ (176.0)

Colorless crystals, ρ 1.989, decomposes on heating; solubility in water: 33.94 g/100 g H_2O at 0°C.

Present in wastes of metallurgical works.

Toxicity: very toxic; has a general toxic, irritating, and allergenic effect; causes vomiting and diarrhea; affects the heart, blood vessels, kidneys, and the nervous system [74,81,113].

Ammonium hydroborate, (ammonium tetrahydroborate); $NH_4HB_4O_7 \cdot 3H_2O$ (228.33)

Colorless crystals, ρ 2.6, decomposes on heating; solubility in water: 10 g/100 g H_2O.

Present in wastes from the production of fire retardants and textiles.

Toxicity: has a general toxic and irritating effect [31,74,113].

Ammonium hydroselenate; NH_4HSeO_4 (162.0)

Colorless crystals, ρ 2.16; decomposes on heating.

Present in wastes from the production of pigments, plastics, photoelectric equipment, and glass; also in wastes of metallurgical works.

Toxicity: very toxic; causes respiratory lesion, damage to the liver and kidneys, and dermatitis [74,81,113].

Ammonium hydroxide, (aqua ammonia); NH_4OH (35.05)

Colorless liquid, m.p. -77°C, ρ 0.9.

Present in wastes from the production of ammonia salts, detergents, aniline dyes, glass, and some plastics.

Toxicity: causes spasms of the throat, asphyxia; irritation of the respiratory tracts, mucous membranes, and the skin [74,81]. LD_{50} for rats and cats - 350 and 250 mg/kg respectively [74,80,81].

Removal: with the aid of filters made of nylon, polyesters, polyethylene, teflon, or polypropylene [57].

Ammonium iodide; NH_4I (144.94)

Colorless crystals, m.p. 405°C(subl.), ρ 2.51; solubility in water: 172.3 g/100 g H_2O at 20°C.

Present in wastes from the production of photographic chemicals and some pharmacetical preparations.

Toxicity: causes general depression, headache, asthenia, anemia, cachexia, and irritation of mucous membranes and the skin [74,81,113].

Ammonium molybdate, (ammonium paramolybdate); $(NH_4)_6Mo_7O_{24}\cdot 2H_2O$ (1235.86)

Colorless crystals, m.p. 150°C(d.), ρ 2.49; soluble in water.

Present in wastes from the production of molybdenum, lacquers and paints, dyes for wool and silk, and microfertilizers; also in wastes from organic processes.

Toxicity: very toxic; affects the lungs, liver, and kidneys; causes dyspnea, tremors, and cramps [74,81,113].

Ammonium nitrate, (saltpeter); NH_4NO_3 (80.04)

Colorless crystals, m.p. 169.6°C, decomposes when heated above m.p., ρ 1.725; solubility in water: 65 g/100 g H_2O at 10°C.

Present in wastes from the production of fertilizers, explosives, some pharmaceuticals, and matches.

Toxicity: has a general toxic and allergenic effect; causes irritation of mucous membranes, the respiratory tracts, and the skin [74,81,103]. MPADC - 0.30 g/m^3; toxicity classification - IV [35].

Removal: see [4,8,42,57,76].

Determination: see [47].

Ammonium pentachlorozincate; $(NH_4)_3ZnCl_5$ (296.79)

Colorless crystals, subl. p. 340°C, ρ 1.81; soluble in water.

Present in wastes of metallurgical works.

Toxicity: causes irritation of mucous membranes and the skin [74].

Ammonium perchlorate; NH_4ClO_4 (117.50)

Colorless crystals, m.p. 270°C(d.), ρ 1.95; solubility in water: 19.89 g/100 g H_2O at 25°C.

Present in wastes from the production of explosives, pyrotechnical chemicals, and rocket fuels.

Toxicity: has a general toxic and irritating effect [74,81,113].

Ammonium persulfate, (ammonium peroxydisulfate); $(NH_4)_2S_2O_8$ (228.2)

Colorless crystals, m.p. 120°C(d.), ρ 1.982; solubility in water: 58.2 g/100 g H_2O at 0°C.

Present in wastes from the production of dyes, photographic chemicals, some foodstuffs, soap, and polygraphic inks; also in wastes of some metallurgical and machine-building plants.

Toxicity: causes asthma and irritation of mucous membranes and the skin [74, 81,102,103].

Determination: see [47].

Ammonium selenide; $(NH_4)_2Se$ (115.0)

Brown crystals; decomposes when heated or in contact with water.

Present in wastes from the production of some chemical reagents.

Toxicity: has a general toxic effect; causes damage to the liver and the nervous system, general depression, and dermatitis [81].

Ammonium selenite; $(NH_4)_2SeO_3$ (163.04)

Colorless crystals; soluble in water.

Present in wastes from the production of glass and alkaloid intermediates.

Toxicity: causes damage to the liver and the nervous system, dermotitis, and general depression [74,81,103,113].

Ammonium sulfamate, (ammate); $NH_4SO_3NH_2$ (114.12)

Colorless crystals, m.p. 125°C, b.p. 160°C(d.); solubility in water: 166.6 g/100 g H_2O at 10°C.

Present in wastes from the production of pesticides, fertilizers, and textiles.

Toxicity: causes irritation of mucous membranes and the skin. $TC_{chronic}$- 0.25-0.5 mg/l [74,81,103].

Ammonium sulfate; $(NH_4)_2SO_4$ (132.09)

Colorless crystals, m.p. 210°C(d.), ρ 1.77; solubility in water: 43.4 g/100 g H_2O at 25°C.

Present in wastes from the production of fertilizers, viscose fibers, ammonium alum, and corundum.

Toxicity: has a general toxic and irritating effect.

Ammonium sulfide; $(NH_4)_2S$ (68.14)

Yellow crystals; decomposes on heating; dissolves in water accompanied by decomposition.

Present in wastes from the production of phosphoric chemicals and textiles; also in wastes of metallurgical works.

Toxicity: has a general toxic and irritating effect [74,81,103,113].

Ammonium tetrachlorocuprate; $(NH_4)_2[CuCl_4]$ (241.45)

Yellow crystals, m.p. 110°C(d.), ρ 1.993; soluble in water.

Present in wastes from the production of reagents for chemical analysis in metallurgy.

Toxicity: has a general irritating and allergenic effect; causes hemolysis, and damage to the lungs, liver, and pancreas [74,81,113].

Ammonium thiocyanate, (ammonium rhodanate); NH_4SCN (76.11)

Colorless crystals, m.p. 149.6°C, b.p. 170°C(d.), ρ 1.305; solubility in water: 62.3 g/100 g H_2O at 10°C.

Present in wastes from the production of matches, dyes, photographic chemicals, plastics, urea, pesticides, and some analytical reagents.

Toxicity: causes irritation of mucous membranes and the skin, and damage to the liver, kidneys, pancreas, and the nervous system [74,81,113].

Ammonium vanadate, (ammonium metavanadate); NH_4VO_3 (116.99)

Colorless crystals, m.p. 100-150°C(d.), ρ 2.33; slightly soluble in water: 0.44 g/100 g H_2O at 12.5°C.

Present in wastes from the production of cotton fabrics, silk, aniline black (an azine dye), inks, printing inks, wool, photographic materials, and some analytical reagents; also in wastes of wood-processing plants.

Toxicity: very toxic; causes anemia and damage to the respiratory tracts, the liver, kidneys, and adrenal glands [74,81,103,113].

Nitric acid; HNO_3 (63.01)

Colorless liquid with a pungent odor; m.p. -41.6°C, b.p. 82.6°C(d.), ρ 1.522; soluble in water, forms azeotropic mixtures with water and crystal hydrates.

Present in wastes from the production of fertilizers, nitrates, ammonia, explosives, lacquers, rayon, and motion-picture films; also in wastes of metallurgical, pharmaceutical, and polygraphic plants.

Toxicity: causes pneumonia, pulmonary edema, tooth decay, ulceration of the nasal septum, damage to the cornea, conjunctivitis, bronchitis, and skin burns. SC_{odor} and TC_{odor} - 0.60 and 0.70 mg/m^3 respectively [148]; $SC_{irritation}$ - 0.0113 mg/m^3 [102,103]; MPSDC and MPADC - 0.4 and 0.15 mg/m^3 respectively; toxicity classification - II [117]. MPSDC and MPADC established in Bulgaria and Yugoslavia - 0.006 mg/m^3; in GDR - 0.14 and 0.06 mg/m^3 respectively; MPADC established in Czechoslovakia - 0.01 mg/m^3 [72]; MPC established in Czechoslovakia and Australia - 0.1 kg/hr and 100 mg/m^3 respectively [88].

Removal: with the aid of filters made of polyesters (including polyacrylates), polyethylene, teflon, or polypropylene [57].

Determination: colorimetric analysis (sensitivity 0.05 μg) [35].

Nitrogen; N_2 (28.01)

Colorless gas, m.p. -210°C, b.p.(760 mm Hg) -195.8°C, ρ 1.250 g/l; slightly soluble in water.

Present in wastes from the production of ammonia, nitric acid, nitrates, cyanides, explosives, and ferroalloys.

Toxicity: has a toxic effect in the event of a sharp decease in oxygen concentration; causes asphyxia and incoordination of muscular movements.

Nitrogen chloride; NCl_3 (120.38)

Yellow liquid with a pungent odor, m.p. -40°C, b.p. 71°C, ρ 1.653; insoluble in water.

Present in wastes from the production of chlorine, and sodium hydroxide.

Toxicity: very toxic; has a general toxic and irritating effect [74,81,102,103].

Nitrogen fluoride, (nitrogen trifluoride); NF_3 (71.01)

Colorless gas, m.p. -206.78°C, b.p. -206.78°C, insoluble in water.

Present in wastes from organic processes.

Toxicity: very toxic; causes headache, vomiting, cachexia, visual disturbance, and irritation of mucous membranes, the respiratory tracts, and the skin; on prolonged exposure - mottled enamel and sclerotic changes in the skeleton [74,81,103]. $TSDC_{acute}$ - 0.29mg/m^3 [81,103].

Nitrogen(I) oxide, (nitrogen hemioxide, nitrous oxide, laughing gas); N_2O (44.01)

Colorless gas with a pleasant odor, m.p. -91°C, b.p. -88.5°C, ρ 1.977; solubility in water: 54.4 ml/100 g H_2O at 25°C.

Present in wastes from the production of rayon, sodium azide, some pharmaceuticals, propylene, and dimethyl ether.

Toxicity: at high concentrations causes asphyxia. $TC_{irritation}$ - 0.06 mg/l [74, 103,148,149].

Nitrogen(II) oxide, (nitrous oxide); NO (30.01)

Colorless gas (when liquefied - blue in color), m.p. -163.6°C, b.p. -151.5°C, ρ 1.340 g/l; slightly soluble in water.

Present in wastes from the production of nitric acid, rayon, propylene, and diethyl ether; also present in air from the combustion of coal at high temperatures [74,103].

Toxicity: very toxic; causes asthemia, dizziness, numbness of limbs, and methemoglobinemia. $SSDC_{acute}$ - 0.03 mg/l, $TSDC_{acute}$ - 0.05 mg/l; LC_{100} for rats - 1.09 mg/l [103]; MPSDC and MPADC - 0.6 and 0.04 mg/m^3 respectively; toxicity classification - II [34]; MPSDC and MPADC established in FRG - 0.8 and 0.4 mg/m^3 respectively [72]; MPC established in Czechoslovakia - 3 kg/hr, in FRG - 1.8 g/m^3, in Great Britain - 1.8 g/m^3, and in Spain - 1.5 mg/m^3 [72,88].

Removal: by combustion [87,95]; adsorption; or catalysis (efficiency 80-85%) [76].

Determination: the luminescence method [151]; measurements with automatic equipment [67].

Nitrogen(IV) oxide, (nitrogen oxide, nitrogen dioxide); NO_2 (46.01)

Yellow liquid, m.p. -11.2°C, b.p. -21°C, ρ 1.49.

Present in wastes from the production of cellulose, nitric acid, sulfuric acid, aniline dyes, explosives, rayon, fertilizers, lacquers, and glycerol; also in wastes from metalworking, organic synthesis processes, and autogenous welding.

Toxicity: very toxic; causes dizziness, bronchopneumonia, cramps, palpitation of the heart, irritation of mucous membranes, the skin, and the respiratory tracts, asthma; affects the gonads [92,102,103]. SC_{odor} - 0.11 mg/m^3, SC and TC sensitivity to light - 0.087 and 0.14 mg/m^3 respectively; $SC_{chronic}$ and $TC_{chronic}$ - 0.15 and 0.6 mg/m^3 respectively [152,203]; $TSDC_{acute}$ for plants - 0.1 mg per cu.m. [152,153]; MPSDC and MPADC - 0.04 mg/m^3; toxicity classification - II [35]. MPSDC and MPADC recommended - 0.085 mg/m^3 [152]; MPSDC and MPADC established in Bulgaria, Hungary, and Yugoslavia - 0.085 mg/m^3; in Romania - 0.3 mg/m^3, and in Czechoslovakia - 0.1 mg/m^3 [72,88].

Removal: by adsoprtion (efficiency 80-85%) [76,112], or combustion [7,43,48,57, 78,87,95].

Determination: colorimetric analysis [103], spectrophotometric analysis [44,155], the luminescence method [154,155], or photocolorimetric analysis [6,92,98,153,154].

Nitrogen(V) oxide, (nitrogen pentoxide); N_2O_5 (108.01)

Colorless crystals, m.p. 32.3°C(subl.), ρ 1.64.

Present in wastes from the production of ozone, nitric acid, and chloroform.

Toxicity: causes irritation of the respiratory tracts and mucous membranes. $TC_{irritation}$ - 0.0025 mg/l [81,103]; MPSDC and MPADC established in Yugoslavia - 0.1 mg/m^3 [72].

OSMIUM

Osmium, Os (190.2)

Bluish silvery metal, m.p. ≅3030°C, b.p. ≅5000°C, ρ 22.5; insoluble in water.

Present in wastes from the production of some catalysts, fountain pens, electric heating elements, radio equipment, research instruments, and porcelain pigments; also in wastes of metallurgical works.

Toxicity: has a general toxic and irritating effect; causes visual disturbance, asthma, dermatitis, and skin ulcers [74,81,103].

Osmium tetraoxide, (perosmic acid); OsO_4 (254.2)

Light yellow crystals, m.p. 41°C, b.p. 131°C, ρ 4.9; soluble in water.

Present in wastes from the production of osmium alloys, some catalysts, and histological reagents.

Toxicity: very toxic; causes pneumonia, anemia, pain in the eyes, conjunctivitis, lacrimation, dermatitis, and skin ulcers [74,81,102,103].

OZONE

Ozone, O_3 (48.00)

Blue gas, m.p. -192°C, b.p. -112°C, ρ 2.14 g/l; slightly soluble in water.

Present in wastes from the production of hydrogen peroxide, some foodstuffs, paper, and some organic compounds; also in wastes of metal-working (welding) plants.

Toxicity: has a general toxic, irritating, carcinogenic, and mutagenic effect; causes fatigue, headache, dizziness, vomiting; irritation of the skin and mucous membranes, coughing, respiratory impairment, chronic bronchitis, pulmonary emphyzema, attacks of asthma, pulmonary edema, and hemolytic anemia. $SC_{chronic}$ and $TC_{chronic}$ - 0.0002 and 0.0003 mg/l respectively; TC_{odor} - 0.0004-0.98 μg/l; LC_{50} - 0.00068-0.003 mg/l [63,75,93,97,103]; has a harmful effect on plants: inhibits growth of tomatoes [181]; lowers tomato yield, and decreases the flesh of tomatoes and their starch, sugar and protein content [182]; the toxic effect on plants becomes evident at a concentration of 0.03 mg/l after a 4-hour exposure [55]; MPSDC and MPADC - 0.16 and 0.03 mg/m^3 respectively; toxicity classification - I [34].

Removal: adsorption on silica gel and alumogel [103].

Determination: gas analysis (automatic method; sensitivity 0.2 mg/m^3) [92]; photocolorimetric method (sensitivity 0.05 mg/m^3) [44,92,183-187].

PALLADIUM

Palladium, Pd (106.4)

Silvery white metal, m.p. 1554°C, b.p. ≅2940°C, ρ 12.02; insoluble in water.

Present in wastes from the production of alloys, precious metals, astronomical equipment, catalysts used in the synthesis of sulfuric acid; communication equipment, and thermoregulators.

Toxicity: slightly toxic [74,81,103].

Palladium chloride; $PdCl_2$ (177.3)

Red crystals, m.p. 680°C, ρ 4.0; soluble in water.

Present in wastes from the production of photographic materials, porcelain, inks, and some catalysts; also in wastes from some electroplating processes.

Toxicity: has a general toxic effect; causes skin irritation [81,113].

Palladium monoxide; PdO (122.4)

Black crystals, m.p. 750°C(d.), ρ 8.3; insoluble in water.

Present in wastes of some chemical plants.

Toxicity: same as for Pd [74,103,113].

PHOSPHORUS

Phosphine; PH_3 (34.00)

Colorless gas, m.p. -133.8°C, b.p. -87.42°C, ρ 1.52 g/l; soluble in water.

Present in wastes from the production of white and red phosphorus, and from the synthesis of some organic compounds; also in wastes of metal-working plants.

Toxicity: very toxic; causes irritation of the respiratory tracts, damage to the lungs and pulmonary edema, impairment of the central nervous system, damage to internal organs, brain edema, cramps, death; in the case of prolonged exposure at low concentrations causes anemia, bronchitis, impairment of vision and speech, and locomotor incoordination. $SC_{irritation}$ and $TC_{irritation}$- 0.0014-0.0035 and 0.007 mg/l respectively; LC_{100}- 0.01 mg/l [74,81,103].

Phosphonium iodide; PH_4I (161.93)

Colorless crystals, b.p. 80°C, ρ 2.8; soluble in water.

Present in wastes from the production of extracting, emulsifying, and flotation agents.

Toxicity: has a general toxic and irritating effect; in the case of severe poisoning causes anxiety, depression of the central nervous system, irritation of the lungs, cramps, nausea, and vomiting; in the event of chronic exposure at low concentrations causes anemia, bronchitis, impairment of vision and speech, irritation of mucous membranes and the skin, and locomotor incoordination [74,81].

Phosphoric acid, meta; HPO_3 (79.98)

Colorless deliquescent mass, ρ 2.2-2.25; sublimes on heating; reacts with water.

Present in wastes from the production of phosphoric fertilizers and phosphorus compounds, some foodstuffs, cement, dental protheses, matches, corrosion inhibitors, fire retardants, motion-picture films, photographic chemicals, some catalysts, and detergents; also in wastes from electrolytic plating of metals.

Toxicity: causes athrophy of the nasal mucous membrane and perforation of the nasal septum, pneumonia, irritation of mucous membranes and the skin, and damage to the liver and kidneys [74,81,103].

Phosphoric acid, ortho; H_3PO_4 (98.00)

Colorless crystals, m.p. 42.35°C, ρ 1.87; soluble in water.

Present in wastes from the production of phosphoric fertilizers and phosphorus compounds, some foodstuffs, cement, dental protheses, matches, corrosion inhibitors, fire retardants, motion-picture films, photographic chemicals, some catalysts, and detergents; also in wastes from electrolytic plating of metals.

Toxicity: causes athrophy of the nasal mucous membrane and perforation of the nasal septum, pneumonia, irritation of mucous membranes and the skin, and damage to the liver and kidneys. $SC_{chronic}$ and $TC_{chronic}$- 2.3 and 10.6 mg/m^3 respectively [74,81,103].

Phosphorus, red; P_4 (123.90)

Reddish brown crystals, m.p. 593°C, ρ 2.0-2.4; insoluble in water.

Present in wastes from the production of some fertilizers, organic compounds and pesticides; matches, pyrotechnical compositions; phosphoric acid, and phosphorus oxide.

Toxicity: causes destruction of hemoglobin erythrocytes, and leucocytes in the blood, and pathological changes in the liver and kidneys. $TC_{chronic}$- 0.04 mg/l [74, 81,103].

Phosphorus, yellow; P_4 (123.90)

Yellowish waxy substance, m.p. 44.14°C, b.p. 257°C, ρ 1.83; very slightly soluble in water.

Present in wastes from the production of phosphorus and its compounds, some pesticides, smoke-forming preparations, electronic equipment, and semiconductors; also in wastes of metallurgical works.

Toxicity: very toxic; causes loss of appetite and weight, anemia, damage to mucous membranes, the respiratory tracts, and bones, functional disturbances of internal organs, retinal apoplexy, and necrosis of jawbone tissues. $TC_{chronic}$- 0.0002-0.0012 mg/l; LD_{50}- 0.05-0.15 g/kg [74,81,102,103,113].

Phosphorus oxybromide; $POBr_3$ (286.72)

Pale yellow crystals, m.p. 55°C, b.p. 192°C, ρ 2.8; reacts with water.

Present in wastes from the production of organophosphorus compounds.

Toxicity: has a general toxic and irritating effect [74,81].

Phosphorus oxychloride; $POCl_3$ (153.33)

Colorless liquid, m.p. 1.18°C, b.p. 107.2°C, ρ 1.67; reacts with water.

Present in wastes from the production of some plastics and pharmaceuticals, dyes, and esters of phosphoric acids.

Toxicity: has a general toxic, irritating, and mutagenic effect; causes irritation of the eyes, the respiratory tracts, and mucous membranes; damage to the liver, kidneys, nerve cells, and the skin resulting in skin ulceration; laryngospasm, and pulmonary edema. $TC_{irritation}$, $TSDC_{acute}$, and $TC_{chronic}$ - 0.00134, 0.07, and 0.01-0.02 mg/l respectively [78,81,103].

Phosphorus oxyfluoride; POF_3 (103.97)

Colorless gas, m.p. -40.15°C, b.p. -39.7°C, ρ 4.8 g/l; reacts with water.

Present in wastes from the production of fluoroorganophosphorus compounds.

Toxicity: very toxic; causes severe irritation of the eyes, mucous membranes, and the respiratory tracts; on prolonged exposure at low concentrations causes fluorosis [81]. MPSDC and MPADC - 0.15 and 0.05 mg/m^3 respectively; toxicity classification - II [34]; established MPSDC and MPADC in both GDR and Yugoslavia - 0.15 and 0.05 mg/m^3 respectively [88].

Determination: colorimetric analysis (sensitivity 0.5 μg) [44,46,47].

Phosphorus pentabromide; PBr_5 (430.49)

Reddish yellow crystals, b.p. 106°C(d.); reacts with water.

Present in wastes from the production of bromine derivatives and some organic compounds.

Toxicity: very toxic; has a general toxic and irritating effect; causes general depression, mental disorders, skin eruption, and furunculosis [74,81].

Phosphorus pentachloride; PCl_5 (208.24)

Greenish crystals, m.p. 160°C, ρ 2.1; reacts with water.

Present in wastes from the synthesis of some organic compounds, phosphorus pentachloride and phosphorus oxychloride, chlorinated hydrocarbons, some alcohols, and acetylcellulose; also in wastes of metal-working works.

Toxicity: very toxic; has a general toxic, irritating effect; causes anxiety, ptyalism, opacity of the cornea, coughing, rhinitis, respiratory impairment, and irritation of the respiratory tracts, mucous membranes, and the skin. $TC_{irritation}$ -

0.01-0.02 mg/l; at high concentrations - causes death due to pulmonary edema [74, 81,103].

Phosphorus pentafluoride; PF_5 (125.97)

Colorless gas, m.p. -93.75°C, b.p. -84.55°C, ρ 5.80 g/l; reacts with water.

Present in wastes from the production of fluorophosphates; some catalysts, organic compounds, and inhibitors; also in wastes of metallurgical works.

Toxicity: very toxic; causes irritation of the eyes, mucous membranes, and the respiratory tracts; on prolonged exposure at low concentrations - fluorosis [74,81].

Phosphorus pentaoxide; P_4H_{10} (283.89)

Colorless crystals, m.p. 563°C, ρ 2.7; reacts with water.

Present in wastes from the production of some catalysts and organic compounds, phosphoric acids, polyisobutylene, and phosphate glass.

Toxicity: causes chemical burns; has a general toxic and irritating effect. MPSDC and MPADC - 0.15 and 0.05 mg/m^3 respectively [34]; MPSDC and MPADC established in both Yugoslavia and GDR - 0.15 and 0.05 mg/m^3 respectively [88].

Determination: see [44,47].

Phosphorus thiochloride; $PSCl_3$ (169.41)

Colorless liquid, m.p. -36.2-(-40.8)°C, b.p. 125°C; reacts with water.

Present in wastes from the production of esters of phosphoric acids.

Toxicity: has an irritating effect [74].

Phosphorus tribromide; PBr_3 (270.69)

Colorless liquid, m.p. -40.5°C, b.p. 175.3°C, ρ 2.87; reacts with water.

Present in wastes from the production of bromine derivatives and some organic compounds.

Toxicity: very toxic; has a general toxic and irritating effect; causes general depression, mental disorders, skin rash, and furunculosis [74,81].

Phosphorus trichloride; PCl_3 (137.33)

Colorless liquid, m.p. -90.34°C, b.p. 75.3°C, ρ 1.57; reacts with water.

Present in wastes from the synthesis of some organic compounds, phosphorus pentachloride and phosphorus oxychloride, chlorine-containing hydrocarbons, some alcohols, and acetylcellulose; also in wastes of metal-working plants.

Toxicity: very toxic; has a general toxic and irritating effect; causes anxiety, ptyalism, opacity of the cornea, coughing, rhinitis, respiratory impairment, and irritation of the respiratory tracts, mucous membranes; and the skin. $TC_{irritation}$ - 0.01-0.02 mg/l; at high concentrations can cause death due to pulmonary edema [74, 81,103].

Phosphorus trifluoride; PF_3 (87.97)

Colorless gas, m.p. -151.5°C, b.p. -101.4°C, ρ 3.9; reacts with water.

Present in wastes from the production of fluorophosphates; some catalysts, organic compounds and inhibitors; also in wastes of metallurgical works.

Toxicity: very toxic; causes irritation of the eyes, mucous membranes, and the respiratory tracts; on prolonged exposure at low concentrations - fluorosis [74,81].

Phosphorus trioxide; P_4O_6 (219.89)

Colorless crystals, m.p. 23.9°C, b.p. 175.4°C, ρ 2.13; reacts with water.

Present in wastes from the production of some catalysts and organic compounds, phosphoric acids, polyisobutylene, and phosphate glass.

Toxicity: causes chemical burns; has a general toxic and irritating effect.

PLATINUM

Chloroplatinic acid; $H_2[PtCl_6]\cdot 6H_2O$ (517.94)

Yellow orange crystals, m.p. 100°C(d.), ρ 2.43.

Present in wastes from the production of glass, jewelry, photographic materials, porcelain, acetic acid, and inks; also in wastes of metallurgical, electronic and electronic and electrolytic plants.

Toxicity: has a general toxic and irritating effect; causes allergy (the latent period may last several years); lacrimation, sneezing, rhinitis, coughing, respiratory impairment, chest congestion, asthma, reddening and peeling of skin, conjunctivis, urticaria, dermatitis, and eczema [81,102,103].

Platinic chloride, (platinum tetrachloride); $PtCl_4$ (336.90)

Red-brown crystals, m.p. 370°C(d.), ρ 2.43; solubility in water: 142.1 g/100 g H_2O at 25°C.

Toxicity: has a general toxic and irritating effect [81,113].

Platinic iodide; PtI_4 (702.91)

Dark brown crystals, m.p. 370°C(d.), ρ 6.06.

Present in wastes from the production of some organic compounds and nonferrous metals.

Toxicity: has a general toxic and irritating effect; causes headache, weakness, anemia, loss of weight, and general depression; irritation of the skin and mucous membranes; and skin eruption [74,81,113].

Platinous chloride, (platinum dichloride); $PtCl_2$ (233.2)

Yellowish green crystals, m.p. 581°C(d.), ρ 5.87; insoluble in water.

Present in wastes of some metallurgical plants and from the production of some catalysts.

Toxicity: has a general toxic and irritating effect [81,113].

Platinous iodide; PtI_2 (448.90)

Black crystals, m.p.>300°C(d.), ρ 6.40; insoluble in water.

Present in wastes from the production of some organic compounds and nonferrous metals.

Toxicity: has a general toxic and irritating effect; causes headache, weakness, anemia, loss of weight, and general depression; irritation of the skin and mucous membranes; and skin eruption [74,81,113].

Platinum; Pt (195.09)

Silvery white metal, m.p. 1772°C, b.p. 3900°C, ρ 21.46.

Present in wastes from the production of some chemical laboratory utensils; corrosion-resistant equipment, and electrodes; Pt thermometers, catalysts used in the synthesis of acetic, nitric, and sulfuric acids; synthetic fibers, jewelry, and dental protheses; and from electroplating.

Toxicity: metallic Pt is practically nonpoisonous; compounds of Pt cause difficulties in breathing, a feeling of chest congestion, asthma, photophobia, conjunctivitis, irritation of the respiratory tracts, rhinitis, sneezing, coughing, cyanosis, dermatitis, urticaria, and eczema. For aerosols of complex compounds of Pt the toxic effect lies in the range of 5-70 mg/m^3 [74,102,103].

PLUTONIUM

Plutonium; Pu (232-246)

Silvery white metal, m.p. 640°C, b.p. 3350°C, ρ 19.80.

Present in wastes from the production of nuclear sources of electric energy; also in wastes of nuclear power stations.

Toxicity: very toxic; has a general toxic and carcinogenic effect; affects the hemopoietic organs, the liver, kidneys, other internal organs; causes irritation of the eyes and mucous membranes, nosebleed, and cirrhosis of the liver [74,81,103].

POTASSIUM

Potassium; K (39.102)

Silvery white metal, m.p. 63.55°C, ρ 0.862; reacts with water.

Present in wastes from the production of inorganic derivatives of potassium, some organic chemicals, photoelectric elements, titanium, and some catalysts.

Toxicity: has a general toxic and irritating effect [74,81,103].

Potassium acid fluoride, (potassium bifluoride; potassium hydrodifluoride); KHF_2 (78.11)

Colorless crystals, m.p. 238.7°C(d.), ρ 2.37; soluble in water (392 g/l).

Present in wastes from the production of pure potassium fluoride, glass, some catalysts, and fluorine.

Toxicity: has a general toxic and irritating effect; on prolonged exposure at low concentrations causes mottled enamel, sclerosis and osteomalacia, and fluorosis [74,81].

Potassium aluminum sulfate; $KAl(SO_4)_2$ (253.20)

Colorless crystals; soluble in water.

Present in wastes from the production of dyes, printing lacquers, paper, plant gums, cements for marble and porcelain, explosives, tanning agents, gelatin, and ammonia; also in wastes of metal-working plants.

Toxicity: at high concentrations causes burns inside the mouth and throat, vomiting, and diarrhea [74,81,113].

Potassium aluminum sulfate, (alumen); $KAl(SO_4)_2 \cdot 12H_2O$ (474.39)

Colorless crystals, decomp. p. 82°C; soluble in water.

Present in wastes from the production of paints, lacquers, paper, plant gums, cement for marble and porcelain; baking powder, compositions for engraving, and some catalysts.

Toxicity: has a general toxic, irritating, and allergenic effect [74,81,103,113].

Potassium arsenate; KH_2AsO_4 (180.02)

Colorless crystals, m.p. 250°C(d.), ρ 2.8; solubility in water: 23 g/100 g H_2O at 20°C.

Present in wastes from the production of textiles, leather, and paper.

Toxicity: very toxic; causes changes in the blood composition; affects the kidneys; and has an irritating and allergenic effect [74,81].

Potassium beryllium tetrafluoride; K_2BeF_4 (163.21)

Colorless crystals; decomposes on heating.

Present in wastes of some metallurgical works.

Toxicity: very toxic; has a general toxic and irritating effect; on prolonged exposure at low concentrations causes fluorosis [74,81,113].

Potassium bismuth iodide; K_2BiI_7 (1253.82)

Red crystals; decomposes on heating; decomposes in water.

Present in wastes from the production of vitamins and antibiotics.

Toxicity: causes irritation of the skin, stomatitis, and damages to the kidneys [74].

Potassium bisulfate, (potassium hydrosulfate); $KHSO_4$ (136.17)

Colorless crystals, m.p. 218.6°C, ρ 2.24; soluble in water.

Present in wastes from the production of analytical reagents used in essaying ores, silicon compounds, and some pharmaceuticals; also in wastes of metallurgical works.

Toxicity: insufficiently investigated.

Removal: with the aid of filters made of nylon, polyesters, polyethylene, poly(vinyl chloride), teflon, or polypropylene [57].

Potassium bromide; KBr (119.00)

Colorless crystals, m.p. 730°C, b.p. 1380°C, ρ 2.75; solubility in water: 68.1 g/100 g H_2O at 25°C.

Present in wastes from the production of photographic materials, some pharmaceuticals, and compositions for engraving.

Toxicity: has a general toxic and irritating effect; at high concentrations causes depression of the central nervous system; on prolonged exposure at low concentrations causes mental disorders and skin eruption [74,81].

Potassium carbonate; K_2CO_3 (138.20)

Colorless crystals, m.p. 891°C, ρ 2.29; solubility in water: 52.4 g/100 g H_2O at 20°C.

Present in wastes from the production of soap, glass, ceramics, potassium salts, engraving and lithographic inks, and leather.

Toxicity: has a general irritating and searing effect [74].

Potassium chlorate; $KClO_3$ (122.53)

Colorless crystals, m.p. 356°C(d.), ρ 2.32; solubility in water: 12.1 g/100 g H_2O at 40°C.

Present in wastes from the production of some pharmaceuticals, explosives, matches, pyrotechnical compositions, dyes, cotton fabrics, some organic compounds, and pesticides.

Toxicity: very toxic; has a general toxic and irritating effect; causes hemolysis of the erythrocytes and methemoglobinemia; irritation of the gastroenteric tracts, damage to the liver and kidneys, jaundice, bilious vomiting, high blood pressure; irritation and ulceration of the skin [74,81,103].

Potassium chloride; KCl (74.56)

Colorless crystals, m.p. 770°C, b.p. 1407°C, ρ 1.98; solubility in water: 34.4 g/100 g H_2O at 20°C.

Present in wastes from the production of fertilizers, some pharmaceuticals, and potassium salts; also in wastes of some instrument-making plants.

Toxicity: has a general toxic and irritating effect; affects the peripheral nervous system; causes irritation of the respiratory tracts and mucous membranes. $TC_{chronic}$ - 10 mg/m^3 [74,103]; ASL - 0.1 mg/m^3 [117].

Potassium chromate, (tarapacaite); K_2CrO_4 (194.20)

Yellow crystals, m.p. 968.3°C, ρ 2.73; solubility in water: 63 g/100 g H_2O at 20°C.

Present in wastes of metal-working and machine-building plants; also in wastes from the production of some organic compounds, leather, textiles, lacquers and paints, some pharmaceuticals, ceramics, matches, photographic materials, enamels, and some analytical reagents.

Toxicity: has a general toxic, irritating, and allergenic effect; causes irritation of mucous membranes and the skin, perforation of the nasal septum, and damage to the respiratory tracts, the liver, and kidneys. $TSDC_{acute}$ - 20-30 mg/m^3 [74,81,103].

Potassium cyanide; KCN (65.11)

Colorless crystals, m.p. 634.5°C, ρ 1.52; solubility in water: 71.6 g/100 g H_2O at 25°C.

Present in wastes from the production of nitriles; also in wastes of gold- and silver-plating plants.

Toxicity: very toxic; has a general toxic effect; at low concentrations causes headache, nausea, vomiting, fatigue, and asthemia; at high concentrations - respiratory impairment, cramps, death. LD_{100} for dogs - 1.6 mg/kg [74,81,103].

Potassium dichromate; $K_2Cr_2O_7$ (294.19)

Red crystals, m.p. 396°C, b.p. 610°C(d.), ρ 2.69; soluble in water: 13 g/100 g H_2O at 25°C.

Present in wastes from organic synthesis processes and from the production of pyrotechnical compositions, matches, leather, dyes, textiles, lacquers and paints, some pharmaceuticals, and ceramics.

Toxicity: has a general toxic, irritating, and carcinogenic effect. LD_{50}, introduced under the skin of rabbits - 7.8 mg/kg, introduced into the lungs of rabbits - 7 mg/kg (causes damage to all the organs); toxic effect on mice becomes evident on prolonged exposure at concentrations of 0.05 mg/kg [81,103].

Potassium fluoride; KF (53.10)

Colorless crystals, m.p. 1074°C, b.p. 1500°C, ρ 2.50; solubility in water: 94.9 g/100 g H_2O at 20°C.

Present in wastes from the production of fluorinated organic compounds, pesticides, and glass.

Toxicity: has a general toxic and irritating effect; at high concentrations causes irritation of the eyes, mucous membranes, and the skin; on prolonged exposure at low concentrations - mottled enamel, sclerosis and osteomalacia, and fluorosis [74,81].

Potassium hydroxide; KOH (56.10)

Colorless crystals, m.p. 404°C, b.p. 1324°C, ρ 2.04; solubility in water: 49.4 g/100 g H_2O at 0°C.

Present in wastes from the production of liquid soaps, lacquers, photographic materials, cotton textiles, lithographic inks, some analytical reagents, and some products of organic synthesis; also in wastes of wood-working plants.

Toxicity: causes severe irritation and searing of the respiratory tracts, mucous membranes, and the skin; damage to the conjunctiva cornea, and pneumonia [74,81, 102,103].

Removal: with the aid of filters made of nylon, polyesters, polyethylene, poly-(vinyl chloride), teflon, or polypropylene [57].

Potassium iodide; KI (166.02)

Colorless crystals, m.p. 686°C, b.p. 1324°C, ρ 3.12; solubility in water: 144.5 g/100 g H_2O at 20°C.

Present in wastes from the production of photographic materials, some analytical reagents, and some pharmaceuticals.

Toxicity: has a general toxic and irritating effect [74,81].

Potassium nitrate, (saltpeter; niter); KNO_3 (101.11)

Colorless crystals, m.p. 336°C, b.p. ≅400°C, ρ 2.11; solubility in water: 31.6 g/100 g H_2O at 20°C.

Present in wastes from the production of fertilizers, gunpowder, pyrotechnical preparations, some foodstuffs, and glass.

Toxicity: has a general toxic and irritating effect; on prolonged exposure at low concentrations causes anemia, methemoglobinemia, nephritis, weakness, general depression, and headache [74,81,103].

Potassium nitrite; KNO_2 (85.11)

Colorless crystals, m.p. 440°C, ρ 1.91; soluble in water.

Present in wastes from the production of dyes and some organic compounds.

Toxicity: has a general toxic and irritating effect; at low concentrations causes a drop in blood pressure, rapid pulse, headache, and dysopia; at high concentrations - nausea, vomiting, cyanosis, collapse, and coma [74,81,103].

Potassium pentaborate; $KB_5O_8 \cdot 4H_2O$ (365.11)

Colorless crystals, b.p. 780°C; slightly soluble in water.

Present in wastes from the production of glass fibers, glaze, plastics, articles made of rubber, and textiles; also in wastes of metallurgical works.

Toxicity: has a general toxic and irritating effect; affects the central nervous system and kidneys; is absorbed by healthy skin, and has a cumulative effect. LD_{50} for rats - 1.7 g/kg [81,103].

Potassium permanganate; $KMnO_4$ (158.09)

Dark purple crystals, m.p. 240°C(d.), ρ 2.70; solubility in water: 6.36 g/100 g H_2O at 20°C.

Present in wastes from the production of some pharmaceuticals, dyes, paper, resins, wax, fats, oils, cotton, silk, organic chemicals, and photographic materials; also in wastes of wood-working and some metallurgical plants.

Toxicity: has a general toxic and irritating effect; affects the gonads and embryos [74,81,102,103].

Potassium persulfate; $K_2S_2O_8$ (158.04)

Colorless crystals, m.p. ≅100°C(d.); solubility in water: 4.49 g/100 g H_2O at 20°C.

Present in wastes from the production of fats, soap, flour, photographic materials, some analytical reagents, explosives, and pesticides.

Toxicity: has a general toxic, irritating, and allergenic effect [74,81,103].

Potassium zirconium fluoride; $K_2[ZrF_6]$ (283.41)

Colorless crystals; decomposes on heating; slightly soluble in water.

Present in wastes from the production of metallic zirconium.

Toxicity: has a general toxic and irritating effect; causes headache, fatigue, chest pains; and irritation of mucous membranes and the skin. $TC_{chronic}$ - 10 mg/m^3 [74,103].

RHENIUM

Rhenium; Re (196.20)

Silvery white metal, m.p. 3190°C, b.p. 5600°C, ρ 21.03.

Present in wastes from the production of cathode-ray tubes, semiconductors, alloys for electric contacts, some catalysts, jewelry, and medical instruments; also in wastes of some metallurgical works.

Toxicity: see [74,81].

Rhenium hexafluoride; ReF_6 (336.90)

Dark red crystals, m.p. 61.3°C, b.p. 69.2°C.

Present in wastes of some metallurgical works.

Toxicity: very toxic; has a general toxic and irritating effect; on prolonged exposure at low concentrations causes fluorosis [74,81].

Rhenium tetrafluoride; ReF_4 (262.31)

Colorless crystals, m.p. 124.5°C.

Present in wastes of some metallurgical works.

Toxicity: very toxic; has a general toxic and irritating effect; on prolonged exposure at low concentrations causes fluorosis [74,81].

RHODIUM

Rhodium; Rh (102.90)

Silvery white metal, m.p. 1963°C, b.p. ≅3700°C, ρ 12.41; insoluble in water.

Present in wastes from the production of some catalysts, viscose, porcelain, and alloys; also from the manufacture of instruments, movie projectors, and thermocouples.

Toxicity: see [74,81,103].

RUBIDIUM

Rubidium; Rb (85.47)

Silvery white metal, m.p. 39.5°C, b.p. 685°C, ρ 1.52; reacts with water.

Present in wastes from the production of ceramics, glass; some catalysts and pharmaceuticals, rubidium salts; and x-ray, electrical, and electronic equipment.

Toxicity: has an irritating effect [74,81,113].

Rubidium carbonate; Rb_2CO_3 (230.95)

Colorless crystals, m.p. 835°C; soluble in water.

Present in wastes from the production of special types of glass, some catalysts, and electrolytes.

Toxicity: has an irritating effect; causes chemical burn [74,81].

Rubidium chloride; RbCl (120.94)

Colorless crystals, m.p. 717°C, b.p. 1390°C, ρ 2.76; soluble in water.

Present in wastes of some metallurgical works and from electrolytic processes.

Toxicity: LD_{50} for rats - 1.2 g/kg [74].

Rubidium chromate; Rb_2CrO_4 (286.97)

Yellow crystals, m.p. 994°C, ρ 3.51; solubility in water: 43.3 g/100 g H_2O at 25°C.

Present in wastes from the production of photoelectronic and photoelectric equipment.

Toxicity: causes irritation of respiratory tracts, mucous membranes, and the skin; perforation of the nasal septum, skin ulcer; and damage to the liver and kidneys [81].

Rubidium fluoride; RbF (104.48)

Colorless crystals, m.p. 775°C, b.p. 1410°C, ρ 2.76; soluble in water.

Present in wastes from the production of optical materials, thermal storage systems, and special types of glass.

Toxicity: at high concentrations has a general toxic and irritating effect; on prolonged exposure at low concentrations causes fluorosis [81].

Rubidium hydroxide; RbOH (102.48)

Colorless crystals, m.p. 380°C, ρ 3.20; solubility in water: 69.61 g/100 g H_2O at 25°C.

Present in wastes from the production of some catalysts, rubidium derivatives, and storage batteries.

Toxicity: causes chemical burns of the skin and mucous membranes [74,81,103].

RUTHENIUM

Ruthenium; Ru (101.70)

Silvery white metal, m.p. 2250°C, b.p. ≅4200°C, ρ 12.37; insoluble in water.

Present in wastes from instrument-making and electrotechnical plants; also in wastes from the production of dyes, porcelain, some organic compounds, jewelry, and textiles.

Toxicity: causes coughing, irritation of the throat, and ulceration of the mucous nasal septum [74,81,103].

Ruthenium tetraoxide; RuO_4 (165.07)

Golden yellow crystals, m.p. 25.5°C, ρ 3.29; soluble in water.

Present in wastes from the production of some reagents.

Toxicity: causes irritation of the respiratory tracts; coughing, and ulceration of the mucous nasal septum [81].

SELENIUM

Selenious acid; H_2SeO_3 (128.97)

Colorless crystals, m.p. >70°C(d.), ρ 3.0; soluble in water.

Present in wastes from the production of selenium and its derivatives.

Toxicity: very toxic; has a general toxic and irritating effect [74,81].

Selenium; Se (78.96)

Dark red, brown, or gray powder; m.p. 221°C, b.p. 685.3°C, ρ 4.80; insoluble in water.

Present in wastes from the production of pigments for coloring glass, photographic materials, semiconductors, electronic equipment, ceramics, enamel, articles made of rubber, typographic inks, lubricants, catalysts for making fire-resistant materials, luminescent compositions, and some electrical instruments and equipment.

Toxicity: has a general toxic and irritating effect; causes paleness, flaccidity, nervous disorders, cramps, general depression, hurried breathing, coughing, pneumonia, pulmonary edema, fatty degeneration of the liver, degeneration of the kidneys, and dermititis [74,81,103].

Determination: see [18,44].

Selenium dioxide; SeO_2 (110.96)

Colorless crystals, m.p. 340°C, ρ 3.25; solubility in water: 68.8 g/100 g H_2O at 25°C.

Present in wastes from the production of selenium compounds, some alkaloids and organic compounds, and luminophors.

Toxicity: very toxic; causes anemia, damage to the mucous membrane of the upper respiratory tract, the lungs, liver, and kidneys; on contact with skin causes severe pain [74,81,103]. TC_{odor} - 0.0002 mg/m^3 [74,81]; MPSDC and MPADC - 0.1 and 0.55 $\mu g/m^3$ respectively; toxicity classification - I [34].

Selenium hexafluoride; SeF_6 (192.96)

Colorless gas, m.p. -50.8°C, b.p. -34.5°C, ρ 3.2 g/l.

Present in wastes from the production of electric insulators and fluorinating agents.

Toxicity: very toxic; has a general toxic and irritating effect; causes irritation of the upper respiratory tracts; affects the lungs; on prolonged exposure at low concentrations causes fluorosis [74,81,103].

Selenium hydride; SeH_2 (80.98)

Gas with an unpleasant odor, m.p. -65.72°C, b.p. -41.5°C, ρ 3.66 g/l; soluble in water (0.096 mole/l).

Present in wastes from the production of selenium and its organic derivatives.

Toxicity: causes allergy and irritation of the eyes and mucous membranes, sneezing, sore throat, dry coughs, severe headache, lacrimation, and damage to the lungs and liver [74,81,103].

Selenium monobromide; Se_2Br_2 (317.75)

Dark red liquid, b.p. 227°C, ρ 3.6.

Present in wastes from some metallurgical plants and chemical processes.

Toxicity: very toxic; has a general toxic and irritating effect; causes mental disorders [74,81].

Selenium monochloride; Se_2Cl_2 (228.83)

Brown liquid, m.p. -8.5°C, b.p. ≅127°C(d.), ρ 2.90; decomposes in water.

Present in wastes from the synthesis of some organic compounds.

Toxicity: very toxic; has a general toxic and irritating effect [74,81].

Selenium nitride; Se_4N_4 (371.87)

Orange-red powder; insoluble in water.

Present in wastes from the production of selenium.

Toxicity: causes irritation of the respiratory tracts, coughing, nausea, and pulmonary edema [74,81].

Selenium oxychloride; $SeOCl_2$ (165.87)

Colorless liquid, m.p. 10.8°C, b.p. 177.6°C, ρ 2.44; decomposes in water.

Present in wastes from the production of some plastics, plasticizers, and chemical compounds, and lasers.

Toxicity: has an irritating effect; causes chemical burns of the skin, skin abscesses, pulmonary edema, and deformation of nails; and is absorbed by healthy skin [74,81,103].

Selenium tetrabromide; $SeBr_4$ (398.62)

Dark red crystals; decomposes on heating.

Present in wastes from some metallurgical and chemical plants.

Toxicity: very toxic; has a general irritating effect; causes mental disorders [74,81].

Selenium tetrachloride; $SeCl_4$ (220.79)

Colorless crystals, m.p. 305°C, ρ 3.78; decomposes in water.

Present in wastes from the synthesis of some organic compounds.

Toxicity: very toxic; has a general toxic and irritating effect [74,81].

Selenium tetrafluoride; SeF_4 (154.96)

Colorless liquid, m.p. -9.5°C, b.p. 107.7°C, ρ 2.7; decomposes in water.

Present in wastes from the production of electric insulators and fluorinating agents.

Toxicity: very toxic; causes irritation of the upper respiratory tracts; affects the lungs; on prolonged exposure at low concentrations causes fluorosis [74,81,103].

Selenium trioxide; SeO_3 (126.96)

Colorless crystals, m.p. 121°C, ρ 3.6; soluble in water [81].

Present in wastes from the production of selenium compounds, some alkaloids and organic compounds, and luminophors.

Toxicity: very toxic; causes anemia, damage to the mucous membrane of the upper respiratory tracts, the lungs, liver, and kidneys; on contact with skin causes severe pain [74,81,103]. TC_{odor} - 0.0002 mg/m^3 [74,81]; MPSDC and MPADC - 0.1 and 0.55 $\mu g/m^3$ respectively; toxicity classification -I [34].

SILICON

Hydrofluosilicic acid; $H_2SiF_6 \cdot 2H_2O$ (180.09)

Present in wastes from the production of fluorides and silicon fluoride. pesticides, cement, wood preservatives, and desinfectants; and from glass polishing and glass etching processes; also in wastes of metal-working and metallurgical plants.

Toxicity: very toxic; lethal dose for humans - 1 g; causes damage to the skin and mucous membranes, irritation of the respiratory tracts, and fluorosis [74,81,103].

Silicon; Si (28.09)

Dark gray crystals, m.p. 1415°C, b.p. ≅3250°C, ρ 2.32; insoluble in water.

Present in wastes from the production of organic derivatives of Si, scientific instruments, and transistors; also in wastes of metallurgical works.

Toxicity: has a general toxic, irritating, and carcinogenic effect; causes chronic catarrh of the upper respiratory tracts; chronic bronchitis, pneumonia, pulmonary emphysema, and coughing [4,42,45].

Silicon carbide; SiC (40.10)

Colorless crystals, m.p. 2830°C(d.), ρ 3.2; insoluble in water.

Present in wastes from the production of semiconductors, abrasives, and electric furnaces; also in wastes of metallurgical works.

Toxicity: has an irritating effect; causes bronchitis and sclerotic changes in the respiratory tracts [74,81,103].

Silicon dioxide; SiO_2 (60.08)

Colorless crystals, m.p. 1728°C, b.p. 2590°C, ρ 2.6; insoluble in water.

Present in wastes from the production of abrasives, glass, ceramics, enamel, and refractory materials.

Toxicity: causes pneumosclerosis; increases the tendency to contract pulmonary tuberculosis [74,81,103].

Determination: colorimetric analysis (sensitivity 1μg) [44].

Silicon fluoride; SiF_4 (104.03)

Colorless gas, m.p. -86.8°C, ρ 4.69 g/l.

Present in wastes from the processing of natural phosphates; and from the production of fluorides, barium fluorosilicate, and aluminum.

Toxicity: very toxic; has a general toxic and irritating effect [74,81,103].

Removal: by using cyclones or electrofilters (efficiency 98.5-99.0%) [74,105].

Silicon hydride, (silane); SiH_4 (32.12)

Colorless gas, m.p. -185°C, b.p. -119.9°C , ρ 2.85 g/l; reacts with water.

Present in wastes from the production of reagents and semiconductors.

Toxicity: causes headache, dizziness, semi-syncopal state [103].

Silicon tetrachloride; $SiCl_4$ (169.90)

Colorless liquid, m.p. -68.8°C, b.p. 57.0°C, ρ 1.48.

Present in wastes from the production of pure silicon and organosilicon compounds.

Toxicity: has an irritating effect [74,81,103].

SILVER

Silver; Ag (107.86)

White metal, m.p. 261.9°C, b.p. 2170°C, ρ 10.5; insoluble in water.

Present in wastes from the production of photographic materials, jewelry; electrotechnical and radio equipment, scientific instruments and apparatus, mirrors, and dental protheses; and from the synthesis of some organic compounds,; also in wastes of some metallurgical works.

Toxicity: slightly toxic; on prolonged exposure causes discoloration of the skin and mucous membranes.

Silver arsenite; Ag_3AsO_3 (446.55)

Light yellow crystals, m.p. 150°C(d.); very slightly soluble in water.

Toxicity: has a general toxic effect; affects the liver and kidneys [74,81].

Silver bromide, (bromyrite); AgBr (187.80)

Pale yellow crystals, m.p. 424°C, b.p. 1505°C, ρ 6.5; insoluble in water.

Present in wastes from the production of photographic materials and some pharmaceuticals.

Toxicity: causes coughing, rhinitis, and lacrimation [74,103].

Silver chlorate; $AgClO_3$ (191.34)

Colorless crystals, m.p. 230°C, ρ 4.4; soluble in water.

Present in wastes of some chemical plants.

Toxicity: causes discoloration of the skin and mucous membranes; irritation of the respiratory tracts and mucous membranes, methemoglobinemia, and destruction of erythrocytes [74,81].

Silver chloride, (ceragyrite); AgCl (143.32)

Colorless crystals, m.p. 455°C, b.p. 1550°C, ρ 5.56; insoluble in water.

Present in wastes from the preparation of some antiseptic compositions, photographic materials, detectors of cosmic radiations, and infrared spectrometers; also in wastes from electrolytic plating with silver.

Toxicity: causes discoloration of the skin and mucous membranes; and irritation of the respiratory tracts and mucous membranes [74,81].

Silver chromate; Ag_2CrO_4 (331.77)

Dark red crystals; slightly soluble in water.

Present in wastes from the production of pigments and ceramics.

Toxicity: very toxic; causes irritation of the respiratory tracts, mucous membranes, and the skin; perforation of the nasal septum, and ulceration and pigmentation of the skin [74,81].

Silver cyanide; AgCN (133.89)

Colorless crystals, m.p. 350°C, ρ 3.95; insoluble in water.

Present in wastes from some chemical processes and from electrolytic plating with silver.

Toxicity: very toxic; causes death from asphyxia; on prolonged exposure at low concentrations causes irritation of the respiratory tracts and the eyes; general weakness, headache, and nausea; decreases appetite [74,81].

Silver dichromate; $Ag_2Cr_2O_7$ (431.78)

Dark red crystals; decomposes on heating; ρ 4.7; insoluble in water.

Present in wastes from the production of pigments and ceramics.

Toxicity: has a general toxic, irritating, and carcinogenic effect; causes irritation of the respiratory tracts and mucous membranes, perforation of the nasal septum, and ulceration of the skin [74,81].

Silver difluoride; AgF_2 (145.86)

Brown crystals, m.p. 690°C, ρ 4.5-4.7; reacts with water.

Present in wastes from the production of organic derivatives of bromine and chlorine, and of some pharmaceuticals.

Toxicity: very toxic; has a general toxic and irritating effect; causes impairment of the lungs, heart, and respiratory tracts; on prolonged exposure at low concentrations causes fluorosis, anemia and loss of weight [74,81].

Silver fluoride; AgF(126.87)

Light yellow crystals, m.p. 435°C, b.p. ≅1000°C, ρ 5.8; solubility in water: 172 g/100 g H_2O at 20°C.

Present in wastes from the production of organic derivatives of bromine and chlorine, and of some pharmaceuticals.

Toxicity: very toxic; has a general toxic and irritating effect; causes impairment of the lungs, heart, and respiratory tracts; on prolonged exposure at low concentrations causes fluorosis, anemia, and loss of weight [74,81].

Silver iodate; $AgIO_3$ (282.80)

Colorless powder, m.p. 200°C, ρ 5.5; slightly soluble in water.

Present in wastes from the production of some reagents.

Toxicity: causes pigmentation of the skin and mucous membranes; and irritation of the respiratory tracts [74,81].

Silver iodide; AgI (234.77)

Light yellow crystals, m.p. 554°C, b.p. 1500°C, ρ 5.6; insoluble in water.

Present in wastes from the production of photographic materials, solid electrolytes, pyrotechnical compositions, and electrotechnical transducers; also in wastes from galvanic processes.

Toxicity: causes pigmentation of the skin and mucous membranes; and irritation of the respiratory tracts [74,81].

Silver nitrate, (lunar caustic); $AgNO_3$ (169.88)

Light yellow crystals, m.p. 209°C, b.p. 300°C(d.); slightly soluble in water.

Present in wastes from the production of photographic materials, some pharmaceuticals and analytical reagents, glass, mirrors, silver compounds, cosmetics, and special inks; also in wastes from galvanic processes.

Toxicity: has a general toxic and irritating effect; causes asthemia, general depression, headache, mental disorders, and pigmentation of the skin [74,81,103].

Silver nitrite; $AgNO_2$ (153.87)

Light yellow crystals, decomp. p. 140°C, ρ 4.49; solubility in water: 0.34 g/100 g H_2O at 20°C.

Present in wastes from the production of some chemical reagents.

Toxicity: has a general toxic and irritating effect; causes rapid pulse, a drop in blood pressure, visual disturbance, headache, and skin pigmentation [74,81].

Silver orthophosphate; Ag_3PO_4 (418.62)

Yellow crystals, m.p. 849°C, ρ 6.4; slightly soluble in water.

Present in wastes from the production of photographic materials.

Toxicity: affects the liver and kidneys; and causes pigmentation of the skin and mucous membranes [74,81].

Silver oxide; Ag_2O (231.74)

Brown-black crystals, m.p. ≅815°C, b.p. >300°C(d.), ρ 7.2; insoluble in water.

Present in wastes from the production of some catalysts and pharmaceuticals, glass, and electric batteries.

Toxicity: causes irritation of the respiratory tracts, the eyes, mucous membranes, and the skin; and pigmentation of the skin and mucous membranes [74,81, 103].

Silver peroxide; Ag_2O_2 (247.74)

Dark gray crystals, decomp. p. 100°C, ρ 7.4; insoluble in water.

Present in wastes from the production of some catalysts and pharmaceuticals, glass, and electric batteries.

Toxicity: causes irritation of the respiratory tracts, the eyes, and mucous membranes, and the skin; and pigmentation of the skin and mucous membranes [74, 81,103].

Silver sulfate; Ag_2SO_4 (311.79)

Colorless crystals, m.p. 660°C, decomp. p. 1085°C, ρ 5.4; solubility in water: 0.8 g/100 g H_2O at 20°C.

Present in wastes of some chemical plants.

Toxicity: causes pigmentation of the skin and mucous membranes [74,81].

Silver sulfide; Ag_2S (247.80)

Black crystals, m.p. 833°C, ρ 7.2; insoluble in water.

Present in wastes from the production of ceramics and solid electrodes.

Toxicity: slightly toxic; causes pigmentation of the skin and mucous membranes; coughing, a tickling in the throat, rhinitis, and lacrimation [74,81,103].

Silver thioarsenite; Ag_3AsS_3 (494.73)

Red crystals, m.p. 488°C, ρ 5.4; insoluble in water.

Present in wastes from the production of electrooptical and acoustic equipment.

Toxicity: has a general toxic, irritating, and allergenic effect [81].

SODIUM

Sodium; Na (22.898)

Silvery white metal, m.p. 97.83°C, b.p. 882.9°C, ρ 0.968; reacts with water.

Present in wastes from the production of tetraethyl lead, sodium oxide, synthetic rubber, dyes, and some medicinal preparations and organic compounds; also in wastes of some metallurgical works.

Toxicity: causes nausea, heartburn, irritation of mucous membranes and the respiratory tracts; and impairs the nervous system and the digestive system [74,81, 103].

Sodium acid carbonate, (bicarbonate; baking soda); $NaHCO_3$ (84.00)

Colorless crystals, m.p. 100-150°C(d.); soluble in water.

Present in wastes from the production of sodium salts, carbon dioxide, baking powder, carbonated beverages, fire extinguishers, and some pharmaceuticals.

Toxicity: slightly toxic; LD_{50} for rats - 4220 mg/kg [74,113].

Sodium acid fluoride; $NaHF_2$ (62.01)

Colorless crystals, m.p. 270°°C(d.); solubility in water: 3.25 g/100 g H_2O at 20°C.

Present in wastes from the production of some catalysts, ceramics, glass, and pesticides; also in wastes of steelworks, wood-working and metallurgical plants, and laundries.

Toxicity: very toxic; causes irritation of the eyes, respiratory tracts, and the skin, coughing, ulceration of the respiratory tracts, nephritis, and dermatitis; on prolonged exposure at low concentrations causes fluorosis [74,81,102]. LD_{50} for guinea pigs - 2000 mg/kg [103].

Sodium acid sulfate, (bisulfate); $NaHSO_4$ (120.07)

Colorless crystals, m.p. 186°C, ρ 2.74; solubility in water: 28.6 g/100 g H_2O at 25°C.

Present in wastes from the production of leather, magnesium cement, and wool; and in wastes of metal-working plants.

Toxicity: has a very toxic and an irritating effect [74,81].

Removal: with the aid of filters made of nylon, polyesters, polyethylene, poly-(vinyl chloride), teflon, or polypropylene [57].

Sodium acid sulfite, (bisulfite); $NaHSO_4$ (104.07)

Colorless crystals, ρ 1.48; soluble in water.

Present in wastes from the processing of wool (as a desinfectant and bleaching agent); also in wastes from the production of dyes, paper, fibers, articles made of rubber, and some pharmaceuticals.

Toxicity: at high concentrations causes irritation of the skin and mucous membranes; atrophy of the bone marrow; general depression and paralysis [74,81,113].

Sodium aluminate; $NaAlO_2$ (81.97)

Amorphous white powder, m.p. 1800°C; decomposes in water.

Present in wastes from the production of dyes, paper, opal glass, soap, and some construction materials.

Toxicity: causes irritation of the eyes, mucous membranes, and the skin [74, 113].

Sodium aluminum sulfate; $NaAl(SO_4)_2$ (242.10)

Colorless crystals, m.p. 60°C, ρ 1.61; soluble in water.

Present in wastes from the production of aluminum, potash alum, some pharmaceuticals, dyes, lacquers, paper, plant gums, cement, gelatin, and ceramics.

Toxicity: has an overall toxic, irritating, and allergenic effect; causes skin irritation and dermatitis [74,81,113].

Sodium arsenate; $Na_2HAsO_4 \cdot 12H_2O$ (424.07)

Colorless crystals, m.p. 86.3°C, ρ 1.76; soluble in water.

Present in wastes from the production of dyes, arsenates, and some pharmaceuticals.

Toxicity: very toxic; has an overall toxic, irritating, mutagenic, and teratogenic effect; causes nausea, vomiting, polyneuritis; degenerative changes in the liver and kidneys; and irritation of the respiratory tracts [74,81,93,103,113].

Sodium azide; NaN_3 (65.01)

Colorless crystals, m.p. >200°C, decomp. p. ≅300°C, ρ 1.85; solubility in water: 40.8 g/100 g H_2O at 20°C.

Present in wastes from the production of hydrazoic acid, lead azide, pure sodium, and some pharmaceuticals.

Toxicity: has an overall toxic, irritating, and searing effect; causes apathy, tremors, cramps, cyanosis; extreme difficulty in breathing, and death. LD_{50} for rats - 45 mg/kg [74,81,103,113].

Sodium boron hydride; $NaBH_4$ (37.75)

Colorless crystals, m.p. 505°C(d.), ρ 1.07; soluble in water.

Present in wastes from the production of aldehydes, ketones, organic acids, esters, chlorides, disulfides, and nitriles.

Toxicity: has an overall toxic and irritating effect; causes burns of mucous membranes, the skin, and the respiratory tracts; and damage to the liver, kidneys and the central nervous system. LD_{50} for rats - 160 mg/kg [74,81,113].

Sodium bromate; $NaBrO_3$ (150.91)

Colorless crystals, m.p. 384°C(d.), ρ 3.34; soluble in water.

Present in wastes from the preparation of some reagents; also in wastes from the purification of gold.

Toxicity: has an irritating effect. LD_{50} for mice - 140 mg/kg [74,81,113].

Sodium bromide; NaBr (102.89)

Colorless crystals, m.p. 760°C, b.p. 1390°C, ρ 3.21; soluble in water.

Present in wastes from the production of photographic materials and some pharmaceuticals and organic compounds.

Toxicity: at high concentrations causes general depression and mental disorders; on prolonged inhalation at low concentrations causes skin eruption (especially on the face) and furunculosis [74,81]. LD_{50} for rats - 3.5 g/kg [74].

Sodium carbonate; Na_2CO_3 (105.99)

Colorless crystals, m.p. 853°C, ρ 2.53; soluble in water.

Present in wastes from the production of glass, aluminum, soap, soda, sodium salts, surface-active agents, dyes, photographic materials, petrochemicals, textiles, and some organic chemicals; also in wastes of metal-working plants and ironworks.

Toxicity: has an overall toxic and irritating effect; causes irritation of the respiratory tracts, the nasal mucous membrane, and the eyes, conjunctivitis, perforation of the nasal septum; nausea, vomiting; skin irritation, eczema, and ulcers on the hands and wrists [74,81,102,103,113]. $TSDC_{acute}$ - 4.9 mg/m^3; ASL - 0.04 mg/m^3 [117].

Sodium chlorate; $NaClO_3$ (106.44)

Colorless crystals, m.p. 248°C, b.p. 630°C(d.), ρ 2.49; solubility in water: 50.2 g/100 g H_2O at 20°C.

Present in wastes from the production of chlorates, dyes, explosives, matches, tanning agents, leather, pesticides, and some pharmaceuticals and organic compounds.

Toxicity: has an overall toxic and irritating effect; causes nausea, vomiting, cyanosis, hemolytic anemia, methemoglobinemia, necrosis of the tubules of the kidney; irritation of the respiratory tracts, mucous membranes, and the skin. LD_{100} for rats and LD_{50} for mice -1200 and 596 mg/kg respectively [74,81,102,103,113].

Sodium chloride, (halite; common salt); NaCl (58.45)

Colorless crystals, m.p. 801°C, b.p. 1465°C, ρ 3.50; soluble in water.

Present in wastes from the production of chlorides, sodium carbonate, sodium hydroxide, some foodstuffs, soap, dyes, ceramics, and leather; also in wastes of some metallurgical works.

Toxicity: slightly toxic. $TC_{chronic}$ - 10 mg/m^3 [74,81,103].

Sodium chromate; Na_2CrO_4 (161.97)

Yellow crystals, ρ 2.72; soluble in water.

Present in wastes from the production of leather, textiles, some chemicals, lacquers and paints, some pharmaceuticals, ceramics, matches, and photographic materials; also in wastes of metal-working plants.

Toxicity: has an overall toxic, irritating, carcinogenic, and allergenic effect; causes irritation of the eyes and mucous membranes; perforation of the nasal septum, ulceration of the skin; and functional disturbances of internal organs [74,81,103,113].

Sodium cyanide; NaCN (49.02)

Colorless crystals, m.p. 563.7°C, b.p. 1497°C, ρ 1.59; solubility in water: 36.8 g/100 g H_2O at 20°C.

Present in wastes from the production of photographic materials, cyanides, some pharmaceuticals, and pesticides, and from lithographic processes and gold-plating and silver-plating processes; also in wastes of steelworks and metallurgical and woodworking plants.

Toxicity: very toxic; in the case of acute poisoning causes respiratory standstill, paralysis, cramps, death; in the case of chronic poisoning - paralysis of the lower extremities, encephalomalacia, dysfunctioning of the central nervous system, and damage to the spinal cord [74,81,103,113].

Sodium dichromate; $Na_2Cr_2O_7$ (252.0)

Red crystals, m.p. 320°C, b.p. 400°C(d.); solubility in water: 187 g/100 g H_2O at 25°C.

Present in wastes from the production of dyes, some organic compounds, inks, leather, electric batteries, fats and oils, resins, petrochemicals, chromic acid, chromates, pigments, and some pharmaceuticals; also in wastes of metal-working and machine-building plants.

Toxicity: has an overall toxic, irritating, allergenic, and carcinogenic effect [74,81,103]. MPSDC and MPADC - 0.0015 mg/m^3; toxicity classification - I [34].

Sodium fluoaluminate; $Na_3[AlF_6]$ (209.96)

Colorless crystals, m.p. 1009°C, ρ 2.9; soluble in water (0.39 g/l).

Present in wastes from the production of aluminum, ceramics, and pesticides; also in wastes of metallurgical and wood-working plants and steelworks.

Toxicity: very toxic; has an overall toxic and irritating effect; causes nausea, vomiting, irritation of the respiratory tracts, asphyxia; inflammation of the kidneys and the skin; pneumosclerosis, and bone sclerosis. $SC_{irritation}$ and $TC_{irritation}$ - 0.5 and 1.0 mg/m^3 respectively [81,102,103]; MPSDC and MPADC - 0.2 and 0.3 mg/m^3 respectively; toxicity classification - II [34].

Sodium fluoride, (villiaumite); NaF (41.99)

Colorless crystals, m.p. 992°C, ρ 2.79; solubility in water: 4 g/100 g H_2O at 15°C.

Present in wastes from the production of glass, ceramics, glues, and pesticides; also in wastes of metallurgical and wood-working plants.

Toxicity: very toxic; considered a protoplasmic poison; causes weakness and debility, lacrimation, ptyalism, difficulty in breathing, fluorosis, inflammation of the eyes, the respiratory tracts, and the skin, rhinitis, nosebleed, mottled enamel, osteosclerosis, and dysfunctioning of the central nervous system. SC- and TC -photosensitivity - 0.03 and 0.05 mg/m^3 respectively [74,81,103]; MPSDC and MPADC - 0.03 and 0.01 mg/m^3 respectively; toxicity classification - II [34].

Sodium fluosilicate; $Na_2[SiF_6]$ (188.05)

Colorless crystals, m.p. ≅570°C(d.), ρ 2.68; soluble in water.

Present in wastes from the production of enamel, glass, acid-proof cements and putties, textiles, pesticides, concrete, fats and oils, and leather; also in wastes of metallurgical and wood-working plants.

Toxicity: has an overall toxic and irritating effect; causes ptyalism, lacrimation, rhinitis, nosebleed, tracheobronchitis, dermatitis, chemical burn of the eyes; pulmonary hemorrhage, bronchial edema, and systolic slowdown. LD_{50} for mice, rats, and rabbits - 50, 158, and 225 mg/kg respectively [74,103,113]; MPSDC and MPADC - 0.03 and 0.01 mg/m^3; toxicity classification - II [34].

Sodium hydride; NaH (24.00)

Colorless crystals, m.p. >400°C(d.), ρ 1.38; decomposes in water.

Present in wastes from the production of fibers, paper, soap, and some chemical compounds; also in wastes of metallurgical works.

Toxicity: has an overall toxic and irritating effect; causes irritation of the cornea and the posterior region of the eyes, which may result in blindness [74,81, 103].

Sodium hydrosulfide; NaSH (56.06)

Colorless crystals, m.p. 350°C, ρ 1.79; solubility in water: 43 g/100 g H_2O at 20°C.

Present in wastes from the production of hides, artificial silk, dyes, sulfur-containing compounds, and some organic compounds.

Toxicity: very toxic; has an overall toxic and irritating effect; in the case of acute poisoning at high concentrations causes excitement followed by depression of the central nervous system; conjunctivitis, formation of blisters on the cornea; dizziness, a staggering gait, asphyxia; irritation of the respiratory tracts, pulmonary edema, respiratory paralysis, death; in the case of chronic poisoning at low concentrations - loss of weight, mental disorders, headache, irritation of the eyes and edema of the eyelids [74,81,113].

Sodium hydroxide; NaOH (40.00)

Colorless crystals, m.p. 322°C, b.p. 1385°C, ρ 2.10; solubility in water: 108.7 g/100 g H_2O at 20°C.

Present in wastes from the production of some pharmaceuticals and chemical compounds, sodium salts, petrochemicals, synthetic fibers, paper, textiles, soap, plastics, cellophane, articles made of rubber, chlorine, dyes, leather, and fats and oils; also in wastes of metal-working plants.

Toxicity: has an overall toxic and irritating effect; causes burns of tissues; damage to the eyes, conjunctivitis, inflammation of the cornea, blindness; irritation of the respiratory tracts, laryngeal edema, asphyxia; chemical burns of the skin, eczema, ulcers, and formation of scales; damage to the cardiovascular system, and coma [74,81,102,103]; ASL - 0.01 mg/m^3 [117].

Removal: with the aid of scrubbers [76]. and of filters made of nylon, polyethylene, teflon, or polypropylene [51].

Sodium iodide; NaI (149.92)

Colorless crystals, m.p. 662°C, b.p. 1304°C, ρ 3.67; soluble in water.

Present in wastes from the production of some pharmaceuticals; also in wastes of electrotechnical plants.

Toxicity: has an overall toxic and irritating effect; causes weakness and debility, headache, general depression, irritation of mucous membranes, the eyes, and the skin [74,81,113].

Sodium metaarsenate; $NaAsO_2$ (129.90)

Light gray crystals, ρ 1.87; soluble in water.

Present in wastes from the production of soap for leather goods, and pesticides; also in wastes of wineries.

Toxicity: very toxic; in the case of acute poisoning causes nausea, vomiting, shock, death; in the case of chronic poisoning - polyneuritis, hemopathy, degenerative changes in the liver and kidneys; skin eruption and pigmentation; and chemical burns of tissues. LD_{50} for mice - 1.17 mg/kg [74,81,103,113].

Sodium molybdate; Na_2MoO_4 (205.92)

Colorless powder, m.p. 687°C, ρ 3.28; soluble in water.

Present in wastes from the production of pigments and growth stimulants; also in wastes of metal-working plants.

Toxicity: slightly toxic; has an irritating effect [74,113].

Sodium monosulfide; Na_2S (78.05)

Colorless crystals, m.p. 1180°C, ρ 1.86; soluble in water.

Present in wastes from the production of leather, wool, viscose, articles made of rubber, dyes, textiles, and some reagents; also in wastes from metal-working, flotation, and etching processes.

Toxicity: very toxic; causes damage to the eyes and skin, and the formation of small ulcers around joints. LD_{100} for rabbits - 6 mg/kg [74,84,103].

Sodium nitrate, (soda niter); $NaNO_3$ (84.99)

Colorless crystals, m.p. 308°C, ρ 2.26; solubility in water: 88 g/100 g H_2O at 20°C.

Present in wastes from the production of fertilizers, some foodstuffs, glass, explosives, nitric and sulfuric acids, sodium nitrite, enamel, matches, tobacco, and meat preservatives; also in wastes of metal-working plants.

Toxicity: has an overall toxic and irritating effect; at large concentrations causes dizziness, vomiting, weakness, cramps, death; at low concentrations - weakness, general depression, headache, mental disorders, and a thickening of the skin of the palms of hands and of soles [74,81,103,113].

Sodium nitrite; $NaNO_2$ (69.00)

Colorless or yellowish crystals, m.p. 271°C, b.p. >320°C(d.), ρ 2.17; solubility in water: 82.9 g/100 g H_2O at 20°C.

Present in wastes from the production of dyes, some foodstuffs, articles made of rubber, textiles, photographic materials, and some pharmaceuticals and analytical reagents; also in wastes of steelworks.

Toxicity: $SC_{chronic}$, $TC_{chronic}$, and LC_{50} - 0.000025, 0.000125, and 0.0055 mg/l respectively [74,81,103,113].

Sodium oxide; Na_2O (61.98)

Colorless crystals, subl.p°C, ρ 2.27.

Present in wastes from the production of some organic chemicals.

Toxicity: very toxic; irritates the eyes, mucous membranes, and the respiratory tracts; on contact causes chemical burns of tissues [81,102,103,113].

Sodium perborate; $NaBO_3 \cdot 4H_2O$ (153.88)

Colorless crystals, m.p. >60°C(d.); soluble in water.

Present in wastes from the bleaching of matchwood, fibers, bristles, spongy materials, wax, and textiles; and from the production of toothpowder and toothpaste, soap, and antiseptics.

Toxicity: causes pathological changes in the central nervous system and in the kidneys [74,81].

Sodium perchlorate; $NaClO_4$ (122.44)

Colorless crystals, m.p. 482°C(d.); soluble in water.

Present in wastes from the production of explosives.

Toxicity: has an overall toxic and irritating effect; causes irritation of mucous membranes and the skin [74,81].

Sodium peroxide; Na_2O_2 (77.99)

Colorless crystals, m.p. 460°C(d.), ρ 2.6; soluble in water.

Present in wastes from the production of bleaching agents for fibers, feathers and bones, ivory, wood, wax, sponges, coral, and textiles; and from the production of dyes and some reagents.

Toxicity: very toxic; has an irritating and searing effect; causes chemical burns of the skin [74,81,103].

Sodium selenate; Na_2SeO_4 (188.94)

Colorless crystals, m.p. 500°C(d.), ρ 3.09; soluble in water.

Present in wastes from the production of pesticides and photographic and lithographic materials.

Toxicity: causes irritation of the central nervous system, the liver, kidneys, the respiratory tracts, and mucous membranes; and dermititis. LD_{100} for rabbits - 4 mg/kg [74,81,103].

Sodium selenite; Na_2SeO_3 (172.94)

Colorless crystals, m.p. 710°C; soluble in water.

Present in wastes from the production of glass and some organic compounds.

Toxicity: very toxic; has an overall toxic and irritating effect; causes irritation of the respiratory tracts, mucous membranes, and the skin. LD_{50} for rats - 3 mg/kg [74,81,103].

Sodium sulfite; Na_2SO_3 (126.06)

Colorless crystals, ρ 2.63; decomposes on heating; soluble in water.

Present in wastes from the production of photographic materials; bleaching agents for wool, matchwood, and silk; hydrogen sulfide, dyes, textiles, paper, and presevatives; also in wastes from the silver-plating of glass.

Toxicity: at high concentrations causes retardation of growth; irritation of nerves, atrophy of the bone marrow, general depression, and paralysis [74,81,113].

Sodium tetraborate; $Na_2B_4O_7$ (201.27)

Colorless crystals, m.p. 742°C, b.p. 1675°C, ρ 2.36; soluble in water.

Present in wastes from the production of antiseptics, glass, cosmetics, pigments, enamel, paper, soap, and some pharmaceuticals; also in wastes of wood-working plants.

Toxicity: has an overall toxic and irritating effect; causes headache, nausea, vomiting, brain edema; fatty degeneration of the liver and kidneys; and baldness. LD_{50} for rats - 2660 mg/kg [74,81,102,113].

Sodium thiosulfate; $Na_2S_2O_3 \cdot 5H_2O$ (248.17)

Colorless crystals, m.p. 48.5°C, ρ 1.71; soluble in water.

Present in wastes from the production of cellulose sulfate, photographic materials, dyes, textiles, leather, and some chemical reagents; and from the bleaching of bones, matchwood, and ivory; also in wastes of mining works (in the extraction of silver).

Toxicity: slightly toxic [74,81].

Sodium tungstate; Na_2WO_4 (293.83)

Colorless crystals, m.p. 698°C, ρ 4.18; soluble in water.

Present in wastes from the production of fireproof and waterproof materials;

some complex compounds, alkaloids, and pigments.

Toxicity: has an overall toxic and irritating effect. LD_{50} for guinea pigs and rats - 1152 and 1190 mg/kg respectively [74,81,103,113]; MPADC (calculated for W) - 0.10 mg/m^3; toxicity classification - III [35].

Sodium uranate; $Na_2U_2O_7 \cdot H_2O$ (652.15)

Yellow powder, slightly soluble in water.

Present in wastes from the production of green-yellow fluorescent glass, pigments for porcelain, and enamel.

Toxicity: very toxic; has an overall toxic, irritating, and carcinogenic effect; causes dermatitis, damage to the kidneys, necrotic changes in the blood vessels, death [74,81].

STRONTIUM

Strontium; Sr (87.62)

Golden yellow metal, m.p. 768°C, b.p. 1390°C, ρ 2.6; reacts with water.

Present in wastes from the production of tracer bullets, storage batteries, pyrotechnical preparations, and red signal lanterns; also in wastes of some metallurgical and electrotechnical plants.

Toxicity: has an overall toxic and irritating effect [74,81,103].

Removal: see [4,42].

Strontium bromide; $SrBr_2$ (247.43)

Colorless crystals, m.p. 643°C, b.p. 1972°C, ρ 4.21; solubility in water: 102.4 g/100 g H_2O at 20°C.

Present in wastes from the production of some monocrystals and pharmaceuticals.

Toxicity: causes methemoglobinemia, damage to the central nervous system, and general depression. LD_{50} for rats - 1 g/kg [74,81].

Strontium carbonate; $SrCO_3$ (147.62)

Colorless crystals, m.p. ≅1500°C(in atmosphere of CO_2), ρ 3.7; slightly soluble in water.

Present in wastes from the production of varnishes, candles, glass, and pyrotechnical compositions; also in wastes of sugar refineries.

Toxicity: has an overall toxic effect [74,103].

Strontium chlorate; $Sr(ClO_3)_2$ (254.54)

Colorless crystals, m.p. 120°C, ρ 3.15; soluble in water.

Present in wastes from the production of pyrotechnical compositions.

Toxicity: very toxic; has an overall toxic and irritating effect; causes methemoglobinemia, destruction of erythrocytes and damage to the kidneys [74,81].

Strontium chloride; $SrCl_2$ (158.53)

Colorless crystals, m.p. 873°C, b.p. ≅2040°C, ρ 3.05; solubility in water: 59.9 g/100 g H_2O at 20°C.

Present in wastes from the production of pyrotechnical compositions, strontium salts, cosmetics, and some pharmaceuticals [103].

Toxicity: LD_{50} for rats, mice, guinea pigs, and rabbits - 1.79, 1.03, 2.84, and 1.51 g/kg respectively [103].

Strontium chromate; $SrCrO_4$ (203.62)

Yellow crystals, m.p. 1283°C(d.), ρ 3.8; slightly soluble in water.

Present in wastes from the production of dyes, pigments, lacquers, anticorrosive enamels and primings.

Toxicity: has an overall toxic and irritating effect; causes irritation of the respiratory tracts, mucous membranes, and the skin. LD_{50} for rats - 3.11 g/kg [74, 81,103].

Strontium hydroxide; $Sr(OH)_2$ (121.62)

Colorless crystals, m.p. 460°C, decomp.p. >500°C, ρ 3.6; solubility in water: 0.8 g/100 g H_2O at 20°C.

Present in wastes from the production of strontium ointment; also in wastes of sugar-beet processing and ore-processing plants.

Toxicity: causes mild chemical burns of the eyes, mucous membranes, and the skin [74,81,103].

Strontium iodide; SrI_2 (341.43)

Colorless crystals, m.p. 538°C, b.p. ≅1900°C, ρ 4.5; solubility in water: 177.7 g/100 g H_2O at 20°C.

Present in wastes from the production of scintillation counters, some pharmaceuticals, and adsorbents.

Toxicity: has an irritating effect [74].

Strontium nitrate; $Sr(NO_3)_2$ (211.63)

Colorless crystals, m.p. 645°C, ρ 2.9; solubility in water: 70.8 g/100 g H_2O at 20°C.

Present in wastes from the production of pyrotechnical compositions, strontium compounds, matches, flares, and signal rockets.

Toxicity: has an overall toxic effect (causes cramps, collapse, death); at low concentrations causes general weakness and depression, headache, and mental disorders [74,81,103].

Strontium peroxide; SrO_7 (119.63)

White powder, m.p. >900°C(d.), ρ 4.5; decomposes in water.

Present in wastes from the production of some antiseptics.

Toxicity: has an irritating effect [74,81].

Strontium selenate; $SrSeO_4$ (230.59)

Colorless crystals, ρ 4.25; reacts with water.

Present in wastes from the production of glass, dyes, some plastics, photoelectric equipment, and fireproof cables; also in wastes of some metallurgical works.

Toxicity: has an overall toxic, irritating, and carcinogenic effect; causes pneumonia, fatty degeneration of the liver; degenerative changes in the kidneys; irritation of the respiratory tracts, coughing, and pulmonary edema [74,81].

SULFUR

Hydrogen sulfide; H_2S (34.8)

Colorless gas with a pungent odor, m.p. -85.44°C, b.p. -60.35°C; solubility in water: 0.378 g/100 g H_2O at 20°C.

Present in wastes from the production of sulfur, sulfuric acid, sulfites, sulfurous dyes, barium sulfide, Prussian blue (by chemical process), ultramarine, hydrogen sulfide, sulfur chloride, soda, some pesticides and pharmaceuticals, articles made of rubber, viscose, textiles, leather, gelatin, and glues; also in wastes of metallurgical works and petroleum refineries.

Toxicity: very toxic; has an overall toxic and irritating effect; is absorbed by healthy skin; causes dizziness, headache, nausea, vomiting, lacrimation, photophobia, conjunctivitis, irritation of the respiratory tracts, pneumonia and pulmonary edema, cyanosis, damage to the cardiac muscle, cramps, collapse, death. $MPSDC_{odor}$ - 0.002-0.01 mg/m^3 [221], 0.031-0.035 mg/m^3 [222]; TC_{odor} - 0.01-0.05 mg/m^3 [203]; MPSDC and MPADC - 0.008 mg/m^3; toxicity classification - II [34]; MPSDC and MPADC established in Bulgaria, Hungary, Poland, Czechoslovakia, and Yugoslavia - 0.0008 mg per m^3, in GDR - 0.008 and 0.015 mg/m^3 respectively, in Romania - 0.03 and 0.01 mg/m^3 respectively [72,88]; MPC established in Czechoslovakia, Great Britain, Sweden, and USA - 0.08 kg/hr, 7.5 mg/m^3, and 10 mg/m^3, and 230 mg/m^3 respectively [72,88].

Removal: by adsorption and then catalytic oxydation [111]; scrubbing (efficiency 89%) [76]; deodorization (efficiency 99%) [208]; multiple-stage adsorption [209]; combustion [7,43,48,95].

Determination: colorimetric analysis [47], photocolorimetric analysis [6,18,44, 69,78,96,98,210].

Sulfur; S (32.06)

Yellow crystals or powder; m.p.112.8-119°C, b.p. 144.6°C; ρ 1.96-2.07; insoluble in water.

Present in wastes from the production of fertilizers; some pharmaceuticals, and organic compounds, pesticides, plastics, synthetic fibers, and explosives; paper, articles made of rubber, sulfuric acid, sulfites, hydrogen sulfide, matches, enamel, cement, dyes, wool, silk, felt, and textiles; also in wastes of metallurgical and woodworking plants and from the manufacture of straw hats.

Toxicity: has an irritating effect; causes bronchitis, pulmonary emphysema, conjunctivitis, damage to the cornea, photophobia, lacrimation, heightened susceptibility to fatigue, abnormal irritability, dizziness, and chest pains. $TC_{irritation}$ - 4.5 mg/m^3 [74,81,102,103]; MPSDC recommended - ≦0.010 mg/m^3 for a period of 1 hr; MPC recommended - ≦10t/year [104].

Sulfur dichloride; SCl_2 (102.97)

Dark red liquid, m.p. -123°C, b.p. 59°C, ρ 1.6; reacts with water.

Present in wastes from the production of articles made of rubber, carbon tetrachloride, lacquers, pesticides, cement, and some pharmaceuticals; also in wastes of some metallurgical and chemical works.

Toxicity: has an overall toxic and irritating effect; causes irritation of the eyes, lungs, and mucous membranes [81].

Sulfur dioxide; SO_2 (64.07)

Colorless gas, m.p. -75.46°C, b.p. -10.1°C, ρ 2.93 g/l; solubility in water: 9.61 g/100 g H_2O at 20°C.

Present in wastes from the production of sulfuric acid, ammonium sulfate, ceramics, caprolactam, linoleum, roof-sheeting materals, tar paper, artificial parchment paper, foam plastics, mineral-fiber plates, some foodstuffs, textiles, and paper; also in wastes of plants operating on solid fuels, metallurgical plants, thermal power plants, and sugar-beet processing plants.

Toxicity: has an overall toxic and irritating effect; irritates the eyes and the respiratory tracts; has a harmful effect on embryos and on the hemopoietic organs; causes lacrimation, coughing, headache, dyspnea, general weakness, bronchial spasms, and methemoglobinemia [93,103]; affects plants [107], metals, construction materials, paints, leather, paper, and textiles [88]. TC_{odor} - 0.08 [55] and 0.25 mg/m^3 [211]; $TC_{chronic}$ - 0.1 mg/m^3 [5]; $TC_{c.n.s.}$ - 0.6 mg/m^3 [5]; destroys plants at concentrations of 1.0 - 2.0 mg/m^3 [116]; MPSDC and MPADC - 0.5 and 0.05 mg/m^3 respectively; toxicity classification - III [34]; MPSDC and MPADC established in Bulgaria, GDR, Hungary, Czechoslovakia, and Yugoslavia - 0.5 and 0.15 mg/m^3 respectively [88], in Poland - 0.25 and 0.075 mg/m^3 respectively, and in Romania - 0.75 and 0.25 mg/m^3 respectively [88].

Removal: by technological methods [5,212]; with the aid of electrostatic precipitators [212]; by adsorption [7,212-217]; and combustion [88,95,111,112,218,219].

Determination: colorimetric, nephelometric [18,47], or gas analysis [18], or photometric analysis [6,24,44,69,76,96,220].

Sulfur hexafluoride; SF_6 (146.05)

Colorless gas, m.p. -51°C, subl. p. 64°C, ρ 6.5 g/l; slightly soluble in water [74,81].

Present in wastes from the production of some pesticides; and electrotechnical and electronic equipment.

Toxicity: has an overall toxic and irritating effect; causes pulmonary hemorrhage and methemoglobinemia; on prolonged exposure at low concentrations - fluorosis [74,81].

Sulfur monochloride; S_2Cl_2 (135.03)

Yellow liquid, m.p. -82°C, b.p. 137.1°C, ρ 1.67; reacts with water.

Present in wastes from the production of articles made of rubber, carbon tetrachloride, lacquers, pesticides, cement, and some pharmaceuticals; also in wastes of some metallurgical and chemical works.

Toxicity: has an overall toxic and irritating effect; causes irritation of the eyes, lungs, and mucous membranes [81].

Sulfur monofluoride; S_2F_2 (102.12)

Colorless gas, m.p. -133°C, b.p. 15°C; reacts with water.

Present in wastes from the production of pesticides and electrochemical and electronic equipment.

Toxicity: has an overall toxic and irritating effect; causes pulmonary hemorrhage and methemoglobinemia; on prolonged exposure at low concentrations - fluorosis [74,81].

Sulfur tetrachloride; SCl_4 (173.87)

Yellow-brown liquid, m.p. -30°C; reacts with water.

Present in wastes from the production of articles made of rubber, carbon tetrachloride, lacquers, pesticides, cement, and some pharmaceuticals; also in wastes of some metallurgical and chemical works.

Toxicity: has an overall toxic and irritating effect; causes irritation of the eyes, lungs, and mucous membranes [81].

Sulfur tetrafluoride; SF_4 (108.05)

Colorless gas, m.p. 121°C, b.p. -38°C, ρ 1.9; reacts with water [74,81].

Present in wastes from the production of some pesticides, and electrotechnical and electronic equipment.

Toxicity: has an overall toxic and irritating effect; causes pulmonary hemorrhage and methemoglobinemia; on prolonged exposure at low concentrations - fluorosis [74, 81].

Sulfur α-trioxide; SO_3 (80.07)

Colorless crystals, m.p. 16.8°, b.p. 44.8°C, ρ 2.75; soluble in water.

Present in wastes from the production of sulfuric acid, vegetable oil, and soap; also in wastes of coal-gasification plants.

Toxicity: has an overall toxic and irritating effect [74,81]. MPS established in Sweden - 0.5 kg/t H_2SO_4 [88].

Removal: see [7,10,43,45].

Sulfuric acid; H_2SO_4 (98.08)

Colorless liquid, m.p. 10.3°C, b.p. 296.2°C(d.), ρ 1.92; soluble in water.

Present in wastes from the production of sulfuric and sulfamic acids, hydrogen sulfide; sulfites and ammonium sulfate; some fertilizers, organic compounds, pesticides, plastics and pharmaceuticals; enamel, glass, articles made of rubber, dyes, matches, paper and pulp, explosives and smoke-forming preparations, textiles, leather, and viscose; and in wastes of metallurgical and metal-working plants and petroleum refineries; also formed in air under natural conditions from sulfur dioxide and nitrogen oxides [10].

Toxicity: has an irritating effect; causes chemical burns; pharyngospasm, damage to the lungs, bronchitis; ulceration of the skin, and dermatitis; a persistent fog in the air which has a harmful effect on humans, animals, plants and the soil, polutes water resrvoirs, and causes damage to buildings. $MPADC_{odor}$ and TC_{odor} - 0.43 and 0.60 mg/m^3 respectively [222]; $TC_{c.n.s.}$ - 0.4 mg/m^3 [5]; $TC_{irritation}$ and TC (sensitivity to light) - 0.5 and 2.4 mg per cu. m respectively [103]; MPSDC and MPADC - 0.3 and 0.1 mg/m^3 respectively; toxicity classification - II [34]; MPC (Czechoslovakia) - 0.1 kg/hr [88].

Removal: with the aid of scrubbers [76]; electrostatic precipitators [5], or filters made of polyesters, polyethylene, teflon, or polypropylene [5]; by burning [95].

Determination: see [18,44,223,224].

Sulfuric oxychloride (sulfuryl chloride); SO_2Cl (134.96)

Colorless liquid, m.p. -54°C, b.p. 69.5°C, ρ 1.66; soluble in water.

Present in wastes from the production of chlorophenol and chlorothymol, and some catalysts and solvents.

Toxicity: causes irritation of mucous membranes, the respiratory tracts, and the skin [74,81].

Sulfuric oxyfluoride, (sulfuryl fluoride); SO_2F_2 (102.07)

Colorless gas, m.p. -135.81°C, b.p. -55.37°C, ρ 3.72 g/l; slightly soluble in water.

Present in wastes from the production of fumigants and some pesticides.

Toxicity: has an overall toxic and irritating effect; causes irritation of the eyes and the respiratory tracts; on prolonged exposure - fluorosis [74,81].

TANTALUM

Tantalum; Ta (180.94)

Gray metal, m.p. 2014°C, b.p. ≅5500°C, ρ 16.6; insoluble in water.

Present in wastes from the production of some medical instruments and electric vacuum equipment; also in wastes of some metallurgical works.

Toxicity: see [74,81,103].

Tantalum chloride; $TaCl_5$ (358.21)

Light yellow crystals, m.p. 216.5°C, b.p. 236°C, ρ 3.68; reacts with water.

Present in wastes of some metallurgical works and also from the production of tantalum.

Toxicity: has an overall toxic and irritating effect [74,81].

Tantalum fluoride; TaF_5 (275.95)

Colorless crystals, m.p. 97.0°C, b.p. 229.2°C, ρ 4.67; reacts with water.

Present in wastes of some metallurgical works.

Toxicity: very toxic; has an overall toxic and irritating effect; on prolonged exposure at low concentrations causes fluorosis [74,81].

Tantalum pentaoxide; Ta_2O_5 (441.89)

Colorless crystals, m.p. 1872°C, ρ 8.5-8.7; insoluble in water.

Present in wastes from the production of special types of glass, some catalysts, and synthetic rubber.

Toxicity: causes bronchitis, pulmonary emphysema, and pneumonia [74,103].

TELLURIUM

Tellurium; Te (127.60)

Silvery gray metal, m.p. 450°C, b.p. 990°C, ρ 6.2; insoluble in water.

Present in wastes from the production of tellurium compounds and its alloys, semiconductors, articles made of rubber, dyes, glass, ceramics, enamel, and some pharmaceuticals.

Toxicity: has an overall toxic effect; causes nausea, vomiting, and sleepiness; depresses the central nervous system; is absorbed by healthy skin. $TSDC_{acute}$ - 0.01-0.05 mg/m^3 [74,81,103].

Determination: spectrophotometric analysis [44,69].

Tellurium dioxide; TeO_2 (159.60)

Colorless crystals, m.p. 732°C, b.p. 1257°C, ρ 5.8; slightly soluble in water.

Present in wastes from the electrolytic purification of metals, and from the production of laser equipment and optical glass.

Toxicity: has an overall toxic effect; causes loss of weight, paralysis, and damage to the central and peripheral nervous systems and the liver. $TC_{chronic}$ - 10 mg per cu. m [74,103]; MPSDC and MPADC - 0.5 μg/m^3; toxicity classification - I [34].

Tellurium hydride; TeH_2 (129.62)

Colorless gas, m.p. -51°C, b.p. -2°C, ρ 5.81 g/l; slightly soluble in water.

Present in wastes from the electrolytic purification of metals.

Toxicity: very toxic; causes irritation of the respiratory tracts, the eyes, and mucous membranes, vomiting, and dizziness [74,81].

THALLIUM

Thallium; Tl (204.37)

Silvery white metal, m.p. 303.6°C, b.p. 1457°C, ρ 11.85; insoluble in water.

Present in wastes from the production of some pesticides, semiconductors, and thermometers; also in wastes of metallurgical works and instrument-making plants.

Toxicity: very toxic; causes impairment of the central and peripheral nervous systems, cramps, paralysis, polyneuratis, pain of joints, interrupted sleep, nausea, vomiting, skin atrophy, and shedding of hair [74,81,102,103].

Removal: see [7,42,43].

Determination: spectrophotometric or polarographic analysis [69].

Thallium carbonate; Tl_2CO_3 (468.75)

Colorless crystals, m.p. 269°C(d.), ρ 7.2; soluble in water.

Present in wastes from the production of optical glass, jewelry, pyrotechnical compositions, and thallium compounds.

Toxicity: very toxic; causes irritation of the respiratory tracts, the eyes, and mucous membranes; polyneuritis, cramps, and damage to the kidneys [74,81,103].

Thallium dichromate; $Tl_2Cr_2O_7$ (624.73)

Red crystals; insolible in water.

Present in wastes from the production of some pesticides; also in wastes of instrument-making plants.

Toxicity: very toxic; has an overall toxic, irritating and carcinogenic effect; causes irritation of the respiratory tracts, mucous membranes, and the skin; ulceration of the skin; psychic disturbances, cramps, and paralysis [81].

Thallium monobromide; TlBr (284.27)

Greenish yellow crystals, m.p. 460°C, ρ 7.5; slightly soluble in water.

Present in wastes from the production of some monocrystals and thermometers; also in wastes of metallurgical works and instrument-making plants.

Toxicity: very toxic; causes damage to the central and peripheral nervous systems, incoordination of movements, cramps, paralysis, psychic disturbances, insomnia, and impairment of vision [74,81,103,113].

Thallium monochloride; TlCl (239.82)

Colorless crystals, m.p. 431°C, b.p. 820°C, ρ 7.0; slightly soluble in water: 0.32 g/100 g H_2O at 20°C.

Present in wastes from the production of infrared optical glass and some catalysts and monocrystals.

Toxicity: very toxic; has an overall toxic and irritating effect; causes polyneuritis and impairment of vision [74,81,103].

Thallium monofluoride; TlF (223.37)

Colorless crystals, m.p. 322°C, b.p. 840°C, ρ 8.36; soluble in water.

Present in wastes from the production of fluorinated ethers; also in wastes of some metallurgical works.

Toxicity: has an overall toxic and irritating effect [74,81].

Thallium monoiodide; TlI (331.27)

Bright yellow crystals, m.p. 442°C, b.p. 833°C, ρ 7.09; slightly soluble in water.

Present in wastes from the production of some monocrystals, halogen lamps, and infrared optical materials.

Toxicity: very toxic; causes irritation of the respiratory tracts, mucous membranes, and the skin, skin eruption; impairment of the central and peripheral nervous systems, cramps [74,81,103].

Thallous hydroxide; TlOH (221.38)

Light yellow crystals, m.p. 125°C(d.); solubility in water: 34.3 g/100 g H_2O at 10°C.

Present in waste from the production of thallium compounds.

Toxicity: very toxic; causes irritation of the respiratory tracts; damage to the liver and kidneys, cramps, polyneuritis, paralysis; and impairment of vision [74,81, 103].

Thallous nitrate; $TlNO_3$ (266.37)

Colorless crystals, m.p. 208°C, decomp. p. >300°C; solubility in water: 9.55 g/100 g H_2O at 20°C.

Present in wastes from the production of some analytical reagents and plastics.

Toxicity: very toxic; causes irritation of the respiratory tracts and mucous membranes; damage to the liver, kidneys, and the brain; cramps, paralysis; general depression, headache, and mental disorders [74,81].

Thallous oxide; Tl_2O (424.78)

Black crystals, m.p. 303°C, b.p. ≅1100°C, ρ 9.5; reacts with water.

Present in wastes from the production of glass with a large refractive index, synthetic precious stones, and thallium compounds.

Toxicity: very toxic; has an overall toxic and irritating effect; causes irritation of the respiratory tracts, mucous membranes, and the eyes [74,81].

Thallous sulfate; Tl_2SO_4 (504.80)

Colorless crystals, m.p. 645°C, ρ 6.75; solubility in water: 4.87 g/100 g H_2O at 20°C.

Present in wastes from the production of thallium salts, some pesticides, some analytical reagents, and cosmetics.

Toxicity: very toxic; causes impairment of the respiratory tracts; cramps, paralysis, collapse, death; at low concentrations has an overall toxic and irritating effect. LD_{50} for mice - 24-35 mg/kg [74,81,103].

Thallous sulfide; Tl_2S (440.80)

Black crystals, m.p. 448.9°C, b.p. 1177°C, ρ 8.4; insoluble in water.

Present in wastes from the production of special types of glass and semiconductors.

Toxicity: has an overall toxic and irritating effect; causes irritation of the respiratory tracts, mucous membranes, and the eyes, cramps, and paralysis [74,81].

THORIUM

Thorium; Th (232.03)

Silvery white metal, m.p. 1750°C, b.p. 4200°C, ρ 11.72; insoluble in water.

Present in wastes from the manufacture of nuclear reactors and watches; and from the production of some catalysts and organic compounds; also in wastes of some metallurgical works.

Toxicity: has a toxic effect due to radioactivity [74,81,103].

Thorium dioxide, (thorianite); ThO_2 (264.04)

Colorless crystals or white powder, m.p. 3350°C, b.p. 4400°C, ρ 9.7; insoluble in water.

Present in wastes from the production of thorium, refractory materials, some catalysts, silicat-free optical glass, electric vacuum equipment, and some pharmaaceuticals.

Toxicity: has an overall toxic effect. $TC_{chronic}$ - 10 mg/m^3 [74,103].

Thorium nitrate; $Th(NO_3)_4 \cdot 4H_2O$ (552.15)

Colorless crystals, d.p. >400°C; soluble in water.

Present in wastes from the production of some reagents and pharmaceuticals and electrotechnical equipment.

Toxicity: has an overall toxic effect. $TC_{chronic}$ - 10 mg/m^3 [74,103].

Thorium tetrachloride; $ThCl_4$ (373.85)

Colorless crystals, m.p. 770°C, b.p. 922°C, ρ 4.5; solubility in water: 55.61 g/100 g H_2O at 0°C.

Present in wastes from the production of thorium.

Toxicity: causes irritation of the eyes, the respiratory tracts, and mucous membranes [74].

TIN

Tin; Sn (118.69)

Gray metal, m.p. 231.9°C, b.p. 2620°C, ρ 5.85; insoluble in water.

Present in wastes from the production of tin plate, bronze, brass, and babbit (alloy for bearings); also in wastes of metallurgical and some chemical works.

Toxicity: has an overall toxic effect [103].

Tin dibromide; $SnBr_2$ (278.50)

Yellow crystals, m.p. 232°C, b.p. 641°C, ρ 5.18; reacts with water.

Present in wastes of metallurgical works.

Toxicity: has an overall toxic and irritating effect; causes general depression and mental disorders, skin eruption, and furunculosis [74,81].

Tin dichloride; $SnCl_2$ (189.60)

Colorless crystals, m.p. 247°C, b.p. 652°C, ρ 3.95; soluble in water.

Present in wastes from the production of textiles, dyes, wires, glass, plastics, tin compounds, some pharmaceuticals, paper, lubricants, leather, catalysts, some analytical reagents, perfumes, soap, ceramics, and silk; also in wastes from metal-working plants.

Toxicity: has an overall toxic and irritating effect; causes chemical burns irritation of the eyes and mucous membranes, and ulceration of the skin [74,81,103].

Tin difluoride; SnF_2 (156.69)

Colorless crystals, m.p. 215.05°C, b.p. 853°C; soluble in water.

Present in wastes from the production of toothpastes and some pharmaceuticals; also in wastes of electrolytic plants.

Toxicity: has an overall toxic and irritating effect [74].

Tin dioxide; SnO_2 (150.69)

Colorless crystals, m.p. 2000°C, b.p. ≅2500°C, ρ 7.01; insoluble in water.

Present in wastes from the production of silicate enamels, glazes, opal glass, ceramics, abrasives, and pigments.

Toxicity: causes pneumoconiosis and pulmonary emphysema [74,103].

Tin disulfide; SnS_2 (156.69)

Golden crystals, m.p. 500°C(d.), ρ 4.5; insoluble in water.

Present in wastes from the production of paper, lacquers and paints, and gypsum; also in wastes of nonferrous metallurgical and wood-working plants.

Toxicity: slightly toxic [74].

Tin hydride; SnH_4 (122.72)

Colorless gas, b.p. -52°C, ρ 4.3.

Toxicity: very toxic [103].

Tin iodide; SnI_4 (626.31)

Brownish yellow crystals, m.p 144.5°C, b.p. 348.6°C, ρ 4.47; reacts with water.

Present in wastes of some chemical works.

Toxicity: has an overall toxic effect [74].

Tin monoxide; SnO (134.70)

Black crystals; when heated changes into tin dioxide, ρ 6.44; insoluble in water.

Present in wastes from the production of silicate enamels, glazes, opal glass, ceramics, abrasives, and pigments.

Toxicity: causes pneumoconiosis and pulmonary emphysema [74,103].

Tin sulfate; $SnSO_4$ (214.75)

Colorless crystals, m.p. >360°C(d.); soluble in water.

Present in wastes from electrolytic plating processes.

Toxicity: has an irritating effect [74].

Tin tetrabromide; $SnBr_4$ (438.31)

Colorless crystals, m.p. 30°C, b.p. 208°C, ρ 3.35; reacts with water.

Present in wastes of metallurgical works.

Toxicity: has an overall toxic and irritating effect; causes general depression, mental disorders, skin eruption, and furunculosis [74,81].

Tin tetrachloride; $SnCl_4$ (260.50)

Colorless liquid, m.p. -33°C, b.p. 114°C, ρ 2.23; reacts with water.

Present in wastes from the production of textiles, dyes, wires, glass, plastics, tin compounds, some pharmaceuticals, paper, lubricants, leather, catalysts, some analytical reagents, perfumes, soap, ceramics, and silk; also in wastes of metal-working plants.

Toxicity: has an overall toxic and irritating effect; causes chemical burns and irritation of the eyes and mucous membranes, and ulceration of the skin [74,81,103].

TITANIUM

Titanium; Ti (47.09)

Present in wastes of some metallurgical and chemical machine-building plants and of steelworks; and from the production of titanium bronze, aluminum, and electric vacuum equipment.

Toxicity: has a carcinogenic effect [74,81,93,103].

Determination: absorption spectroscopy (sensitivity 0.01 $\mu g/m^3$) [80].

Titanium diboride; TiB_2 (49.9)

Gray crystals, m.p. 2850°C, ρ 4.45; insoluble in water.

Present in wastes from the production of refractory materials, muffles, cruicibles, and heat-resisting alloys; also in wastes of some metallurgical works.

Toxicity: slightly toxic; on prolonged exposure at low concentrations causes pathological changes in the lungs; affects the coordination of movements, and causes paralysis [103].

Titanium dioxide, (anatase, rutile); TiO_2 (79.90)

Colorless (anatase) or yellow (rutile) crystals, m.p. 1870°C, ρ 3.6-3.9 (anatase), 4.2-4.3 (rutile); insoluble in water.

Present in wastes from the production of lacquers and paints; plastics, synthetic fibers, articles made of rubber, leather, ceramics, titanium, enamels, inks, and some pharmaceuticals; also in wastes of some metallurgical works.

Toxicity: has an overall toxic and irritating effect; causes sclerotic changes in the lungs [74,81,103].

Titanium hydride; TiH_2 (49.9)

Dark gray powder, ρ 3.79.

Present in wastes of some metallurgical works and also from the production of cathode-ray tubes and ceramics.

Toxicity: very toxic; irritates and causes chemical burns of the respiratory tracts and mucous membranes [74,81].

Titanium phosphate; TiP (78.88)

Present in wastes of metallurgical works and also from the production of semiconducting materials.

Toxicity: impairs the activity of the central nervous system; causes irritation of the lungs and pulmonary edema; functional disturbances of the heart and other internal organs; tremors, fatigue, insomnia, nausea, and vomiting; in the case of prolonged exposure causes anemia, bronchitis, and impairment of vision [81].

Titanium tetrachloride; $TiCl_4$ (189.71)

Light yellow liquid, m.p. -24.1°C, b.p. 136.3°C, ρ 2.8; reacts with water.

Present in wastes from the production of titanium, textiles, leather, glass, and artificial pearls.

Toxicity: has an overall toxic and irritating effect; causes difficulty in breathing, chronic bronchitis, pulmonary edema; and opacity of the cornea [74,81,103].

Titanium trichloride; $TiCl_3$ (154.28)

Reddish purple crystals, m.p. 440°C, d.p. >500°C; reacts with water.

Present in wastes from the production of some organic compounds and catalysts.

Toxicity: causes irritation of the eyes, the respiratory tracts, and mucous membranes [74,81].

TUNGSTEN

Tungsten; W (193.85)

Light gray metal, m.p. 3420°C, b.p. 5700°C, ρ 19.3; insoluble in water.

Present in wastes of metallurgical works; and from the production of electrotechnical and radio equipment, and x-ray and TV equipment.

Toxicity: has an irritating effect. $SSDC_{acute}$ and $TSDC_{acute}$ - 0.1 and 0.5 mg/m^3 respectively [1].

Tungsten dioxide; WO_2 (215.84)

Brown crystals, m.p. ≅ 1500°C, b.p. 1700°C, ρ 12.11; insoluble in water.

Present in wastes from the production of metallic tungsten, tungsten derivatives, ceramics, glass, and some catalysts.

Toxicity: has an overall toxic and irritating effect [103].

Tungsten hexacarbonyl; $W(CO)_6$ (351.98)

Colorless crystals, m.p. 169°C, b.p. 175°C, ρ 2.65; insoluble in water.

Present in wastes from the production of tungsten.

Toxicity: has an overall toxic effect; affects the lungs. $TSDC_{acute}$ - 0.35-0.45 mg/l [81,103].

Tungsten hexachloride; WCl_6 (396.57)

Bluish black crystals, m.p. 283°C, b.p. 340°C, ρ 3.52; insoluble in water.

Present in wastes of metal-working and machine-building plants.

Toxicity: causes retardation of growth, trachco-bronchitis, and necrosis of the skin; also has an irritating effect. $TC_{chronic}$ - 6 mg/m^3 [103].

Tungsten hexafluoride; WF_6 (297.86)

Colorless gas or slightly yellow liquid, m.p. 2.0°C, b.p. 17.3°C, ρ 3.44.

Present in wastes from the production of tungsten.

Toxicity: has an irritating effect; also causes changes in bone tissues and ligaments, and fluorosis [74,103].

Tungsten trioxide; WO_3 (231.85)

Orange yellow crystals, m.p. ≅1473°C, b.p. ≅1850°C, ρ 7.16; insoluble in water.

Present in wastes from the production of tungstates, pigments, ceramics, and some catalysts.

Toxicity: has an overall toxic effect. $SC_{irritation}$ and $TC_{irritation}$ - 0.020 and 0.49 mg/m^3 respectively [2].

Tungstic acid; H_2WO_4 (249.86)

Yellow powder, m.p. 180°C(d.), ρ 5.5; slightly soluble in water.

Present in wastes from the production of tungsten, textiles, and some catalysts.

Toxicity: affects the respiratory tracts and mucous membranes [103].

URANIUM

Uranium; U (238.03)

Silvery white metal, m.p. 1134°C, b.p. ≅4200°C, ρ 19.04; reacts with water.

Present in wastes from the production of uranium and its derivatives, and nuclear fuels; also in wastes of nuclear power plants.

Toxicity: very toxic; has an overall toxic and carcinogenic effect; causes necrosis of the blood vessels and damage to the kidneys [74,81,103].

Uranium carbide; UC_2 (262.09)

Reddish gray crystals, m.p. ≅2500°C(d.), b.p. 4370°C, ρ 11.2; decomposes in water.

Present in wastes from the production of ceramic components of nuclear fuel elements; and in wastes of nuclear power plants.

Toxicity: has an overall toxic and carcinogenic effect; causes damage to the blood vessels and kidneys [81].

Uranium diboride; UB_2 (248.9)

Silvery gray crystals, m.p. ≅2380°C, ρ 12.70; insoluble in water.

Present in wastes from the production of refractory materials, heat-resistant alloys, resistors, and nuclear equipment.

Toxicity: very toxic; has an overall toxic and irritating effect; is carcinogenic; causes necrosis of the blood vessels and pulmonary edema; does damage to the kidneys, the skin, and the nervous system; impairs vision [81].

Uranium dioxide, (uraninite); UO_2 (270.03)

Dark brown crystals, m.p. 2850°C, b.p. 3450°C, ρ 10.9; insoluble in water.

Present in wastes from the production of nuclear fuel and ceramic components of nuclear fuel elements.

Toxicity: has an overall toxic and carcinogenic effect; causes damage to the blood vessels and kidneys [81].

Uranium disulfide; US_2 (302.20)

Grayish black crystals, m.p. 1680°C(d.); decomposes in water.

Present in wastes from the production of solid lubricants.

Toxicity: very toxic; has an overall toxic and carcinogenic effect; causes damage to the blood vessels, the kidneys, and the skin [81].

Uranium hexafluoride; UF_6 (352.06)

Colorless crystals, m.p. 64.0°C, ρ 5.06; reacts with water.

Present in wastes from the production of uranium compounds.

Toxicity: very toxic; has an overall toxic, irritating, and carcinogenic effect; causes necrosis of the blood vessels; irritation of mucous membranes, the respiratory tracts, the eyes, and the nasopharynx; pulmonary edema; and degenerative changes in the kidneys. LC_{100}- 1.22 mg/m^3 [74,81,103].

Uranium pentachloride; UCl_5 (415.29)

Reddish brown crystals, m.p. 320(d.), ρ 3.18; reacts with water.

Present in wastes from the production of uranium compounds.

Toxicity: very toxic; has an overall toxic and carcinogenic effect; causes necrosis of the blood vessels and damage to the kidneys [74,81,103].

Uranium peroxide; $UO_4 \cdot 2H_2O$ (338.10)

Light yellow crystals, d.p. 260°C, ρ 4.31-4.66; slightly soluble in water [74,81, 103].

Uranium tetrabromide; UBr_4 (557.73)

Brown crystals, m.p. 519°C, b.p. 761(d.), ρ 5.3; soluble in water.

Present in wastes from the production of uranium compounds.

Toxicity: very toxic; has an overall toxic and carcinogenic effect; causes damage to the walls of blood vessels and to the kidneys [81].

Uranium tetrachloride; UCl_4 (379.84)

Dark green crystals, m.p. 590°C, b.p. 792°C, ρ 4.8; reacts with water.

Present in wastes from the production of uranium compounds.

Toxicity: very toxic; has an overall toxic and carcinogenic effect; causes necrosis of the blood vessels and damage to the kidneys [74,81,103].

Uranium tetrafluoride; UF_4 (314.07)

Green crystals, m.p. 1036°C, b.p. 1730°C; soluble in water.

Present in wastes from the production of uranium compounds.

Toxicity: very toxic; has an overall toxic and carcinogenic effect; causes necrosis of the blood vessels; irritation of mucous membranes, the respiratory tracts, the eyes, and the nasopharynx; pulmonary edema; and degenerative changes in the kidneys. LC_{100}- 1-22 mg/m^3 [74,81,103].

Uranium tribromide; UBr_3 (477.74)

Dark red crystals, m.p. 730°C, ρ 5.98; reacts with water [81].

Present in wastes from the production of uranium compounds.

Toxicity: very toxic; has an overall toxic and carcinogenic effect; causes damage to the walls of blood vessels and to the kidneys [81].

Uranium trichloride; UCl_3 (344.39)

Red crystals, m.p. 842°C, b.p. 1780°C, ρ 5.35; reacts with water.

Present in wastes from the production of uranium compounds.

Toxicity: very toxic; has an overall toxic and carcinogenic effect; causes necrosis of the blood vessels and damage to the kidneys [74,81,103].

Uranium trioxide; UO_3 (286.03)

Yellow crystals, ρ 7.29; insoluble in water.

Present in wastes from the production of ceramic components of nuclear fuel elements; also in wastes of nuclear power plants.

Toxicity: has an overall toxic and carcinogenic effect; causes damage to the blood vessels and kidneys [81].

Uranous sulfate; $U(SO_4)_2 \cdot 4H_2O$ (502.27)

Green crystals, d.p. >300°C; soluble in water.

Present in wastes from the production of uranium fluoride.

Toxicity: very toxic; has an overall toxic and carcinogenic effect; causes necrosis of the blood vessels, and damage to the kidneys and the skin [81].

Uranyl nitrate; $UO_2(NO_3)_2 \cdot 6H_2O$ (502.13)

Yellow crystals, m.p. 59.5°C, ρ 2.8; soluble in water.

Present in wastes from the production of uranium, photographic materials, glaze, porcelain, and some analytical reagents.

Toxicity: very toxic; has an overall toxic and carcinogenic effect; causes damage to the blood vessels, kidneys, and the skin; internal bleeding, and death. LD_{50} for dogs and rabbits - 5 mg/kg [74,103].

VANADIUM

Vanadium; V (50.95)

Light yellow metal, m.p. 1920°C, b.p. 3400°C, ρ 5.96; insoluble in water.

Present in wastes of metallurgical works.

Toxicity: has an overall toxic effect on internal organs, the nervous system, and the circulatory system; also an irritating and carcinogenic effect. $TSDC_{acute}$ -

0.06 mg/m^3 [74,81,93,103]; MPADC - 0.02 mg/m^3; SC recommended in the USA - 0.005 mg/m^3 [76]; toxicity classification - I [76].

Determination: colorimetric analysis [44,47]; absorption or emission spectroscopic analysis [99,115].

Vanadium dichloride; VCl_2 (121.85)

Green crystals, m.p. ≅ 1350°C, ρ 3.23; soluble in water.

Present in wastes of metallurgical works.

Toxicity: has an overall toxic and irritating effect [81].

Vanadium pentaoxide; V_2O_5 (181.88)

Orange crystals, m.p. 680°C, ρ 3.36; slightly soluble in water.

Present in wastes from the production of ferrovanadium, sulfuric acid, glass, and some organic compounds.

Toxicity: causes bronchitis, bronchopneumonia, pulmonary edema, dystrophic changes in the internal organs, impairment of circulatory and respiratory organs and the nervous system, and of metabolism; and irritation of the eyes and mucous membranes. $SC_{irritation}$ - 0.002 mg/l; $TSDC_{acute}$ - 0.003-0.005 mg/l; TC_{acute} - 0.01 mg/l; $TC_{irritation}$ - 0.006 mg/m^3 [238]; MPADC - 0.002 mg/m^3; toxicity classification - I [34]; MPADC established in Bulgaria, GDR, Czechoslovakia, and Yugoslavia - 0.002, 0.002, 0.003, and 0.003 mg/m^3 respectively [72].

Vanadium trioxide; V_2O_3 (149.88)

Black crystals, m.p. 1970°C, b.p. 3000°C, ρ 4.87; insoluble in water.

Present in wastes from the production of textiles.

Toxicity: has an overall toxic and irritating effect. $TSDC_{acute}$ - 40-75 mg/m^3 [103].

Vanadyl dichloride, (vanadium oxychloride); $VOCl_2$ (137.86)

Green crystals, ρ 2.88.

Present in wastes from the production of dyes.

Toxicity: has an irritating effect [74].

Vanadyl trichloride, (vanadium oxytrichloride); $VOCl_3$ (173.32)

Yellow liquid, m.p. -78°C, b.p. 126.7°C, ρ 1.830.

Present in wastes from the production of semiconductors and films.

Toxicity: has an overall toxic and irritating effect [74,81].

ZINC

Zinc; Zn (63.37)

Silvery white metal, m.p. 419.5°C, b.p. 906°C, ρ 7.13; insoluble in water.

Present in wastes from the production of electrodes in chemical batteries; also in wastes of metallurgical plants, steel mills, and metal-working and machine-building plants.

Toxicity: has an overall toxic, irritating, and carcinogenic effect [69,103]; causes a sweet taste in the mouth, dryness of throat, coughing, nausea, vomiting; irritation of the skin and mucous membranes; insomnia, loss of weight, impairment of memory, sweating, anemia, hemorrhage, and pulmonary edema. $TC_{chronic}$ - 5 mg/m^3 [74,81,113]; MPSDC recommended in the USA - 0.1 mg/m^3 [76].

Removal: with the aid of electrostatic filters [76].

Determination: colorimetric analysis (sensitivity 1 μg) [47]; absorption spectroscopic analysis (sensitivity 0.0002 mg/m^3); or emission spectroscopic analysis (sensitivity 0.12 $\mu g/m^3$) [44,99,115].

Zinc arsenide; Zn_3As_2 (345.96)

Gray crystals, m.p. 1015°C; insoluble in water.

Present in wastes from the production of arsenous hydride (arsine).

Toxicity: has an overall toxic, irritating, and allergenic effect; causes destruction of erythrocytes and hemoglobin; at high concentrations causes damage to the kidneys, pulmonary edema, cramps, collapse, death; at low concentrations - anemia, headache, weakness, nausea, and vomiting [81].

Zinc arsenite, (zinc metarsenite); $Zn(AsO_2)_2$ (279.2)

Colorless crystals; insoluble in water.

Present in wastes from the production of some pesticides; also in wastes of wood-working plants.

Toxicity: very toxic; has an overall toxic and allergenic effect; destroys hemoglobin and erythrocytes; causes pulmonary edema, cramps, collapse, death; at low concentrations causes anemia, headache, weakness, nausea, and vomiting [81].

Zinc bromide; $ZnBr_2$ (225.2)

Colorless crystals, m.p. 394°C, b.p. 670°C, ρ 4.22; solubility in water: 470 g/100 g H_2O at 25°C.

Present in wastes from the production of photographic materials and nuclear fuel [74].

Toxicity: causes general depression, mental disorders, skin eruption, furunculosis, irritation of mucous membranes, coughing, insomnia, and impairment of memory [74,81].

Zinc chlorate; $Zn(ClO_3)_3 \cdot 4H_2O$ (304.36)

Colorless crystals, ρ 2.15; decomposes on heating; soluble in water.

Present in wastes from the production of some herbicides, defoliants, desiccants, and pyrotechnical compositions.

Toxicity: very toxic; causes irritation of mucous membranes and the skin; loss of weight, impairment of memory, destruction of hemoglobin and erythrocytes, and damage to the cardiac muscle [81].

Zinc chloride; $ZnCl_2$ (136.28)

Colorless crystals, m.p. 318°C, b.p. 732°C, ρ 2.9; solubility in water: 367 g/100 g H_2O at 20°C.

Present in wastes from the production of wood preservatives, paper, viscose, zinc paints, activated carbon, articles made of rubber, desinfectants, impregnating compositions for crossties, cement, and textiles; also in wastes of metallurgical, woodworking and machine-building plants, petroleum refineries, and some chemical plants.

Toxicity: very toxic; has an overall toxic, irritating, and carcinogenic effect; causes irritation and chemical burns of the respiratory tracts, mucous membranes, and the skin; coughing, and dyspnea. LD_{50} for rats - 75 mg/kg [74,81,103].

Removal: with the aid of filters made of polyesters, polypropylene, poly(vinyl chloride), or teflon [57].

Zinc chromate; $ZnCrO_4$ (181.36)

Lemon-colored crystals, ρ 5.3; slightly soluble in water.

Present in wastes from the production of tanning agents, preservatives, and some analytical reagents.

Toxicity: very toxic; causes irritation of the eyes, the respiratory tracts, the skin(ulceration), and mucous membranes [81].

Zinc cyanide; $Zn(CN)_2$ (117.42)

Colorless crystals, m.p. 800°C(d.); slightly soluble in water.

Present in wastes from electrolytic plating of metals and from the production of some pharmaceuticals.

Toxicity: very toxic; causes difficult breathing, cramps, and paralysis of the respiratory center; at low concentrations - headache, nausea, vomiting, dizziness, irritation of mucous membranes, insomnia, and impairment of memory [74,81].

Zinc dichromate; $ZnCr_2O_7 \cdot 3H_2O$ (335.40)

Reddish brown crystals; soluble in water.

Present in wastes from the production of explosives, some organic compounds, tanning agents, and wood preservatives.

Toxicity: very toxic; has an overall toxic and irritating effect; causes irritation of the eyes, mucous membranes, and the respiratory tracts, perforation of the nasal septum; eczema, skin ulceration, and functional disturbances of internal organs [74, 81].

Zinc fluoride; ZnF_2 (103.88)

Colorless crystals, m.p. 875°C, b.p. 1505°C, ρ 4.8; soluble in water.

Present in wastes from the fluorination of organic compounds and from galvanic processes; also in wastes from the production of phosphor for fluorescent lamps, porcelain, glass, enamels, wood preservatives, luminophors, and lasers.

Toxicity: has an overall toxic and irritating effect; causes irritation of mucous membranes and the skin; on prolonged exposure at low concentrations causes fluorosis [74,81].

Zinc fluosilicate; $Zn[SiF_6]\cdot 6H_2O$ (315.5)

Colorless crystals, ρ 2.104; decomposes on heating; solubility in water: 36.16 g/100 g H_2O at 20°C.

Present in wastes from the production of wood preservatives, glass, cement, and ceramics.

Toxicity: has an overall toxic and irritating effect; causes irritation of mucous membranes and the skin; on prolonged exposure at low concentrations - fluorosis [74, 81].

Zinc hexaborate; $Zn_2B_6O_{11}$ (371.62)

Colorless crystals, m.p. 610°C, ρ 4.2; slightly soluble in water.

Present in wastes from the production of glazes, glass, enamels, ceramics, pigments, antiseptics, and fungicides.

Toxicity: has an overall toxic effect, and also a cumulative effect; is harmful to the central nervous system, and causes general depression [81].

Zinc hydroxide; $Zn(OH)_2$ (92.4)

Colorless crystals, m.p. 125°C(d.), ρ 3.05; insoluble in water.

Present in wastes of some chemical plants.

Toxicity: causes a sweet taste in the mouth; xerosis of the throat; irritation of mucous membranes and the skin; and insomnia [81].

Zinc iodate; $Zn(IO_3)_2$ (415.17)

Colorless crystals, ρ 4.98; decomposes on heating; slightly soluble in water.

Present in wastes from the production of antiseptics.

Toxicity: has an overall toxic and irritating effect; causes irritation of the skin and mucous membranes; coughing, formation of methemoglobin in the blood, and paralysis of the central nervous system [74,81].

Zinc iodide; ZnI_2 (319.22)

Colorless crystals, m.p. 446°C, b.p. 624°C, ρ 4.67; solubility in water: 432 g/100 g H_2O at 18°C.

Present in wastes from the synthesis of some pharmaceuticals.

Toxicity: causes irritation of mucous membranes and the skin, skin eruption, coughing, formation of methemoglobin in the blood, paralysis of the central nervous system, headache, anemia, loss of weight, and general depression [74,81].

Zinc nitrate; $Zn(NO_3)_2 \cdot 6H_2O$ (297.49)

Colorless crystals, m.p. 36.4°C, ρ 2.06; solubility in water: 127.3 g/100 g H_2O at 26°C.

Present in wastes from the production of some dyes.

Toxicity: causes irriatation of mucous membranes, coughing, dizziness, weakness, cramps, collapse; at low concentrations - general depression, headache, and mental disorders [81].

Zinc oxide; ZnO (81.38)

Colorless crystals, m.p. 1975°C, ρ 5.67; slightly soluble in water.

Present in wastes from the production of glass, dyes, some pharmaceuticals, articles made of rubber, matches, textiles, ceramics, porcelain, cement, dental cement, enamel, celluloid, typographic paints, cosmetics, and inks; also in waste of metallurgical and engineering works.

Toxicity: causes a sweet taste in the mouth, fatigue, insomnia, dry cough, irritation of the upper respiratory tracts, headache, zinc-fume fever, immobility, difficult breathing, purulent bronchitis, cyanosis, and pneumonia [74,81,102,103]. MPADC - 0.05 mg/m^3; toxicity classification - III [34].

Zinc permanganate; $Zn(MnO_4)_2 \cdot 6H_2O$ (411.33)

Dark purple crystals, m.p. 100°C, ρ 2.47; soluble in water.

Present in wastes from the production of some pharmaceuticals.

Toxicity: affects the central nervous system; causes speech disturbance, insomnia, and irritation of the respiratory tracts [81].

Zinc peroxide; ZnO_2 (97.38)

Yellowish powder, ρ 1.57; soluble in water.

Present in wastes from the production of some pharmaceuticals.

Toxicity: has an overall toxic and irritating effect; causes irritation of mucous membranes, insomnia, loss of weight, impairment of memory [74,81].

Zinc phosphide; Zn_3P_2 (258.06)

Steel gray crystals, m.p. 1193°C, ρ 4.5; soluble in hot water.

Present in wastes from the production of some pesticides.

Toxicity: very toxic; has an overall toxic and irritating effect; causes headache, dizziness, dyspnea, chest pains, respiratory impairment, cyanosis, cramps, damage to the heart, liver and kidneys, pulmonary hemorrhage, and irritation of mucous membranes and the skin [74,81,102,103].

Zinc selenide; ZnSe (144.34)

Lemon-colored crystals, m.p. 1575°C, ρ 5.4; insoluble in water.

Present in waste from the production of lasers and luminophors.

Toxicity: has an overall toxic, irritating, and carcinogenic effect; causes irritation of the respiratory tracts, coughing, pulmonary edema, pneumonia, fatty degeneration of the liver, and degenerative changes in the liver cells [74,81].

Zinc sulfate; $ZnSO_4 \cdot 7H_2O$ (287.54)

Colorless crystals, ρ 1.97; soluble in water.

Present in wastes from the production of viscose, mineral paints, wood preservatives, paper, lithopone, zinc salts, glues, and some analytical reagents; also in wastes of engineering, metal-working, and wood-working plants, and from electrolytic plating of metals.

Toxicity: causes ulceration of the skin [74,103].

Removal: by filtration [57].

Zinc sulfide; ZnS (97.45)

Colorless crystals, m.p. 1775°C, ρ 4.08; insoluble in water.

Present in wastes from the production of pigments, oilcloth, linoleum, leather, and lithopone.

Toxicity: slightly toxic [81].

Zinc telluride; ZnTe (192.99)

Red crystals, m.p. 1238°C, ρ 6.34; insoluble in water.

Present in wastes from the production of semiconductors.

Toxicity: absorbed by undamaged skin; has an overall toxic and irritating effect; causes loss of appetite, fatigue, and insomnia [74,81].

ZIRCONIUM

Zirconium; Zr (91.22)

Silvery white metal, m.p. 1855°C, b.p. ≅4340°C, ρ 6.5; insoluble in water.

Present in wastes from the production of plastics, electrotechnical equipment, pyrotechnical compositions, ceramics, glass, dyes, some catalysts used in organic synthesis, and pure zirconium; also in wastes of steelworks, chemical and engineering plants, and nuclear energy installations.

Toxicity: has an overall toxic and irritating effect; causes granulomatosis, general depression, dermititis, and irritation of the respiratory tracts [74,81,102,103].

Zirconium boride; ZrB_2 (112.84)

Gray crystals, m.p. 3200°C, ρ 5.82; insoluble in water.

Present in wastes from the production of special alloys, cermets, abrasives, and refractory materials; also in wastes of nuclear power plants.

Toxicity: has an overall toxic effect; causes dystrophic changes in the liver, kidneys, and cardiac muscle; and incoordination of movements [103].

Zirconium carbide; ZrC (103.23)

Gray crystals, m.p. ≅3800°C, b.p. ≅5100°C, ρ 6.7; insoluble in water.

Present in wastes from the production of abrasives, ceramics, and refractory coatings for metals; also in wastes of steelworks.

Toxicity: has an overall toxic and irritating effect; causes dizziness and headache; at high concentrations causes asphyxia, granulomatosis, general depression, and dermititis; at low concentrations - irritation of the respiratory tracts [81,103].

Zirconium chloride; $ZrCl_4$ (233.03)

Colorless crystals, m.p. 437°C(in vacuo), ρ 2.8; reacts with water.

Present in wastes from the production of zirconium, and some catalysts used in organic synthesis.

Toxicity: has an overall toxic and irritating effect; causes irritation of mucous membranes, the respiratory tracts, and the eyes, pharyngospasm, and pulmonary edema [74,103,113].

Zirconium dichloride; $ZrCl_2$ (162.13)

Dark crystals, m.p. 350°C(d.); insoluble in water.

Present in wastes from some chemical plants.

Toxicity: has an overall toxic and irritating effect; causes irritation of the respirartory tracts, the eyes, and mucous membranes [81].

Zirconium dioxide, (baddeleyite); ZrO_2 (123.22)

Colorless crystals, m.p. 2700°C, b.p. ≅4300°C, ρ 5.68; insoluble in water.

Present in wastes from the production of refractory materials, ceramics, enamels, special types of glass, and crucibles.

Toxicity: has an overall toxic and irritating effect; causes granulomatosis, general depression, dermatosis, and irritation of the respiratory tracts [74,81,102,103].

Zirconium diphosphide; ZrP_2 (153.17)

Gray crystals, m.p. 750°C(d.), ρ 4.7; insoluble in water.

Present in wastes from the production of semiconductors and antioxidants.

Toxicity: very toxic; causes depression of the central nervous system, irritation of the lungs, pulmonary edema, cramps, fatigue, anemia, bronchitis, and impairment of vision, speech and locomotory functions [81].

Zirconium fluoride; ZrF_4 (167.21)

Colorless crystals, m.p. 910°C, ρ 4.43; slightly soluble in water: 1.5 g/100 g H_2O at 25°C.

Present in wastes from the production of pure zirconium and special types of glass.

Toxicity: has an overall toxic and irritating effect; at low concentrations causes osteosclerosis, calcification of the ligaments, and mottled enamel [74].

Zirconium hydride; ZrH_2 (93.24)

Dark gray crystals, m.p. 800°C(d.); insoluble in water.

Present in wastes of powder metallurgy plants and from the production of ceramics and vacuum equipment.

Toxicity: very toxic; has an overall toxic and irritating effect; causes destruction of erythrocytes and chemical burns [74,81].

Zirconium iodide; ZrI_4 (598.84)

Brown crystals, m.p. 500°C(at 0.92 MPa); reacts with water.

Present in wastes from the production of pure zirconium.

Toxicity: has an overall toxic and irritating effect; causes granulomatosis, general depression, dermatitis, and irritation of the respiratory tracts [74].

Zirconium nitrate; $Zr(NO_3)_4 \cdot 5H_2O$ (429.32)

Colorless crystals, d.p. 75°C; soluble in water.

Present in wastes of some chemical plants.

Toxicity: has an overall toxic and irritating effect; causes dizziness, weakness, cramps, collapse, general depression, headache, and mental disorders [81].

Zirconium oxybromide; $ZrOBr_2$ (216.0)

Colorless crystals, m.p. 120°C; soluble in water.

Present in wastes of nuclear engineering plants; and from the production of coagulants and zirconium compounds.

Toxicity: has an overall toxic and irritating effect [81].

Zirconium oxychloride; $ZrOCl_2 \cdot 8H_2O$ (322.25)

Colorless crystals, ρ 1.55; decomposes on heating; soluble in water.

Present in wastes from the production of zirconium salts and some analytical reagents.

Toxic: has an overall toxic and irritating effect; causes general depression, dermatitis, granulomatosis, and irritation of the respiratory tracts [74,81,102,103].

Zirconium oxysulfide; ZrOS (139.28)

Yellow crystals, ρ 4.87; insoluble in water.

Present in wastes of some chemical plants.

Toxicity: very toxic; has an overall toxic and irritating effect [81].

Zirconium silicate, (zirconium orthosilicate); $ZrSiO_4$ (183.30)

Colorless crystals, m.p. 1650°C(d.), ρ 4.56; insoluble in water.

Present in wastes from the production of glass, ceramics, cement, and cosmetics; also in wastes of foundries.

Toxicity: see [74,103].

Zirconium sulfate; $Zr(SO_4)_2 \cdot 4H_2O$ (355.40)

Colorless crystals, m.p. >450°C(d.); solubility in water: 64 g/100 g H_2O at 18°C.

Present in wastes from the production of some catalysts, amino acids, and proteins; and tanning agents.

Toxicity: has an overall toxic and irritating effect; causes granulomatosis, general depression; dermatitis, and irritation of the respiratory tracts [74].

APPENDIX

Inorganic dust

Inorganic dust is present in the wastes of thermal power stations, metallurgical plants, mining works, machine-building plants, etc; and from the production of construction materials.

Toxicity: causes irritation of the respiratory tracts, conjunctivitis, and keratitis; particles $\geqq 10\mu m$ in diameter are retained in the upper respiratory tracts and cause their irritation; particles $\leqq 10\mu m$ in diameter penetrate the lung tissues and cause pulmonary fibrosis; particles $<5\mu m$ in diameter are especially dangerous [225, 226,229]. SC-chronic and TC-chronic - 0.05 and 0.15 mg/m^3 respectively; MPSDC and MPADC - 0.5 and 0.15 mg/m^3 respectively; toxicity classification - III [34]. In the USSR the MPSDC and MPADC (mg/m^3) for dusts containing silicon oxide are as follows [34]:

% SiO_2 in wastes from the production of	MPSDC	MPADC	Toxicity classification
Silica refractory brick (85-90%)	0.15	0.05	III
Refractory clay (50%)	0.3	0.10	III
Cement (20%)	0.3	0.10	III
Dolomite (8%)	0.3	0.15	III

In the USA and France the maximum permissible average daily concentration for these types of dust is 0.15 mg/m^3 [88,101].

Below are listed the maximum permissible concentrations of inorganic dust present in the air of populated areas which meet the sanitation standards set in different countries [88]:

Industries	Country	MPC (mg/m^3)
All industries	Australia, New Zealand	250
All industries	Czechoslovakia	5 kg/hr
Asphalt	USA	70
Asphalt	Canada	230
Calcium carbide	FRG	150
Ceramics	FRG	150
Coal-tar chemicals	Canada	46
Glass-melting	FRG	150
Coal		
briquetting	FRG	150
burning:		
ash content <20%	France	350

ash content >20%	France	500
combustion in domestic furnaces	Belgium	300
combustion in all types of furnaces	Switzerland	100
in the combustion of garbage	Belgium	150
Sulfite pulp (new plants)	Denmark	250
Sulfite pulp (old plants)	Denmark	500
Sulfite pulp (new plants)	Sweden	250
Sulfite pulp (old plants)	Sweden	500
Cement (new plants)	Great Britain	229
Cement (old plants)	Great Britain	457
Cement (new plants)	Denmark	250
Cement (old plants)	Denmark	500
Cement	FRG	150
Electrodes	FRG	150

Removal: by means of cyclones, filters, electrostatic precipitators, or reverse osmosis, depending on the concentration of the inorganic dust present in industrial wastes and in atmospheric air and on the required degree of removal; the degree of purification varies with different purification installations, ranging from 58.6 to 99.95% [4,8,42,45,57,61,69,76,105,230].

Determination: gravimetric analysis (sensitivity from 0.3 μg to 0.25 μg) [18,47, 96,231,232].

REFERENCES

1. Materials of the XXVII CPSU Congress. Moscow: Politizdat, 1986, 352 pp.
2. USSR Legislation on Preventing Air Pollution. Records of the USSR Supreme Soviet, No. 27 (2049), July 2, 1980, Moscow. Article 528.
3. The Constitution of the USSR. Moscow, 1977, Article 67.
4. M. I. Birger, A. Yu. Valdberg, and B. I. Myagkov. Dust- and Ash-Trapping. A Handbook. Moscow: Energiya, 1975, 296 pp.
5. K. A. Bushtueva (Ed.). Keeping the Air Clean. A Handbook. Moscow: Meditsina, 1976, 416 pp.
6. E. Eder, S. Heise, and K. H. Kelner. Umwelt, 1976, No. 2, pp. 102-107.
7. S. H. Ganz (Ed.). Purification of Industrial Gases. A Handbook. Dnepropetrovsk: Promin, 1977, 116 pp.
8. G. M. Gordon and I. L. Peisakhov. Dust-Trapping and Purification of Gases in Nonferrous Metallurgy (3rd ed.). Moscow: Metallurgiya, 1977, 456 pp.
9. T. N. Zhigalovskaya, E. P. Makhonko, and A. I. Shilina. Microelements in Natural Waters and Atmosphere. (Eds: T. N. Zhigalovskaya and S. G. Malakhova). In: Pollution of Natural Waters. Moscow: Gidrometeoizdat, 1974, 86 pp.
10. I. E. Kuznetsov and T. M. Troitskaya. Air Protection from Harmful Substances Present in Wastes of Chemical Plants. Moscow: Khimiya, 1979, 344 pp.
11. E. N. Levina. General Toxicology of Metals. Moscow: Meditsina, 1972, 184 pp.
12. Nature Conservation. Atmosphere. Compositional Classification of Wastes According to the All-Union State Standard 17.2.101-76. Moscow: State Committee on Standards, 1976, 5 pp.
13. Nature Conservation. Atmosphere. Meteorological Aspects of Pollution and Industrial Wastes. Main Terms and Definitions. All-Union State Standard 17.2.1.-04-77. Moscow: State Committee on Standards, 1977.
14. Nature Conservation. Regulations on Air Quality Control in Populated Areas. All-Union State Standard 17.2.3.01-77. Moscow: State Committee on Standards, 1977, 4 pp.
15. Nature Conservation. Atmosphere. Regulations on Establishing the Permissible Concentration of Harmful Substances in Wastes from Industrial Plants. All-Union State Standard 17.2.3.02-78. Moscow: State Committee on Standards, 1978.
16. Approximate Safe Level (ASL) of Air Pollution in Populated Areas: Supplement No.1 to List No.1430-76 of July 3, 1976. Approved by the State Sanitary Inspectorate, 1978, No.1890-78.
17. Approximate Safe Level (ASL) of Air Pollution in Populated Areas: Supplement No.2 to List No.1430-76 of July 3, 1976. Approved by the State Sanitary In-

spectorate, 1979. No.2062-79.
18. E. A. Peregud. Handbook of Sanitary-Chemical Control of Air. Leningrad: Khimiya, 1980, 336 pp.
19. S. M. Andonev and O. V. Filipov. Aerosol Wastes of Metallurgical Plants. Moscow: Metallurgiya, 1973, 198 pp.
20. L. I. Ishukova, S. N. Zueva, and L. S. Kuritsina. Khim. Prom., 1980, No.7, p. 402.
21. A. Ya Korolchenko (Ed.). Inflammable and Dangerously Explosive Substances. Trudy VNII of Fire Prevention, Moscow, 1979, No. 2.
22. A. A. Zuikov. Protection of the Environment from Industrial Pollution. Leningrad, 1975. Issue 2, pp. 21-25.
23. V. Leite. Determination of Air Pollution Level in the Atmosphere and at the Work Area (translated from German). Leningrad: Khimiya, 1980, 344 pp.
24. M. D. Manita, R. M.-F. Salidzhanova, and S. F. Yavorskaya. Modern Methods for Determining Pollution Level in Populated Areas. Moscow: Meditsina, 1980, 252 pp.
25. L. I. Medved (Ed.). Handbook of Pesticides: Hygienic Use and Toxicology. Kiev: Urozhai, 1974, 448 pp.
26. V. A. Ryazanov (Ed.). Maximum Permissible Concentration of Atmospheric Pollutants. Moscow: Medgiz, 1952, No.1, 150 pp.
27. V. A. Ryazanov (Ed.). Maximum Permissible Concentration of Atmospheric Pollutants. Moscow: Medgiz, 1957, No.3, 170 pp.
28. V. A. Ryazanov (Ed.). Maximum Permissible Concentration of Atmospheric Pollutants. Moscow:Medgiz, 1960, No.4, 158 pp.
29. V. A. Ryazanov (Ed.). Maximum Permissible Concentration of Atmospheric Pollutants. Moscow: Medgiz, 1961, No.5, 180 pp.
30. V. A. Ryazanov (Ed.). Maximum Permissible Concentration of Atmospheric Pollutants. Moscow: Medgiz, 1962, No.6, 220 pp.
31. V. A. Ryazanov (Ed.). Maximum Permissible Concentration of Atmospheric Pollutants. Moscow: Medgiz, 1963, No.7, 124 pp.
32. V. A. Ryazanov and M. S. Goldberg (Eds.). Maximum Permissible Concentration of Atmospheric Pollutants. Moscow: Medgiz, 1964, No.8, 192 pp.
33. Maximum Permissible Concentration of Pollutants in the Air in Populated Areas List No.3086-84 (replacing Lists Nos. 1892-78, 2063-79, 2391-81, 2616-82, 2936-83). Approved by the State Sanitary Inspectorate, 1984.
34. Maximum Permissible Concentration of Pollutants in the Air: Supplement No.1 to List No.3086-84 of August 27, 1984. Approved by the State Sanitary Inspec-

torate, 1985.

35. Calculated Maximum Permissible Concentration of Pesticides in the Air in Populated Areas, No. 1233-75. Approved by the State Sanitary Inspectorate, 1975.

36. I. V. Ryabov (Ed.). Handbook of Fire-Hazardous Substances. Part I. Moscow: Gosstroiizdat, 1966, 242 pp.

37. I. V. Ryabov (Ed.). Handbook of Fire-Hazardous Substances. Part II. Moscow: Gosstroiizdat, 1970, 336 pp.

38. I. V. Ryabov (Ed.). Handbook of Fire-Hazardous Substances in the Chemical Industry. Moscow: Khimiya, 1970, 336 pp.

39. Inflammable and Dangerously Explosive Substances and Materials. Trudy VNII of Fire Prevention. Moscow, 1978, No.1, 184 pp.

40. I. A. Sigal. Protection of Air from Combustion of Fuels. Leningrad: Nedra, 1977, 294 pp.

41. T. A. Semenova and I. L. Leites (Eds.). Purification of Technological Gases. Moscow: Khimiya, 1977, 388 pp.

42. T. V. Soloveva and V. A. Khrustaleva. Methods for Determining Harmful Substances in the Air. A Handbook. Moscow: Meditsina, 1974, 300 pp.

43. S. B. Stark. Dust Trapping and Purification of Gases from Metallurgical Plants. Moscow: Metallurgiya, 1977, 328 pp.

44. Guide to Calculation of the Degree of Dispersion of Harmful Substances from Industrial Plants. SN.369-74. Moscow: Stroiizdat, 1975, 42 pp.

45. G. I. Sidorenko and M. T. Dmitrieva (Eds.). Standardized Methods for Determining Atmospheric Pollutants. Moscow: Khimiya, 1976, 198 pp.

46. E. N. Serpionova. Absorption of Gases and Vapors in Industries. 2nd ed. Moscow: Vysshaya Shkola, 1969, 414 pp.

47. B. M. Fedyushin, S. A. Anurov, and N. V. Leltsev. Khim. Prom., 1977, No.8, 597-598.

48. O. P. Cheinoga. Hygienic Use and the Toxicology of Pesticides and a Clinical View of Poisoning. (Ed.: L. I. Medved). Kiev: VNIIGINTOKS, No.8, pp. 30-40.

49. E. Ahland and H. Marteus. VDI-Ber., 1980, vol. 358, pp. 107-111; Chem. Abstr., 1981, No.4, Ref. 94,19713x.

50. ABC Umweltschutz unter besonderer Berucksichtigung der Umweltschutztechnologie. Leipzig: M. Quarg, 1976, 268 pp.

51. BDR Technische Anleitung zur Reinhaltung der Luft. 1974.

52. A. Bjorseth, G. Lunde, and A. Lindskog. Atm. Environm., 1979, vol. 13, No.1, pp. 45-53.

53. R. G. Bond and C. P. Straul (Eds.). Handbook of Environmental Control. Vol.1.

Air Pollution. Cleveland: CRC Press, Inc., 1972, 576 pp.

54. A. J. Buonicore and L. Theodore. Industrial Control Equipment for Gaseous Pollutants. Vols. 1 and 2. Cleveland: ERR Press, Inc., 1975, 204 pp. and 128 pp. respectively.
55. P. N. Cheremisinoff and R. I. Young. Pollution Engineering Practice Handbook. Michigan: Ann. Arbor Science, 1975, 1073 pp.
56. R. Coleman. Proceed. Intern. Symposium Fort Collins, Colorado, 1973. Washington, 1976, pp. 162-169.
57. J. A. Cooper. J. Air Pollution Control Assoc., 1980, vol. 30, No.8, pp. 855-867.
58. R. W. Coughlin, R. D. Siegel, and Ch. Rai (Eds.). Recent Advances in Air Pollution Control. ALChE Symposium Series No. 137, vol. 70. New York, 1974, 528 pp.
59. J. A. Danielson (Ed.). Air Pollution Engineering Manual; 2nd ed., Washington, 1973, 988 pp.
60. L. Fishbein. Chromatography of Environmental Hazards. Vol. 1. Carcinogens. Amsterdam: Elsevier Publishers, 1972, 499 pp.
61. L. Frieberg and R. Cederlof. Environm. Health Perspectives, 1978, vol. 22, pp. 43-66.
62. F. G. Banit and A. D. Malgin. Dust Trapping and Purification of Gases in the Construction Materials Industry. Moscow: Stroiizdat, 1979, 346 pp.
63. H. E. Hesketh. Understanding and Controlling Air Pollution. 2nd ed. Michigan: Ann Arbor, 1974, 414 pp.
64. C. J. Hilado and H. J. Cumming. Fire Flam., 1979, vol. 10, pp. 252-260.
65. R. B. King, A. C. Altone, and J. S. Fordyce. J. Air Pollution Control Assoc., 1977, vol. 27, No. 9, pp. 867-871.
66. R. E. Lee (Ed.). Air Pollution from Pesticides and Agricultural Processes. Cleveland: CRC Pres, Inc., 1976, 264 pp.
67. B. G. Liptak (Ed.). Environmental Engineers Handbook. Vol. 2. Air Pollution. Radnor: Chilfon Book Co., 1974, 1340 pp.
68. H. F. Lund (Ed.). Industrial Pollution Control Handbook. New York, 1971.
69. J. M. Marchello. Control of Air Pollution Sources. Basel: Dakker, 1976, 630 pp.
70. W. Martin, A. C. Stern. The World's Air Quality Management Standards. Vol. 1. Washington: EPA, 1974, 382 pp.
71. T. H. Maugh. Science, 1975, vol. 201, No. 1200.
72. Merck Index. An Encyclopedia of Chemicals and Drugs. 8th ed. Rahway, 1968, 1714 pp.

73. D. F. Natusch. Environm. Health Perspectives, 1978, vol. 22, pp. 79-90.

74. H. W. Parker. Air Pollution. Englewood: Prentice Hall, 1977, 288 pp.

75. F. A. Patty. Industrial Hygiene and Toxicology. Vol 2. New York: Interscience Publ., 1967.

76. K. D. Anil and B. K. Pal. J. Indian Chem Soc., 1977, vol. 54, Nos. 1-3, pp. 265-273.

77. E. Sawicki. Environmental Pollutants. Detection and Measurement. (Eds.: T. Y. Taribera, J. R. Coleman, B. E. Dahneke, and Y. Feldman). London: Plenum Press, 1977.

78. N. I. Sax (Ed.). Industrial Pollution. New York: Van Nostrand Reinhold, 1974, 702 pp.

79. N. I. Sax. Dangerous Properties in Industrial Materials. 4th ed. New York: Van Nostrand Reinhold, Corp., 1975, 1258 pp.

80. R. W. Serth and T. Hughes. Environmental Sci. Technol., 1980, vol 12, No. 3, pp. 298-301.

81. M. Sittig. How to Remove Pollutants and Toxic Materials from Air and Water: A Practical Guide. New York: Noues Data Corp., 1977, 621 pp.

82. A. C. Stern (Ed.). Air Pollution. 3rd ed. Air Pollutants, their Transformation and Transport. Vol. 1. New York: Academic Press, 1976, 716 pp.

83. A. C. Stern (Ed.). Air Pollution. 3rd ed. Effects of Air Pollution. Vol 2. New York: Academic Press, 1977, 684 pp.

84. A. C. Stern (Ed.). Air Pollution. 3rd. ed. Measuring, Monitoring and Surveillance of Air Pollution. Vol 3. New York: Academic Press, 1976, 800 pp.

85. A. C. Stern (Ed.). Air Pollution. 3rd ed. Engineering Control of Air Pollution. Vol. 4. New York: Academic Press, 1977, 946 pp.

86. A. C. Stern (Ed.). Air Pollution. 3rd ed. Air Quality Management. Vol 5. New York: Academic Press, 1977, 500 pp.

87. J. Tarr and C. Damme. Chem Eng., 1978, No. 10, pp. 86-89.

88. L. Van Vaeck and K. Van Cauwenberghe. Atm. Environm., 1978, vol. 12, No. 11, pp. 2229-2239.

89. H. U. Wanner, A. Deuber, and J. Satish. Atmospheric Pollution Proc. of the 12th Int. Colloquim. Paris, May 5-7, 1976. (Ed.: M. M. Benarie). Amsterdam: Elsevier Sci. Publ. Co., 1976, pp. 99-108.

90. E. A. Peregud, M. S. Bykovskaya, and E. V. Gernet. Methods for a Quick Determination of Harmful Substances in the Air. Moscow: Khimiya, 1970, 358 pp.

91. T. Y. Taribara, J. R. Coleman, B. E. Dahneke, and J. Feldman (Eds.). Environmental Pollution, Detection and Measurements, vol. 13. New York, 1977,

500 pp.

92. D. A. Lynn. Air Pollution. Threats and Response. Reading: Addison Weley, 1976, 388 pp.
93. J. H. Seinfeld. Air Pollution. Physical and Chemical Fundamentals. New York: McGraw Hill, 1975, 524 pp.
94. H. S. Stoker and S. L. Teager. Environmental Chemistry. Air and Water Pollution. 2nd ed. Glenview: Forman, 1976, 234 pp.
95. M. Lipman and R. B. Schlesinger. Chemical Contamination in the Environment. New York: Oxford Univ. Press, 1979, 456 pp.
96. V. I. Krasova (Ed.). Methods and Equipment for Automatic Control of Atmosphere Pollutants. Leningrad: Gidrometeoizdat, 1969, 132 pp.
97. J. W. Scales. Air Quality Instrumentation. Selected Papers from International Symposia Presented by the Instrument Society of America. Pittsburgh, 1972, 188 pp.
98. W. B. Deichmann and H. W. Gerade. Toxicology of Drugs and Chemicals. New York: Academic Press, 1969, 758 pp.
99. M. M. Benaire (Ed.). Atmospheric Pollution. Proc. 12th Int. Colloquim, Paris, May 5-7, 1976. Amsterdam: Elsevier Sci. Publ. Co., 1976, 650 pp.
100. E. R. Plunkett. Handbook of Industrial Toxicology. Industrial Health Service. New York: Chem. Publ. Co., 1976, 552 pp.
101. N. V. Lazarev, I. D. Gadaskina (Edds.). Handbook of Harmful Substances in Industry. Inorganic and Hetero-Organic Compounds, 7th ed. Leningrad: Khimiya, 1977, 608 pp.
102. H. M. Ellis, A. R. Greenway. J. Air Pollution Control Assoc., 1981, vol. 31, No. 2, 136-138.
103. D. B. Meeker. Air Quality Control. National Standards and Goals. Washington: National Assoc. of Manuf. Resources and Technol. Dept., 1975, 94 pp.
104. R. Guderian. Air Pollution: Phytotoxicity of Acidic Gases and its Significance in Air Pollution Control. Berlin, 1977, 125 pp.
105. R. Guderian. Polizei Verkehry, 1976, vol. 14, No. 1, pp. 18-20.
106. J. R. Mudd and T. T. Koslowski (Eds.). Responses of Plants to Air Pollution. Physiological Ecology. New York: Academic Press, 1975, 384 pp.
107. V. A. Zaitsev and A. P. Tsygankov. Zhur. Vsesoyuz. Khim. Obshch. im D. I. Mendeleev, 1979, vol. 24, No. 1, pp. 3-12.
108. Yu. S. Drugov and V. G. Berezkin. Gas Chromatographic Analysis of Air Pollutants. Moscow: Khimiya, 1981, 254 pp.
109. H. Jüntgen. Tech. Mitteilungen, 1977, vol. 70, No. 1, pp. 55-64.

110. R. Bothea. Pollution Eng., 1977, vol. 9, No. 1, pp. 51-52.

111. I. Sax. Dangerous Properties of Industrial Materials. 4th ed. New York: Van Nostrand, Reinhold Co., 1979, 1120 pp.

112. J. R. Pfaffin and E. N. Ziegler. Encyclopedia of Environmental Science and Engineering. Vols. 1 and 2. New York, 1976.

113. W. Gallay, H. Egan, and J. L. Monuman. Environmental Pollutants. Selected Analytical Methods (scope 6). London: Butterworths, 1975, 278 pp.

114. A. P. Stern (Ed.). Air Pollution. Sources of Air Pollution and their Control. 2nd., vol. 3. New York: Academic Press, 1968, 866 pp.

115. Approximate Safe Levels (ASL) of Pollutants in the Air in Populated Areas. List. No. 2847-83. Approved by the State Sanitary Inspectorate. Moscow, 19-83.

116. V. A. Ryazantsev. Sanitary Protection of the Air. Moscow: Medgiz, 1954, 236 pp.

117. L. D. Temermann. Rev. Agriculture (Belg.), 1979, vol. 32, No. 4, pp. 1007-1017.

118. A Semb. Atmospheric Environment, 1978, vol. 12, Nos. 1-3, pp. 455-460.

119. H. C. Perkins. Air Pollution. New York: McGraw-Hill Book Co., 1974, 407 pp.

120. E. C. Fuller. Chemistry and Man's Environment. Boston: Mufflin, 1974, 502 pp.

121. B. A. Chertkov, V. N. Pankratova, and I. M. Tropp. Khim. Prom., 1978, No.6, pp. 441-444.

122. S. Gerkin and W. Schulze. Amer. Econ. Rev., 1981, vol. 71, No. 2, pp. 228-234.

123. L. K. Ember. Chem Eng. News, 1981, vol. 59, No. 6, pp.18-20.

124. R. E. Rush and C. H. Coodm. AIChE, 1980, vol. 76, No. 196, pp. 139-149.

125. H. Trenkler.11th World Energy Conference, London, 1980, vol. 3, pp. 29, 75, 305-324.

126. K. Dolgner, Th. Eikmann, and H. Einbrodt. Staub Reinhalt Luft, 1980, vol. 40, No. 9, pp. 418-425.

127. A. V. Lysak, I. M. Nazarov, and A. G. Ryaboshapko. Zhur. Vsesoyuz. Khim. Obshch. im D. I. Mendeleev, 1979, vol. 24, No. 1, pp. 25-29.

128. M. D. High. J. Air Pollution Control Assoc., 1976, vol. 26, No. 5, pp. 471-479.

129. C. Jouce. New Sci., 1981, vol. 89, No. 1243, p. 541.

130. Combating Environmental Pollution in the Paper and Pulp Industry. Moscow: Znanie, 1976, 59 pp.

131. R. E. Munn. J. Air Pollution Control Assoc., 1977, vol. 27, No. 9, pp. 842-843.

132. C. J. Stairmand. International Clean Air Conference. Melbourne: Charman Hartman H. F., 1972.
133. V. M. Styazhina. Maximum Permissible Concentration of Air Pollutants. Issue 6. Moscow: Medgiz, 1962, pp. 96-108.
134. Sanitary-Protection Zones. Criteria for Selecting Construction Sites. SN-245-71. Moscow: Stroiizdat, 1972.
135. M. E. Berlyand. Sanitation Standards and Control of Industrial Wastes in the Atmosphere. Leningrad: Gidrometeoizdat, 1977, 123 pp.
136. V. S. Gurev, A. V. Ilchenko, and L. M. Kupyshin. Air and Water Purification at Metallurgical Plants. 1977, pp. 3-8.
137. V. A. Zaitsev and A. P. Tsygankov. Zhur. Vsesoyuz. Khim. Obshch. im. D. I. Mendeleev, 1979, vol. 24, No. 1, pp. 3-12.
138. W. Straus. Industrial Gas Cleaning. 2nd ed. Oxford: Pergamon Press, 1975, 621 pp.
139. V. F . Maksimov, V. B. Lesikhin, and L. M. Izyanov. The Timber Industry. 1972, 312 pp.
140. Yu. A Izrael and L. M. Filippova. Monitoring of the Environment. Soviet-US Symposium 1977. Leningrad: Gidrometeoizdat, 1977, pp. 34-40.
141. J. R. Machoney. The Effects of the Air Environment in Industrial and Residential Facilities. New York, 1975, pp. 131-134.
142. Manual of Air Quality Control in Cities. Copenhagen, 1980, 264 pp.
143. Environmental Sci. and Technol., 1971, vol. 5, No. 2, p. 106.
144. V. P. Melekina. Biological Effect and Sanitary Aspects of Atmospheric Pollution. (Eds.: V. A. Ryazantsev and M. S. Goldberg). Moscow: Medgiz, 1966, pp. 133-141.
145. B. M. Smirnov. Priroda, 1977, No. 4, pp. 10-19.
146. R. E. Baumgardner and T. A Clark. Environm. Sci. and Technol., 1975, vol. 9, No. 11, pp. 67-69.
147. I. M. Nazarov, A. N. Nikolaev, and Sh. D. Fridman. Remote-Control Methods for a Quick Determination of Environmental Pollutants. Moscow: Gidrometeoizdat, 1977, 194 pp.
148. P. P. Yakimchuk. Maximum Permissible Concentration of Air Pollutants. Moscow: Medgiz, 1963, pp. 66-67.
149. M. T. Dmitriev, N. A. Kitrosky, and L. G. Maslenkivsky. Gig. i Sanit., 1977, No. 1, pp. 61-64.
150. Ch. Rai and L. A. Spielman (Eds.). Air Pollution Control and Clean Energy. AIChE Symp. New York, 1975, vol. 71, No. 147, 184 pp.

151. V. S. Nikolaevsky and A. T. Miroshnikova. Gig. i Sanit., 1974, No. 4, pp. 16-18.

152. L. P. Dudakov and I. L. Leites. Khim. Prom., 1977, No. 1, pp. 32-34.

153. Technocrat, 1976, vol. 9, No. 9, p. 84.

154. A. N. Shcherban, A. V. Primak, and V. I. Kopeikin. Automated Systems for Air Pollution Control. Kiev: Tekhnika, 1979, 158 pp.

155. I. I. Demin and V. I. Erenchuk. Tsement, 1979, No. 5, p. 12.

156. C. W. Louw. Chemsa, 1975, vol. 1, No. 8, pp. 169-172.

157. E. A. Skrikina and I. A. Pinigina. Gig. i Sanit. , 1985, No. 9, pp. 52-53.

158. V. P. Filonov, E. M. Shpilevsky, and S. M. Sokolov. Gig. i Sanit., 1984, No. 3, pp. 71-73.

159. F. K. Idiatullina. Gig. i Sanit., 1981, No. 9, pp. 79-81.

160. V. P. Voronov. Gig. i Sanit., 1983, No. 7, pp. 71-72.

161. O. Hutzinger (Ed.). Handbook of Environmental Chemistry. Vol. 3, Part A, Antropogenic Compounds. Berlin: Springer Verlag, 1980, pp. 59-107.

162. T. L. Radovskaya and L. A. Khazemova. Gig. i Sanit., 1977, No. 7, pp. 69-72.

163. P. P. Kish, I. S. Bolog, and V. I. Buletsa. Gig. i Sanit., 1981, No. 8, pp. 58-60.

164. A. B. Kamkin. Gig. i Sanit., 1981, No. 10, pp. 13-16.

165. V. F. Dokuchaeva and N. I. Skvartsova. Maxium Permissible Concentration of Atmospheric Pollutants. Issue 6. Moscow: Medgiz, 1962, pp. 173-184.

166. V. A. Litkens. Maximum Permissible Concentration of Atmospheric Pollutants. Issue 2. Moscow: Medgiz, 1955, pp. 47-63.

167. V. M. Pozynich. Gig. i Sanit., 1984, No. 12, pp. 53-55.

168. M. M. Ginonyan. Gig. i Sanit., 1976, No. 6, pp. 8-12.

169. E. M. Roisenblat, T. M. Veryatina, and K. P. Ivanova. Gig. i Sanit., 1984, No. 4, pp. 36-37.

170. Schadstoffe in der Atmosphere aus onkologischen und toxicologischen Sichtw. Berlin: Akadem. Verlag, 1988, 128 pp.

171. I. A. Mazur, V. M. Pazynich, and V. I. Mandrichenko. Gig. i Sanit., 1983, No. 5, pp. 52-53.

172. M .T. Dmitriev and B. I. Fradkin. Gig. i Sanit., 1985, No. 4, pp. 63-64.

173. T. A. Mansfield (Ed.). Effects of Air Pollutants on Plants. Cambridge: University Press, 1976, 210 pp.

174. J. Air Pollution Control Assoc., 1979, vol. 29, No. 12, pp. 1253-1255.

175. M. T. Dmitriev, V. V. Osechkin, and L. D. Pribytkov. Gig. i Sanit., 1975,

No. 8, pp. 46-48.

176. R. E. Pecsar and C. H. Harman. Air Quality Instrumentation. Vol. 1. Selected Papers from International Symposia Presented by the Instrument Society of America (Ed.: J. W. Scales). Pittsburgh, 1972, pp. 89-113.

177. T. M. Royal, C. E. Decker, and J. B. Tommerdahl. Proc. YEEE National Aerospace and Electron. Conf., NAECON, 75. Dauton, 1975. New York, 1975, pp. 697-703.

178. V. N. Kurnosov. Maximum Permissible Concentration of Atmospheric Pollutants. Moscow: Medgiz, 1961, pp. 54-71.

179. R. G. Leites. Maximum Permissible Concentration of Atmospheric Pollutants. Issue 1. Moscow: Medgiz, 1952, pp. 90-99.

180. L. S. Fedorovskaya, V. A. Skripnik, and A. S. Romashov. Khim Prom., 1978, No. 8, pp. 45-48.

181. G. Dowd, G. Corte, and J. L. Monkman. J. Air Pollution Control Assoc., 1976, No. 7, pp. 678-679.

182. I. N. Ermachenko. Gig. i Sanit., 1984, No. 9, pp. 11-13.

183. M. T. Dmitriev and B. I. Fradkin. Gig i Sanit., 1985, No. 12. pp. 47-48.

184. M. T. Dmitriev. Gig. i Sanit., 1985, No. 9, pp. 52-53.

185. L. Frieberg and R. Cederlof. Environm. Health Perspectives, 1978, vol. 22, pp. 45-66.

186. D. Henschler. Angew. Chemie, 1973, vol. 85, p. 317.

187. A. V. Lysak, I. M. Nazarov, and A. G. Ryaboshapko. Zhur. Vsesoyuz. Khim. Obshch. im D. I. Mendeleev, 1979, vol. 24, No. 1, pp. 25-29.

188. P. R. Harrison, H. W. Georgii, and G. L. Fernaar. Analysis of Industrial Air Pollutants. New York, 1974, 185 pp.

189. H. H. Gruhn. Ind. Anzliger, 1978, vol. 100, No. 29, pp. 24-26.

190. N. Izmerov. Social and Sanitary Aspects of Protecting Atmospheric Air and Scientific and Technological Progress. Moscow: Meditsina, 1971, 184 pp.

191. O. P. Shalamberidze. Maximum Permissible Concentration of Atmospheric Pollutants. Issue 5, Moscow: Medgiz, pp. 40-53.

192. I. I. Demin and V. I Eremenchuk. Tsement, 1979, No. 5, p.12.

193. Control of Air Pollution Caused by Technological Wastes. Moscow: MDNTP, 1978, 134 pp.

194. B. P. Gurinov. Maximum Permissible Concentration of Atmospheric Pollutants. Issue 1. Moscow, 1952, pp. 55-62.

195. R. A. Loginova. Maximum Permissible Concentration of Atmospheric Pollutants. Issue 3. Moscow: Medgiz, 1957, pp. 63-84.

196. W. U. Kotting. Tech. Unweltmagazin, 1976, No. 5, pp. 20-22.
197. A. V. Andreeva. Production of Sulfur from Gases. Moscow: Metallurgiya, 1977, 172 pp.
198. B. M. Fedyushin, S. A. Anurov, and N. V. Keltsev. Khim. Prom., 1977, No. 8, pp.597-598.
199. K. Work and S. Worner. Air Pollution. Sources and Control (Translated from English). Moscow: Mir, 1980, 534 pp.
200. Environm. Sci. and Technol., 1976, vol 10, No. 5, pp.416-417.
201. I. P. Mukhlenov, E. Ya. Tarat, and G. N. Buzanova. Khim. i Prom., 1978, No. 8, pp. 37-40.
202. S. A. Anurov, N. V. Keltsev, and V. I. Smola. Khim. i Prom., 1978, No. 8, pp. 37-40.
203. B. A. Chertkov, V. N. Pankratova, and I. M. Tropp. Khim. i Prom., 1978, No. 6, pp. 441-444.
204. I. N. Shokin. Khim. i Prom., 1975, No. 5, pp. 369-370.
205. B. Boeuf. Fluoride, 1977, vol. 10, No. 1, pp. 12-13.
206. M. S. Sadilova. Biological Effect and Sanitary Aspects of Atmospheric Pollution. Issue 11. Moscow: Meditsina, 1967, pp. 186-201.
207. K. A. Bushtueva (Ed.). Manual on Keeping Atmospheric Air Clean. Moscow: Meditsina, 1976, 416 pp.
208. G. R. Winch. Instrumental Clean Air Conference. Melbourn, 1972, pp. 98-105.
209. G. M. Woodwelt, R. K. Sewers, and J. E. Lovelock. Ecological, Biological Effect of Air Pollution. New York: Int. Corp., 1973, 181 pp.
210. Ch. Rai and L. A. Spielman. Air Pollution Control and Clean Energy. AIChE Symp. Ser., New York, 1976, No. 156.
211. J. A. Cox, G. L. Lundquist, and Przyjaznl. Environm. Sci. and Technol., 1978, vol. 12, No. 6, pp. 722-723.
212. V. M. Pazynich. Biological Effect and Sanitary Aspects of Atmospheric Pollution. Moscow: Meditsina, 1967, pp. 201-207.
213. A. Kh. Kamildzhanov and R. U. Ubaidullaeva. Gig. i San., 1985, No. 7, pp. 73-74.
214. O. V. Eliseeva. Biological Effect and Sanitary Aspects of Atmospheric Pollution. (Eds.: V. A. Ryazantsev and M. S. Goldberg). Moscow: Meditsina, 1966, pp. 7-27.

Part II

Organic Substances

Chapter 3

PREVENTIVE MEASURES AGAINST ENVIRONMENTAL POLLUTION BY INDUSTRIAL WASTES

Sources of Air Pollution

Organic substances are a main source of air pollution. It has been estimated [79] that 1.8 million chemical compounds (both organic and inorganic) have been synthesized by man, and 250,00 new ones are devised each year. About 3000 of these are known to be carcinogenic, and 20,000-30,000 are industrially produced. Today the chemical industry worldwide makes more than 50,000 harmful organic compounds, and over 1000 new harmful ones are synthesized each year [184]. Of these only a small number have been investigated for carcinogenic or mutagenic properties. In the United States 6 billion pounds of vinyl chloride and 10 billion pounds of dichloroethylene are produced annually whose carcinogenic effect has been definitely proved [184]. Many pesticides used all over the world are carcinogenic and mutagenic [73]. Up to 80% of the air pollutants at industrial centers are carcinogenic; the number of carcinogenic substances is particularly large in the wastes of petrochemical works [109]. The air over cities in industrial countries is the most polluted [41]. For example, in the air over Los Angeles (U.S.A.) 60 harmful pollutants (alkanes, alkenes, acetylenes, and aromatic derivatives) have been identified, which have not only a toxic but also a photochemical effect [80].

According to the World Health Organization 600,000 toxic chemicals are industrially produced worldwide, and 3000 new ones synthesized each year. In the United States alone 64,000 toxic compounds are used in industry [202].

In the Soviet Union anti-pollution measures are worked out by teams of specialists (chemists, sanitary engineers, hygienists, and climatologists [1,2,9-12,18,21,23,24,28,29].

Toxic Wastes

The toxic effect of an organic substance is determined on the basis of observations of its effect on volunteers and on experimental animals exposed to the substances [6,7]. The results of the investigation are used to establish the maximum permissible single-dose concentration (MPSDC) of the substance in the air. But it is not enough to determine the toxic effect of the substance if inhaled; it is also necessary to determine its harmful effect if ingested, for example, with drinking water and food [106]. In the case of heptachlor (insecticide), for instance, a person exposed to it for 8 hours at the maximum permissible single-dose concentration of 0.5 mg/m^3 as recommended in the United States absorbs 7 mg of the substance; if he drinks 2 liters of water within a 24-hour period at the maximum permissible single-dose concentration of 0.0001 mg/ℓ of the insecticide, he receives 0.0002 mg of it; and if he ingests it with food at the maximum permissible single-dose concentration of 0.0005 mg/kg of the insecticide for 1 kg of weight of a person (with an average weight of 70 kg), he gets 0.035 mg of the insecticide [40]. A similar pattern of toxicity has been observed by the present author with respect to other pesticides.

Some polynuclear aromatic compounds, aromatic amines, resinous substances, aldehydes, and nitrosoamines can cause cancer and blastmycosis if inhaled with air [8,47,52,107-109]. It should be borne in mind when determining the carcinogenic effect of substances that in many cases the effect becomes evident only after a prolonged period. Often it takes 20 years and more before the first clinical symptoms of cancer manifest themselves [185]. It is therefore important that as soon as the carcinogenic properties of a compound are established measures should be taken to destroy the wastes containing the compound or that it should be replaced by a compound which is not toxic.

Some chemical compounds, including aromatic amines, nitrosoamines [38,110], aldehydes [38], halogen derivatives of alkanes and their derivatives [111], and vinyl chloride [112,113] have a mutagenic effect. Active mutagens are formed in reactions between precursors that may be present in the air, such as 1,2-benzopyrene or polycyclic aromatic hydrocarbons, and ozone, nitrogen dioxide, and nitro derivatives [97,114]. There can be no genetic adaptation of man to mutagens from without. Therefore, strict measures should be taken to prevent the latter from entering the environment.

Many organic compounds are allergenic. They can cause bronchial asthma and skin diseases such as dermatitis and eczema [5]. Many organic compounds contained in industrial wastes are harmful to plants and microflora in the soil. The chemicals that are the most stable are also the most toxic [116,170]. Many of them are deposited on the soil and are washed by rain into rivers, polluting the surface and ground waters. They are also harmful to buildings and architectural and cultural monuments in the cities [118].

The pollution of atmospheric air reduces its transparency; decreases the amount of solar radiation reaching the earth; and increases the humidity of the air, which leads to the formation of fogs [6,7,100]. Pollutants in the air often undergo photochemical oxidation yielding products that cause further damage to the environment [119,120]. It has been estimated that from 41 to 126 different volatile organic compounds are present in the air over cities with many industries [186].

When several substances are present in the air whose total toxic effect is the sum of the toxic effect of the individual components, the sum of their concentrations must not exceed unity:

$$(C^1/MPSDC^1) + (C^2/MPSDC^2) + (C^n/MPSDC^n) \leq 1;$$

where, C^1, $C^2 \ldots C^n$ - the actual concentration of the substances in the air; $MPSDC^1$, $MPSDC^2, \ldots MPSDC^n$ - the maximum permissible concentration of the same substances [2,4,28].

It should be kept in mind that in the majority of cases the combined toxic effect of toxic substances that are present as a mixture is the sum total of the toxic effect of the individual components. Below is a list of such mixtures [1,23]: acetone, acrolein, phthalic anhydride; acetone, phenol; acetone, acetophenone; acetaldehyde, vinyl acetate; benzene, acetophenone; valeric acid, caproic acid, butyric acid; ozone, nitrogen dioxide, formaldehyde; carbon monoxide, nitrogen dioxide, formaldehyde, hexane; isopropylbenzene, isopropylbenzene hydroperoxide; sulfur dioxide, phenol; sulfur dioxide, carbon monoxide, phenol, dust from metallurgical plants; 2,3-dichloro-1,4-naphthoquinone, 1,4-naphthoquinone; acetone, furfural, formaldehyde, phenol; acetic acid, acetic anhydride; phenol, acetophenone; furfural, methanol, ethanol; phenol, sulfur dioxide, carbon monoxide, nitrogen dioxide; cyclohexane, benzene; ethylene, propylene, butylene, amylene; 1,2-dichloropropane, 1,2,3-trichloropropane, tetrahydroethylene.

The necessary degree of purification of waste materials containing substances having a combined effect should be calculated on the basis of the MPSDC less the registered value for each of the components multiplied by the number of the components present in the mixture [1,64].

The combined effect of toxic compounds has been determined only for mixtures containing 2 to 3 components [125,126]. But since many more components are often present in the air their synergistic effect should be considered [127].

Combating Atmospheric Pollution

The main methods of combating air pollution consist in the use of waste-free technological processes, the utilization of harmful wastes, and the hermetic sealing of equipment [6-8,10,13,15,17,25,26,62,98,128,196,197]. Other helpful measures include the location of plants on the leeward side of residential districts, establishing the minimum distance between industrial plants and residential districts, and the planting of green belts between them [2].

Methods of Removing Toxic Substances from Industrial Wastes

Organic substances, especially in a wet state and at high concentrations, are destroyed by catalytic oxidation or by being burned in air; at low concentrations - by adsorption; and at very low concentrations - by adsorption on activated carbon and other adsorbents followed by burning [173].

Various mechanical filters, scrubbers, and electrostatic precipitators have been developed and are used in industry for removing dusts and gases from industrial wastes [4,15-17,25,34,158]. The efficiency of dust collectors is as follows (Table 1): cyclones - 84.2%; multivariant cyclones - 93.8%; electrostatic precipitators - 99.0% (for particles 5-50 and 1 μm in diameter the efficiency is 99 and 86% respectively); and filters - 99.7-99.9% (for particles 5-50, 5, and 1 μm in diameter the efficiency is 96, 73, and 27% respectively) [131,132]. According to some estimates [131] the best designed electrostatic precipitators have an 100% efficiency for 50 μm-diameter particles, and a 99 and 98% efficiency respectively for particles 5 and 1μm in diameter.

Table 1 Operating Efficiency of Cyclones [130]

Diameter of particles (μm)	Concentration of dust particles (mg/m^3)	Efficiency (%)
<5	<50	50-80
5-20	50-80	80-95
21-40	80-95	95-99
>40	95-99	95-99

Scrubbers are the most effective (80-99%) for removing components [46]. Fatty acids present in gaseous wastes can be removed by using a two-stage unit [166], or by adsorption and absorption followed by catalytic oxidation or combustion [164]. Amines, mercaptans, alcohols, and ketones are removed with the aid of horizontal scrubbers followed by adsorption of residues on activated carbon [171]. Similar methods are employed for getting rid of ill-smelling substances in gaseous wastes at food processing plants [17,172,173]. In the production of benzene 92% of the organic components in the gaseous wastes are removed by adsorption on activated carbon [133]. However, in a majority of cases activated carbon can remove not more than 50% of the organic pollutants and substances having an unpleasant odor [46]. Toluene and xylene present in wastes from the production of carbon paper have been removed by filtration for 60 minutes with the aid of glass-fiber filters; this makes it possible to recycle 95-98% of these components in the process [134]. Often, as for example at petrochemical works, mercaptans and other toxic and ill-smelling components present in wastes are neutralized by a combination of technological means: adsorption, filtration, and centrifuging with the aid of cyclones [17, 121].

The combustion of organic wastes is carried out at 450-1200°C, and in the presence of catalysts - at 400°C. The thermal units are so designed that they make possible complete combustion within 0.2-1.0 s [188]. This method of destroying wastes is most widely used at petrochemical works [136] and at plants making methylmethacrylate [137], methanol [99], and other organic substances [16,136]. Catalytic combustion is usually employed as the final stage of the burning process, which yields carbon dioxide and water. It has been estimated that in the world today approximately 9000 catalytic combustion units are in operation (Table 2) [89]. The use of precious metals as the catalysts for this purpose is economically justified because of their long service life.

When no waste-free processes are available, or when the recycling of waste products is economically unacceptable and their removal and destruction are too costly, or when the hermetic sealing of the equipment is technically impossible, harmful wastes are allowed to scatter in the atmosphere through tall chimneys (300-350 m in height) [28,129,201]. The use of chimneys helps to reduce the concentration of wastes in the immediate vicinity of the plants concerned, but it does not prevent contamination of the environment. Wastes ejected by chimneys have been detected as far as 200 to 1000 km from their original sites [115]. In Great Britain and Central Europe gaseous wastes reach a height of 2 km from the ground and are subsequently detected 1000 km away in the air over Sweden and Southern Norway [139]. When the wind blows in the direction of Sweden from Great Britain and Central Europe, a chemical analysis of air samples in a Swedish town has shown the presence of 26 different polycyclic aromatic compounds in concentrations 13.8 times greater than when the wind blows in the opposite direction [31].

Tall chimneys facilitate the ejection of gaseous wastes high into atmosphere. There various oxides undergo further changes yielding acids which subsequently precipitate as acid rains, which cause damage to the soil and plants [198,199]. Therefore, the ejection of gaseous wastes high into the atmosphere is not a solution to the problem of pollution. It is better to seek ways of recycling or destroying the wastes before they enter the atmosphere [128,141,142].

Table 2 Some organic components in industrial wastes subjected to catalytic combustion

Substance	Catalyst	Temperature (°C)	Efficiency (%)
Aldehydes	copper chromite + BaO	250-300	90-100
	hopcalite	200-350	90-100
	Co-Mn spinel	300-350	90-100
	Pt	150-400	99-100
	Pt-Al alloys	200-650	90-100
Ketones	Co-Mn spinel	300-350	90-100
	Pt	150-400	99-100
	copper chromite + BaO	250-300	90-100
Mercaptans	$CuO + Cr_2O_3$	280-300	90-100
	cobalt sulfides	70 (p = 0.7 MPa)	90-100
	molybdenum sulfides	350-400	90-100
Alkenes	copper chromite + BaO	250-300	90-100
S-containing compounds	Cr_2O_3	310-630	80-90
	thorium sulfides + ZnO	330	90
	nickel sulfides	490-520	90
	copper chromite + BaO	250-300	90-100
Alcohols	hopcalite	200-350	90-100
	Co-Mn spinel	300-350	90-100
	Pt	150-400	99-100
	Pt-Al alloys	200-650	90-100
	copper chromite + BaO	250-300	90-100
Hydrocarbons	hopcalite	200-350	90-100
	$MnO_2 + CuO + Ag$	200-400	100
	$CuO + Cr_2O_3$	280-300	90-100
	Cr_2O_3	310-630	80-90
	ZnO	600	80
	iron oxides	200-350	90-100
	Pt	150-400	99-100
	Pt-Al alloys	200-650	90-100
	copper chromite + Al_2O_3	250-300	95-100
	copper chromite + BaO	250-300	90-100
Gasoline	Pd	205-230	100
Ethylmercaptan	molybdenum sulfides	350-400	90-95

Methods for Determining Harmful Substances in the Air

Chromatographic analysis of air samples aimed at detecting the presence of organic pollutants represents a rapid, highly sensitive, and precise (±≦10%) method. It can be used in analyzing practically all organic compounds automatically and continuosly [122,146,149-163,181,182,183].

Regulations on air quality control for populated areas set out in All-Union State Standard (Russian abbreviation GOST) 17.2.301-77 constitute a whole program for monitoring atmospheric pollutants [12].

Methods for rendering industrial wastes harmless have been described in many publications:

Industries	References
Chemical	7,17,41,59,60
Metallurgical	7,16,59,61-63
Petroleum refineries	7
Pharmaceutical	59
Cellulose	59,64
Construction	139
For the production of:	
oxalic acid	17
synthetic rubber	166
poly(vinyl chloride)	171
dichloroethane	176
methyl methacrylate	137
paints and varnishes	174,175.

CHAPTER 4

AIR POLLUTANTS

Abietic acid (sylvic acid); $C_{20}H_{30}O_2$ (302.46)

Yellow powder, m.p. 173-175°C, b.p. 248-250°C at p=1.266 kPa; self-ignition p. 842°C; soluble in benzene, methanol, and acetone; insoluble in water; CLE 15 g/m^3 [65].

Present in wastes from the production of esters, emulsifiers, desiccants, paints and varnishes [70].

Toxicity: causes irritation of mucous membranes and the skin [55].

Acenaphthene; $C_{12}H_{10}$ (154.21)

White crystals, m.p. 96°C, b.p. 279°C, ρ 0.831; soluble in ethanol, toluene; insoluble in water.

Present in wastes from fuel-burning stacks; from the production of plastics, petrochemicals, dyes, and pesticides; from shale-processing plants.

Toxicity: has an overall toxic effect [47,55,71]. MPC - 800 μg per 1 g of carbon black or soot [88].

Acetal; $C_6H_{14}O_2$ (118.17)

Liquid with a specific odor, b.p. 102-104°C, ρ 0.831.

Present in wastes from the production of basic organic compounds and some plastics and perfumes.

Toxicity: has an overall toxic and irritating effect [55].

Acetaldehyde (metaldehyde); C_2H_4O (44.05)

Liquid; m.p. -123.5°C, b.p. 20.2°C, CLE 3.97-57%, ρ 0.783; soluble in water and organic solvents.

Present in wastes from the production of some plastics, dyes, wood chemicals, coal-tar chemicals, organic chemicals, and pharmaceuticals; and glycerin.

Toxicity: has an overall toxic and narcotic effect (like ethanol); has also an irritating, mutagenic, and carcinogenic effect [37,56,407,408]. TC_{odor} - 0.012 mg per m^3; $TC_{irritation}$ (according to its effect on the eyes' sensitivity to light) - 0.012 mg/m^3; and SC_{odor} - 0.01 mg/m^3 [409]; MPSDC and MPADC - 0.01 mg/m^3 [20,21];

toxicity classification - III [20,21]; MPSDC established in Bulgaria and Yugoslavia - 0.01 mg/m^3, and in GDR - 0.03 mg/m^3; MPSDC and MPADC recommended for sites of industrial plants - 0.01 mg/m^3 [55]; MPSDC and MPADC established in FRG - 150 and 20 mg/m^3 respectively [51].

Removal: with the aid of filters made of polyesters including polyacrylates, poly(vinylidene chloride), teflon, polypropylene, or polyethylene [34]; by adsorption on activated carbon (efficiency 7%) [50].

Determination: polarographic analysis (sensitivity 0.0008-0.01 mg/m^3) [58].

Acetaldehyde ammonia (1-aminoethanol); C_2H_7NO (61.09)

White powder, m.p. 97°C, b.p. 110°C (with decomposition); soluble in water and ethanol.

Present in wastes from the production of basic organic compounds.

Toxicity: has an overall toxic and irritating effect [55,71].

Acetamide (ethanamide); C_2H_5NO (59.07)

White crystals (technical grade - liquid with a sharp odor), m.p. 82-83°C (on heating evolves cyanides), b.p. 221.2°C, f.p. 154.4°C, ρ 1.159; soluble in chloroform and water.

Present in wastes from the production of some basic organic compounds, plastics, pharmaceuticals, varnishes and paints.

Toxicity: has an overall toxic effect; causes irritation of mucous membranes [70]; suspected of being carcinogenic [5,41]. LD_{50} for rats - 30 g/kg [214].

Determination: polarographic method (sensitivity 0.008 mg/m^3) [62].

S-[(2-Acetamido)ethyl]-O,O-dimethyldithio phosphate; $C_6H_{14}NO_3PS_2$ (243.3)

White crystals, m.p. 22-23°C; soluble in acetone, benzene, and ethanol; slightly soluble in water.

Present in wastes from the production of some pesticides.

Toxicity: $TC_{irritation}$ - 0.8-4.0 mg/kg; LD_{50} for rats, mice, dogs, rabbits, and guinea pigs - 400 mg/kg [47].

Determination: lliquid-gas chromatography [34].

Acetanilide (antifebrin; N-phenylacetamide); C_8H_9NO (135.16)

White leaves, m.p. 114.3°C, b.p. 304°C, ρ 1.026; soluble in organic solvents; slightly soluble in water.

Present in wastes from the production of some dyes and pharmaceuticals, and of delluloid.

Toxicity: has an overall toxic effect (causes changes in the erythrocyte count), an irritating effect (causes eczema), and an allergenic effect. LD_{50} for rats - 80 mg/kg [5].

Acetic acid (ethanoic acid); $C_2H_4O_2$ (60.05)

Liquid with a sharp odor; m.p. 16.75°C, b.p. 118.1°C, f.p. 34°C, s.-i.p. 454°C, CLE 3.3-22%, ρ 1.049; soluble in water.

Present in wastes from the production of acetic acid, cellulose acetate, alkyl acetates, linoleum, ethyl acetate, textiles, synthetic fatty acids, and some foodstuffs.

Toxicity: has an overall toxic effect; causes irritation of mucous membranes, the nose, throat, larynx, and bronchi, and the eyes; chemical burns; impairment of the functioning of liver. TC_{odor}, $TC_{sensitivity\ to\ light}$, $TC_{central\ nervous\ system}$, and $TC_{irritation}$ - 0.0006, 0.00048, 0.00029, and 0.0002 mg/ℓ respectively [5]; MPSDC and MPADC - 0.2 and 0.06 mg/m^3 respectively; toxicity classification - III [20,21]; MPSDC established in GDR and Bulgaria - 0.2 mg/m^3; MPADC established in GDR - 0.06 mg/m^3 [44,54]; MPC established in FRG - 150 mg/m^3 [51].

Removal: with the aid of scrubbers (efficiency 98-99%) [46]; by adsorption on activated carbon [46]; or by filtering [34].

Determination: paper chromatography (sensitivity 5 μg) [29].

Acetic anhydride; $C_4H_6O_3$ (102.09)

Liquid with a sharp odor; m.p. -73.1°C, b.p. 139.9°C, f.p. 40°C, s.-i.p. 389°C, ρ 1.083; soluble in organic solvents and cold water.

Present in wastes from the production of some organic chemicals.

Toxicity: has an irritating effect; causes chemical burns of the skin. TC_{odor}, $TC_{sensitivity\ to\ light}$, and $TC_{c.n.s.}$ - 0.00049, 0.00036, and 0.00018 mg/ℓ respectively [5]; MPSDC and MPADC - 0.1 and 0.03 mg/m^3 respectively; toxicity classification - III [20,21,410]; MPSDC established in GDR and Bulgaria - 0.1 mg/m^3; MPADC established in GDR - 0.03 mg/m^3 [44].

Removal: by adsorption on activated carbon [46].

Determination: polarographic analysis (sensitivity 10 μg/ml) [63].

Acetonanyl (poly(2,2,4-trimethyl-1,2-dihydroquinoline)); (173.0)

Yellowish brown crystals, m.p. ≦120°C, ρ 1.05-1.12; soluble in organic solvents; insoluble in water.

Present in wastes from the production of synthetic rubber and articles made of rubber.

Toxicity: causes brain and pulmonary edema, and degeneration of the liver and kidney. LD_{50} for rats and mice - 2.0 and 1.45 mg/kg [40].

Acetone (propanone); C_3H_6O (58.08)

Liquid with a peculiar odor, m.p. -95.35°C, b.p. 56.24°C, f.p. -18°C, CLE 2.91-12.8% by vol., ρ 0.790; soluble in organic solvents and water.

Present in wastes from the production of pharmaceuticals, wood chemicals, varnishes and paints, plastics, motion picture film, synthetic rubber, acetylene, acetaldehyde, acetic acid, methyl- and butyl acrylates, poly(methyl methacrylate), phenol, and acetone; also in wastes of solid fuel processing plants.

Toxicity: adsorbed by undamaged skin; has an overall toxic effect; causes irritation of the respiratory tracts [47]. SC_{odor} - 0.8 mg/m^3 [229]; TC_{odor} - o.25 [55] 0.26-0.83 [32] and 1.1 mg/m^3 [229]; $SC_{irritation}$ and $TC_{irritation}$ - - 0.35-0.44 and 0.44-0.55 mg/m^3 respectively [229]; MPSDC and MPADC - 0.35 mg/m^3 [20,22]; both MPSDC and MPADC established in Bulgaria, Hungary, and Yugoslavia - 0.35 mg/m^3; MPSDC and MPADC established in GDR - 1.0 and 0.35 mg/m^3 respectively, and in Romania - 5.0 and 2.0 mg/m^3 respectively; MPC established in FRG - 300 mg/m^3 [44,51].

Removal: by catalytic oxidation over a Cu-Cr catalyst [230]; with the aid of scrubbers (efficiency 99%9 [46]; with the aid of filters made of nylon, polyacrylates, polyethylene, polyesters, poly(vinyl chloride), or polypropylene [34]; by adsorption on activated carbon (efficiency 90%) [34,46,87].

Determination: turbidimetric analysis (sensitivity 1 μg) [29]; chromatographic analysis (sensitivity 5 μg) [231]; photometric analysis (sensitivity 2 μg) [58]; or by using a salicylaldehyde reagent (sensitivity 1 μg) [27].

Acetonecyanohydrin (2-hydroxy-2-methylpropionitrile); C_4H_7NO (85.11)

Liquid with an odor of bitter almond, m.p. -19°C, b.p. 120°C, f.p. 73°C, CLE -2.2-12%, ρ 0.930; soluble in water and organic solvents.

Present in wastes from the production of acrylate esters, plastics, and some pharmaceuticals.

Toxicity: adsorbed by undamaged skin; causes irritation of mucous membranes, the skin, and the respiratory organs [5,55,70]. LD_{50} for mice - 15 mg/kg [72].

Acetonitrile (methyl cyanide; cyanomethane); C_2H_3N (41.05)

Colorless liquid with an odor of ether; m.p. -45.72°C, b.p. 81.6°C, f.p. 6°C, ρ 0.786; soluble in water and organic solvents.

Present in wastes from the production of synthetic rubber, poly(methyl acrylate), acrylonitrile, dyes, some pharmaceuticals, and polyisoprene; also in wastes of slate-processing plants.

Toxicity: adsorbed by undamaged skin; has an overall toxic effect. LD_{50} for rats, mice, and guinea pigs - 380, 200, and 140-260 mg/kg respectively; $TC_{irritation}$ - 2.5 mg/ℓ [5]; ASL - 0.10 mg/m^3 [18].

Acetophenone (methylphenyl ketone); C_8H_8O (120.14)

Liquid with an odor of bird cherry, m.p. 19.6°C, b.p. 202.3°C, f.p. 82°C, s.-i.p. 571°C; the lower concentration limit of explosivness -1.1% by volume.

Present in wastes from the production of phenol, acetone, and some perfumes and pharmaceuticals.

Toxicity: $SC_{irritation}$, SC_{odor}, and $SC_{sensitivity\ to\ light}$ - 0.003, 0.008, and 0.007 mg/m^3 respectively; $TC_{irritation}$, TC_{odor}, and $TC_{sensitivity\ to\ light}$ - 0.007, 0.01, and 0.02 mg/m^3 respectively [232]; MPSDC and MPADC - 0.003 mg/m^3; toxicity classification - III [20,21]. The maximum permissible concentration (mg/m^3) established in some countries is as follows:

Country	MPADC	MPADC
Bulgaria	0.35	0.35
GDR	0.01	0.003
Yugoslavia	0.003	0.003

Determination: see [60].

Acetophos (O,O-diethyl-S-(carboethoxymethyl) thiophosphate); $C_4H_{17}O_4PS$ (256.26)

Light yellow liquid with a sharp odor, b.p. 120°C, ρ 1.17; soluble in water and organic solvents.

Present in wastes from the production of some pesticides.

Toxicity: LD_{50} for rats - 300-700 mg/kg [47].

Acetyl acetone (diacetylmethane; 2,4-pentanedione); $C_5H_8O_2$ (100.11)

Colorless liquid with a mild odor, m.p. -23.2°C, b.p. 140.5°C, ρ 0.976; soluble in water and organic solvents.

Present in wastes from the production of some basic organic compounds, varnishes and paints, and some pesticides.

Toxicity: has an overall toxic effect; causes irritation of the skin and mucous membranes [55,70].

Acetyl chloride (ethanoyl chloride); C_2H_3ClO (78.50)

Liquid with a sharp odor; solidification p. -112°C, b.p. 51.8°C, f.p. 4°C, s.-i.p. 390°C; the lower concentration limit of explosivness 5% by vol., ρ 1.105; soluble in organic solvents; in water undergoes hydrolysis.

Present in wastes from the production of some dyes, pharmaceuticals, and basic organic compounds.

Toxicity: causes irritation of the skin and mucous membranes (the eyes and the upper respiratory tracts) [55,71].

Removal: with the aid of filters made of polyesters, including polyacrylates; polyethylene, or polypropylene [34].

Acetylene (ethyne, ethine); C_2H_2 (26.04)

Gas with a garlic odor, b.p. 83.6°C, s.-i.p. 335°C, CLE 2.5-81%, ρ 1.1734; soluble in acetone, slightly soluble in water.

Present in wastes from the production of petrochemicals, acetic acid, acetylene, acetaldehyde, synthetic rubber, trichloroethylene, vinyl chloride, poly(vinyl chloride), and some plastics, pesticides and pharmaceuticals; also in wastes of machine-building and metal- and fuel-processing plants.

Toxicity: has an overall toxic effect; causes dizziness, headache, asphyxia, collapse [56]; has a narcotic effect when mixed with oxygen (40% by vol.) [55,56, 70,86].

Removal: by combustion [26]; catalytic decomposition [89].

Determination: gas chromatographic analysis [64]; various other methods including automatic ones have been described in literature [225-228].

Acetylhydroperoxide (peracetic acid); $C_2H_4O_3$ (76.05)

Liquid with a sharp odor, m.p. -0.1°C, b.p. 105-110°C, ρ 1.22; soluble in water and organic solvents.

Present in wastes from the production of some basic organic chemicals and bleaching agents.

Toxicity: has an overal toxic and irritating effect [47,55,70].

Acrex (dinifen, (6-sec-butyl-2,4-dinitrophenyl) isopropylcarbonate); $C_{14}H_{18}N_2O_7$ (326.3)

Light yellow crystals, m.p. 61-62°C,; soluble in most organic solvents; soluble in water (0.1%).

Present in wastes from the production of some pesticides.

Toxicity: slightly adsorbed by undamaged skin. $LC_{chronic}$ for mice - 4.85 mg/m^3 (for a period of 4 hrs); LD_{50} for rats - 140 mg/kg [47]; recommended MPC - 0.2 mg/m^3 [5].

Acridine; $C_{13}H_9N$ (179.22)

Yellow crystals with a specific odor, m.p. 110-111°C, b.p. 346°C, ρ 1.1005; soluble in benzene and hydrogen sulfide; slightly soluble in water (1:20,000).

Present in wastes of coal-tar chemical plants and natural gas processing plants; also in wastes from the production of paints and varnishes and some pharmaceuticals.

Toxicity: slightly adsorbed by undamaged skin; suspected of being carcinogenic [38]; causes headache, weakness, and irritation of the mucous membranes of the upper respiratory tracts and the eyes. $TC_{chronic}$ for rats - 0.27 mg/m^3.

Acrolein (acrylic aldehyde); C_3H_4O (56.7)

Liquid with a sharp odor, m.p. -86.95°C, b.p. 52.5°C, f.p. -17.8°C, s.-i.p. 277°C [65], ρ 0.8410; soluble in ethanol, diethyl ether, and water (40 g/100 ml of H_2O).

Present in wastes from the production of polyacrylates, varnishes and paints, plastics, linoleum, glycol tristearate, oilcloth, and glycerol; also in wastes of electrical engineering plants, petroleum refineries, foundries, and metallurgical and machine-building works.

Toxicity: has an allergenic and mutagenic effect [37,38,47]; suspected of being carcinogenic [45]; causes irritation of mucous membranes [208]. $TC_{sensitivity\ to\ light}$, $TC_{irritation}$, TC_{odor} - 0.6, 1.75, and 0.8 mg/m^3 respectively [209]; according to other estimates TC_{odor} - 0.52-4.50 mg/m^3 [32]. $TC_{chronic}$ for rats - 0.5 mg/m^3; $SC_{irritation}$ - 0.1 mg/m^3 [210], while according to other estimates $SC_{irritation}$ - 0.63 mg/m^3 [212]; toxicity classification - II [20,21]; MPSDC and MPADC - 0.03 mg/m^3; recommended MPSDC - 0.025 [32] and 0.01 mg/m^3 [46]. The maximum permissible concentration (mg/m^3) established in some countries is as follows:

Country	MPSDC	MPADC
Bulgaria	0.3	0.1
Hungary	0.3	0.1
GDR	0.02	0.01
Romania	0.3	0.1
Czechoslovakia	0.3	0.1
Yugoslavia	0.3	0.1
FRG	0.01	-

Removal: with the aid of filters made of polyacrylates, poly(vinyl chloride), polypropylene, or polyethylene [34]; by adsorption on activated carbon (efficiency 20-25%) [46].

Determination: colorimetric analysis (sensitivity 1 μg) [29]; the fluorescence method (sensitivity 0.05 μg) [211,213]; the spectrophotometric method (sensitivity 0.026 mg/m^3) [78].

Acrylamide; C_3H_5NO (71.08)

White crystalline powder, m.p. 84-85°C, b.p. 215°C, ρ 1.122; soluble in methanol, acetone, and water.

Present in wastes from the production of synthetic rubber, coal, and some organic compounds.

Toxicity: absorbed by undamaged skin; has an overall toxic effect. LD_{50} for rats, guinea pigs, and rabbits - 150-180 mg/kg [5,55,70].

Acrylic acid (propenoic acid); $C_3H_4O_2$ (72.06)

Colorless liquid with a sharp odor, m.p. 14°C, b.p. 141°C, f.p. 440°C, ρ 1.062; soluble in water.

Present in wastes from the production of polymers.

Toxicity: has an overall toxic and irritating effect. LD_{50} for rats - 2.5 g/kg [55,70]; MPSDC established in FRG - 20 mg/m^3 [51].

Removal: with the aid of filters made of polyacrylates, poly(vinyl chloride), teflon, or polypropylene [34]; by adsorption on activated carbon (efficiency - 20-5-%) [46].

Acrylic nitrile (vinyl cyanide); C_3H_3N (53.03)

Liquid with a mild odor (becomes yellowish on storage); m.p. -83.5°C, b.p. 77.5°C, f.p. 0±2.5°C, s.-i.p. 370°C [65], ρ 0.8069, CLE - 3-17% [56].

Present in wastes from the production of some pharmaceuticals, acrylic acid, dyes, surface-active substances, pesticides, plastics, synthetic rubber, and synthetic fabrics.

Toxicity: very toxic; has an irritating effect. $TC_{irritation}$ and TC_{odor} - 0.0008-0.0018 and 0.008 mg/ℓ respectively [5]; according to other estimates [32] $TC_{irrtation}$ - 3.72-51.0 mg/m^3; LD_{50} for mice and rats - 0.35 and 0.47 mg/kg respectively [5]; MPADC - 0.03 mg/m^3; MPC established in FRG - 20 mg/m^3 [51]; toxicity classification - II.

Removal: by adsorption on different adsorbents [48], including activated carbon (efficiency 20-5-%) [46].

Determination: see [206,207].

Adipic acid (hexandioic acid); $C_6H_{10}O_4$ (146.14)

White crystals or powder, m.p. 125-130°C, b.p. 112-115°C, ρ 1.366; solubility in water: 15 g/ℓ H_2O at 15°C; fine powder - highly explosive; ignites on contact with flame, i.p. 320°C, lower concentration limit of explosiveness - 40.3 g/m^3 [65].

Present in wastes from the production of beet sugar, synthetic fibers, and plastics.

Toxicity: slightly toxic for warm-blooded animals [55].

Adipyl dinitrile (tetramethylene dicyanide); $C_6H_8N_2$ (108.56)

Colorless liquid, m.p. 0-1°C, b.p. 295°C; solubility in water: 50 g/ℓ H_2O at 15°C.

Present in wastes from the production of plastics.

Toxicity: causes dizziness and headache; adsorbed by undamaged skin. LD_{50} for rats and mice - 154.8 mg/kg [5].

Aldehydin (2,5) (2-methyl-5-ethyl-pyridine); $C_8H_{11}N$ (121.18)

Liquid with a sharp odor, b.p. 178.3°C, ρ 0.918; soluble in organic solvents; slightly soluble in water (1.2%).

Present in wastes from the production of 2-methyl-5-vinyl pyridine, some pharmaceuticals, textiles, nicotinic acid, and nicotinamide.

Toxicity: has an overall toxic effect (affects the functioning of the central nervous system and causes changes in the blood composition); also causes irritation of mucous membranes, the skin, and the respiratory tracts. LD_{50} for rats and mice - 1.46 and 1.68 g/kg respectively [5,70,86].

Aldol (2-hydroxybutyraldehyde); $C_4H_8O_2$ (88.10)

Colorless (sometime yellowish) viscous liquid; m.p. 0°C, b.p. 83°C at 2.65 kPa, ρ 1.103; soluble in water, ethanol, and diethyl ether.

Present in wastes from the production of some pharmaceuticals, articles made of rubber, and perfumes; also in wastes of metallurgical works.

Toxicity: has an overall toxic and irritating effect. LD_{50} for rats - 2.2 g/kg [55,70].

Aldrin (1,2,3,4,10,10-hexachloro-1,4,4a,5,8,8a-hexahydro-1,4-endo,exo-5,8-dimethanonaphthsline); $C_{12}H_8Cl_6$ (364.94)

White crystals with a mild odor, m.p. 104-105°C; slightly soluble in water (0.07-0.01%).

Present in wastes from the production of some pesticides.

Toxicity: has a carcinogenic effect [73]. $TC_{irritation}$ - 0.25 mg/m^3; MPSDC established in USA - 0.25 mg/m^3 [85]. The use of this pesticide is forbidden in the USSR, Bulgaria, and Poland; in Yugoslavia and Czechoslovakia its use is permitted in limited amounts [85].

Alizarin (1,2-dihydroxy-9,10-anthraquinone); $C_{14}H_8O_4$ (240.23)

Orange-red crystals, m.p. 289-290°C, b.p. 430°C; soluble in ethanol, diethyl ether, and benzene; slightly soluble in water ($2.1 \cdot 10^{-6}$ mole/ℓ).

Present in wastes from the production of some dyes.

Toxicity: causes skin irritation;; may cause edema and eczema [5]; has a mutagenic effect [37,93].

Allethrin (d,ℓ-allyl-3-methyl-2-cyclopenten-1-on-4-yl-d,ℓ-cis-trans-chrysantemate; $C_{19}H_{26}O_3$ (305.41)

Viscous liquid, b.p. 80°C at 0.13 Pa, ρ 1.01; soluble in most organic solvents; insoluble in water.

Present in wastes from the production of some pesticides.

Toxicity: causes irritation of the skin and mucous membranes; has an overall toxic and allergenic effect. LD_{50} for mice - 480 mg/kg [55,70]; recommended MP-SDC - 0.05 mg/m^3 [46].

Allyl alcohol (propen-1-ol-3); C_3H_6O (56.08)

Liquid with a mustard odor; m.p. -129°C, b.p. 96.9°C, f.p. 21°C, s.-i.p. 378°C, CLE - 2.4%, ρ 0.852; readily soluble in most organic solvents and in water.

Present in wastes from the production of some plastics, plasticiers, esters, glycerol, articles made of rubber, and medicinal preparations.

Toxicity: causes irritation of mucous membranes; has an overall toxic and mutagenic effect; is adsorbed by undamaged skin [37,93]. LD_{50} for dogs and rats - 40 and 64 mg/kg respectively [47,55,70,72]; ASL - 0.03 $mg.m^3$ [90].

Allyl amine (2-propenylamine); C_3H_7N (57.09)

Liquid with an odor of ammonia; b.p. 58°C, ρ 0.761; soluble in water, ethanol, diethyl ether, and chloroform.

Present in wastes from the production of plastics and some pharmaceuticals and pesticides.

Toxicity: adsorbed by undamaged skin [5,55]; has an irritating, allergenic effect. TC_{odor} - 0.015 mg/ℓ; LD_{50} for rats - 106 mg/kg [72].

Allyl bromide (3-bromo-propene-1); C_3H_5Br (120.98)

Liquid with a sharp odor, m.p. -119.4°C, b.p. 71.3°C, s.-i.p. 295°C, ρ 1.398, CLE - 4.36-7.25% [56]; soluble in chloriform, carbon disulfide and carbon tetrachloride; insoluble in water.

Present in wastes from the production of perfumes and some pharmaceuticals.

Toxicity: has an overall toxic effect; causes irritation of mucous membranes; is adsorbed by undamaged skin [55].

Allyl chloride (3-chloro-propene-1); C_3H_5Cl (76.53)

Liquid with a sharp odor; m.p. -134.5°C, b.p. 44.96°C, f.p. -29°C, s.-i.p. 420°C, CLE 3-14.8%, ρ 0.938; soluble in organic solvents; slightly soluble in water (0.36%).

Present in wastes from the production of some organic compounds, pharmaceuticals, and insecticides.

Toxicity: causes irritation of mucous membranes; is adsorbed by undamaged skin; has an overall toxic effect [55,70]. LD_{50} for rats - 0.78 g/kg [72]; MPSDC and MPADC - 0.07 and 0.01 mg/m^3 respectively; toxicity classification - II [1].

Allyl cyanide (vinyl acetonitrile); C_4H_5N (67.1)

Colorless liquid, m.p. -86.8°C, b.p. 118.5°C, f.p. 23°C, s.-i.p. 460°C, ρ 0.838; soluble in organic solvents and in water (3.8%).

Present in wastes from the production of some organic compounds.

Toxicity: causes irritation of mucous membranes; is adsorbed by undamaged skin. LD_{50} for mice - 50 mg/kg [5].

Allyl mercaptan; C_3H_6S (74.14)

Liquid with an odor of garlic (becomes darker on storage) b.p. 63-67°C, ρ 0.93; soluble in ethanol and diethyl ether; insoluble in water.

Present in wastes from the production of some polymers and antibiotics, and rubber.

Toxicity: has an irritating effect. TC_{odor} - 0.0009 mg/m^3 [86].

Allyl propionate; $C_6H_{10}O_2$ (114.15)

Colorless liquid, m.p. -100°C, b.p. 153.9°C, ρ 0.968; soluble in water (14.7%) and organic solvents.

Present in wastes from the production of some organic compunds.

Toxicity: TC_{odor} - 47 mg/m^3; LD_{50} for rats and mice - 1.60 and 0.39 g.kg respectively [72,86].

Allyl sulfide; $C_6H_{10}S$ (114.20)

Oily liquid with a garlic odor; m.p. -83°C, b.p. 139°C, ρ 0.888; soluble in organic solvents; slightly soluble in water.

Present in wastes from the production of some organic compounds.

Toxicity: causes severe irritation of the respiratory organs; has an overall toxic effect [55]. TC_{odor} - 0.0007 mg/m^3 [64].

Allyl iso-thiocyanate; C_4H_5NS (99.15)

Oily liquid with a sharp odor; m.p. -80°C, b.p. 152°C, f.p. 46°C, ρ 1.013; soluble in ethanol, diethyl ether, and carbon disulfide; slightly soluble in water (0.2%).

Present in wastes from the production of some foodstuffs, pharmaceuticals, and perfumes.

Toxicity: causes irritation of mucous membranes; has an overall toxic and allergenic effect. TC_{odor} - 0.0006 mg/ℓ [5,55,70].

Altax, (bis(2-benzothiazoloilyl)disulfide); $C_{14}H_8N_2S_4$ (332.50)

White (or yellowish) powder with a sharp odor; m.p. 185°C, s.-i.p. 645°C; lower concentration limit of explosiveness - 37.8 g/m^3.

Present in wastes from the production of synthetic rubber and articles made of rubber.

Toxicity: has an irritating effect (causing inflammation of the respiratory organs) [5].

Amidopyrine (1-phenyl-2,3-dimethyl-4-dimethylaminopyrazol-5-one); $C_{13}H_{17}N_3O$ (213.30)

White (or yellowish crystalline powder; m.p. 107-109°C; soluble in ethanol (50%) and water (5%).

Present in wastes from the production of some pharmaceuticals.

Toxicity: has an overall toxic effect. LD_{50} for mice - 1.8 g/kg; $TC_{chronic}$ for rats - 25 mg/m^3 [5].

Aminazine (rhodoamine; 10-(3-dimethylaminopropyl)-2-chlorophenothazine hydrochloride); $C_4H_8ClN_2S \cdot HCl$ (389.84)

White crystalline powder with a slight garlic odor; m.p. 194-197°C; soluble in water and ethanol.

Present in wastes from the production of some pharmaceuticals.

Toxicity: has an overall toxic effect. LC_{50} for rats and mice - 210-230 mg/kg; some animals died at 40 mg/kg; $TC_{irritation}$ - 0.02-0.1 mg/m^3 [5,70].

Amines, aliphatic (C_{16}-C_{20}).

Solid substances with a strong odor of amine.

Present in wastes from the production of some chemicals, pharmaceuticals, articles made of rubber, textiles, polymeric materials, pesticides, and synthetic rubber.

Toxicity: MPSDC and MPADC - 0.003 mg/m^3; toxicity classification - I [4].

5,6-Amino-2,p-aminophenylbenzimidazole; $C_{13}H_{12}N_4$ (230.00)

White powder; m.p. 235°C; slightly soluble in water (15 mg/ℓ).

Present in wastes from the production of some synthetic fibers.

Toxicity: has a cumulative toxic effect. TC_{odor} - 15 mg/ℓ [218], $SC_{irritation}$ - 0.01 mg/m^3, $TC_{irritation}$ - 0.02 mg/m^3; LD_{50} for rats - 5 g/kg; MPADC - 0.01 mg/m^3; toxicity classification - III [23].

1-Aminoanthraquinone; $C_{14}H_9NO_2$ (223.24)

Reddish purple crystals; m.p. 252°C (sublimes), CLE 38 g/m^3; soluble in ethanol and diethyl ether; insoluble in water.

Present in wastes from the production of some dyes and pharmaceuticals.

Toxicity: $TC_{chronic}$ for rats - 10 mg/m^3 [5]; recommended MPSDC and MPADC - 0.01 mg/m^3 [219].

o-Aminoazotoluene (4-o-tolylazo-o-toluidine); $C_{14}H_{15}N_3$ (225.28)

Orange crystals; m.p. 101-102°C; slightly soluble in ethanol and diethyl ether; insoluble in water.

Present in wastes from the production of some dyes.

Toxicity: adsorbed by undamaged skin; has an overall toxic effect; has been observed to have a carcinogenic effect [5,68] and a mutagenic effect [37].

m-Aminobenzoic acid; $C_7H_7NO_2$ (137.13)

Crystalline powder; m.p. 179.5°C, ρ 1.511; soluble in water (5.9 g/ℓ).

Present in wastes from the production of some dyes, basic organic compounds, pharmaceuticals, and vitamins.

Toxicity: LD_{50} for mice (intra-abdominal injection) - 250-500 mg/kg [70,86].

o-Aminobenzoic acid (anthranilic acid); $C_7H_7NO_2$ (137.13)

Crystalline powder; m.p. 146-147°C, CLE 31.2 g/m^3, ρ 1.412; soluble in benzene; slightly soluble in water (0.34%).

Present in wastes from the production of some dyes, basic organic compounds, pharmaceuticals, and vitamins.

Toxicity: LD_{50} for mice - 2.85 g/kg [55,70,86].

p-Aminobenzoic acid (aminodracylic acid); $C_7H_7NO_2$ (137.13)

Crystalline powder; m.p. 186-187°C; slightly soluble in water (0.34%).

Present in wastes from the production of vitamin H; some dyes, organic compounds, and pharmaceuticals.

Toxicity: LD_{50} for dogs, rats, and mice - 1, 6, and 2.85 g/kg respectively [55,70,86].

m-Aminobenzotrifluoride (α-trifluoro-m-toluidine; 3-aminobenzotrifluoride); $C_7H_6F_3N$ (161.13)

Viscous liquid; m.p. 3°C, b.p. 187.5°C, ρ 1.305; soluble in ethanol and diethyl ether; slightly soluble in water.

Present in wastes from the production of some dyes and pharmaceuticals.

Toxicity: causes skin irritation [55].

1-Aminobutane (butylamine); $C_4H_{11}N$ (73.14)

Amber-colored liquid with an odor of ammonia; m.p. -50.5°C, b.p. 77.8°C, f.p. -12°C, s.-i.p. 312°C, CLE 1.7-9.8%, ρ 0.740; soluble in water, ethanol, and diethyl ether.

Present in wastes from the production of some dyes, pharmaceuticals, and pesticides; of tanning agents and articles made of rubber.

Toxicity: has an overall toxic effect; causes irritation of the skin and mucous membranes. LD_{50} for rats - 500 mg/kg [47,55,70]; MPSDC - 10 mg/m^3 [20,21].

2-Aminobutane (sec-butylamine); $C_4H_{11}N$ (73.14)

Amber-colored liquid with an odor of ammonia; m.p. -104.5°C, b.p. 64°C, f.p. -10°C, ρ 0.725; soluble in water, ethanol, and diethyl ether.

Present in wastes from the production of some dyes, pharmaceuticals, and pesticides; of tanning agents and articles made of rubber.

Toxicity: has an overall toxic effect; causes irritation of the skin and mucous membranes. LD_{50} for rats - 380-500 mg/kg [47,55,70].

4-Amino-3,4'-dimethylazobenzene (dimethyl-p-azobenzene; benzene-azo-dimethylaniline); $C_{14}H_{15}N_3$ (225.30)

Orange-yellow crystals; m.p. 127-128°C; slightly soluble in organic solvents; insoluble in water.

Present in wastes from the production of some chemical reagents, dyes, and basic organic compounds.

Toxicity: adsorbed by undamaged skin; causes depression of the central nervous system, dermatitis; has a carcinogenic and mutagenic effect [5,37,47,70].

Determination: colorimetric analysis (sensitivity 0.2 μg) [292].

4-Aminodiphenylamine (N-phenyl-o-phenylenediamine); $C_{12}H_{12}N_2$ (184.24)

White crystals; m.p. 75°C, b.p. 354°C (in H_2 atmosphere); soluble in ethanol and diethyl ether; slightly soluble in water.

Present in wastes from the production of some dyes, photographic chemicals, antioxidants, and pharmaceuticals.

Toxicity: has an overall toxic effect; causes irritation of the skin. LD_{50} for rats and mice - 447 and 247 mg/kg respectively [5].

bis-(2-Aminoethyl)amine (diethylenetriamine); $C_4H_{13}N_3$ (103.1)

Liquid with an ammonia odor; m.p. -39°C, b.p. 206.7°C, f.p. 94°C, s.-i.p. 365°C, ρ 0.954; soluble in water and organic solvents; slightly soluble in diethyl ether.

Present in wastes from the production of epoxides and polyamides.

Toxicity: adsorbed by undamaged skin; has an overall toxic and irritating effect. LD_{50} for rats - 1.08-2.39 g/kg [5,55].

2-Aminophenol (o-aminophenol); C_6H_7NO (109.14)

Colorless needles; m.p. 174°C, b.p. 214°C, s.-i.p. 390°C; soluble in water (1.7% at 0°C), ethanol and diethyl ether.

Present in wastes from the production of synthetic rubber, dyes, and some pharmaceuticals.

Toxicity: causes dermatitis. LD_{50} for rats, mice, rabbits, and guinea pigs - 500, 600, 1530, and 1900 mg/kg respectively [55]; ASL - 0.26 mg/m^3 [18].

3-Aminophenol (m-aminophenol); C_6H_7NO (109.14)

White crystals; m.p. 123°C, b.p. 164°C (at 1.2 kPa); soluble in ethanol, diethyl ether, and water (2.6%).

Present in wastes from the production of synthetic rubber, dyes, and some pharmaceuticals.

Toxicity: causes dermatitis. LD_{50} for rats and mice - 1 and 0.144 mg/kg respectively [55]; ASL - 0.026 mg/m^3 [18].

4-Aminophenol (p-aminophenol); C_6H_7 NO (109.14)

White crystals; m.p. 186°C, s.-i.p. 500°C; slightly soluble in water (1.1% at 0°C).

Present in wastes from the production of synthetic rubber, dyes, and some pharmaceuticals.

Toxicity: has an overall toxic and irritating effect; causes dermititis and bronchial asthma [55]. ASL - 0.026 mg/m^3 [18].

p-Aminopropiophenone (4-aminopropiophenone); $C_9H_{11}NO$ (149.2)

Yellow needles, m.p. 1±40°C; soluble in water.

Present in wastes from the production of some basic organic compounds.

Toxicity: has a carcinogenic effect [37,93].

2-Aminopyridine; $C_6H_6N_2$ (94.11)

White crystalline powder; m.p. 58.1°C, b.p. 210.6°C; soluble in water and in most organic solvents.

Present in wastes from the production of some dyes and pharmaceuticals.

Toxicity: is adsorbed by undamaged skin; causes headache and cramps. LD_{50} for rats and mice - 0.2 and 0.05 g/kg respectively [5,55,72].

2-Aminothiazole; $C_3H_4N_2S$ (100.15)

Pale brown crystals; b.p. 90°C, b.p. 117°C (at 2 kPa); soluble in ethanol, diethyl ether, and water.

Present in wastes from the production of some pharmaceuticals.

Toxicity: has an overall toxic and irritating effect. LC_{50}-chronic for guinea pigs - 25 mg/m^3; LD_{50} for rats - 480 mg/kg [5,55,70].

Amiton (O,O-diethyl-S-(2-diethylaminoethyl)thiophosphate oxalate); $C_{10}H_{24}NO_3PS$ (269.35)

Liquid; b.p. 110°C; soluble in water and organic solvents.

Present in wastes from the production of some pesticides.

Toxicity: suppresses the action of enzymes; causes nausea, vomiting, ptyalism, cramps, coma [70]. MPSDC - 0.002 mg/m^3; toxicity classification - III [20,21].

Amitrole (3-amino-1,2,4-triazole); $C_2H_4N_4$ (84.08)

White crystals, m.p. 157-159°C; soluble in water (28%).

Present in wastes from the production of some pesticides and photographic chemicals.

Toxicity: has an overall toxic and mutagenic effect. LD_{50} for rats - 14.7 g/kg [5,70,93].

Amyl acetate; $C_7H_{14}O_2$ (130.18)

Liquid with a fruity odor; m.p. -70.8°C, b.p. 149.2°C, f.p. 25-43°C, s.-i.p. 360°C, CLE 1.1-7.8%, ρ 0.875; soluble in ethanol and diethyl ether; slightly soluble in water (18 g/ℓ).

Present in wastes from the production of varnishes and paints, and some pharmaceuticals and foodstuffs.

Toxicity: causes irritation of mucous membranes [55]. SC_{odor} - 0.5 mg/m^3, TC_{odor} - 0.6 mg/m^3, $SC_{sensitivity\ to\ light}$ - 0.12 mg/m^3, $TC_{irritation}$ - 0.3 mg/m^3 [215]; MPSDC and MPADC - 0.1 mg/m^3; toxicity classification - IV [20,21]; MPSDC and MPADC established in Bulgaria, Hungary, and Yugoslavia - 0.1 mg/m^3; and in GDR - 0.3 and 0.1 mg/m^3 respectively [44,51]; MPSDC established in FRG - 150 mg per m^3 [51].

Removal: with the aid of filters made of nylon, polyacrylates, poly(vinyl chloride), polypropylene, or polyethylene [34]; by adsorption on activated carbon (efficiency up to 34%) [46,49].

iso-Amylacetate (common amyl acetate); $C_{17}H_{14}O_2$ (180.19)

Liquid with a fruity odor; m.p. -78.5°C, b.p. 142°C, f.p. 36°C, CLE 1.25-8.0%, ρ 0.872; readily soluble in ethanol; slightly soluble in water (≅2%).

Present in wastes from the production of soap, mineral waters, syrups, some solvents, perfumes, confectionery, leather, textiles, dyes, and photographic films.

Toxicity: causes functional disturbances of internal organs, irritation of mucous membranes and the respiratory tracts; has a narcotic effect at high concentrations. TC_{odor} and $TC_{sensitivity\ to\ light}$ - 0.0006 and 0.0003 mg/ℓ respectively [5,314].

Amyl alcohol (pentanol-1); $C_5H_{12}O$ (88.15)

Yellow liquid with an odor of fusel oil; m.p. -78°C, b.p. 138°C, f.p. 51°C, s.-i.p. 300°C, CLE 1.2-10%, ρ 0.809; soluble in organic solvents; slightly soluble in water (1.9%).

Present in wastes from the production of surface-active substances, varnishes and paints, some organic compounds and foodstuffs, and perfumes.

Toxicity: causes irritation of mucous membranes; is adsorbed by undamaged skin; has an overall toxic effect [56]. MPSDC and MPADC - 0.01 mg/m^3; toxicity classification - III [20,21].

Removal: with the aid of horizontal srubbers and then adsorption on activated carbon [46]; or with the aid of filters [34,216-218].

Amyl amine (1-aminopentane); $C_5H_{13}N$ (87.16)

Liquid; m.p. -55°C, b.p. 104.4°C, f.p. 7°C, ρ 0.755; soluble in water, ethanol, and diethyl ether.

Present in wastes from the production of some dyes, emulsifiers, flotation agents, corrosion inhibitors, and some pesticides and medicinal preparations.

Toxicity: causes irritation of mucous membranes and the skin [50].

Amyl chloride (1-chloropentene); $C_5H_{11}Cl$ (106.60)

Liquid; m.p. -99°C, b.p. 108.2°C, CLE 1.4%, ρ 0.883; soluble in ethanol and diethyl ether; insoluble in water.

Present in wastes of some chemical plants.

Toxicity: has an overall toxic and irritating effect [55,70].

cis-ß-Amylene (cis-pentene-2); C_5H_{10} (70.13)

Liquid with a sharp odor; m.p. -140.24°C, b.p. 36.36°C, ρ 0.648; soluble in organic solvents; insoluble in water.

Present in wastes from the production of some organic chemicals; also in wastes of printing houses.

Toxicity: causes irritation of mucous membranes; has an overall toxic effect [370]. TC_{odor}, $TC_{chronic}$, and $SC_{chronic}$ - 1.8, 9.8, and 1.9 mg/m^3 respectively [49]; MPSDC and MPADC - 1.5 mg/m^3 [20,21]; MPSDC and MPADC established in Bulgaria and Yugoslavia - 1.5 mg/m^3, and in GDR - 1.5 and 1.0 mg/m^3 respectively.

n-Amylene (pentene-1); C_5H_{10} (70.13)

Liquid with a sharp odor; m.p. -165.22°C, b.p. 29.97°C, ρ 0.641; soluble in organic solvents; insoluble in water.

Present in wastes from the production of some organic chemicals; also in wastes of printing houses.

Toxicity: causes irritation of mucous membranes; has an overall toxic effect [370]. TC_{odor}, $TC_{chronic}$, $SC_{chronic}$ - 1.8, 9.8, and 1.9 mg/m^3 respectively [49]; MPSDC and MPADC - 1.5 mg/m^3 [20,21]; MPSDC and MPADC established in Bulgaria and Yugoslavia - 1.5 mg/m^3, and in GDR - 1.5 and 1.0 mg/m^3 respectively.

trans-ß-Amylene (trans-pentene-2); C_5H_{10} (70.13)

Liquid with a sharp odor; m.p. -124°C, b.p. 37.8°C, f.p. -18°C, s.-i.p. 273°C, CLE 1.4-8.7%; soluble in ethanol and diethyl ether; insoluble in water.

Toxicity: causes irritation of mucous membranes; has an overall toxic effect [370]. TC_{odor}, $TC_{chronic}$, and $SC_{chronic}$ - 1.8, 9.8, and 1.9 mg/m^3 respectively [49]; MPSDC and MPADC - 1.5 mg/m^3 [20,21]; MPSDC and MPADC established in Bulgaria and Yugoslavia - 1.5 mg/m^3, and in GDR - 1.5 and 1.0 mg/m^3 respectively.

Anabasine sulfate (ℓ-2-(2-pyridyl)piperidine sulfate); $C_{10}H_{14}N_2 \cdot H_2SO_4$ (422.5)

Dark red oily liquid; b.p. 276°C, ρ 1.046; soluble in water.

Present in wastes from the production of some pesticides.

Toxicity: causes headache, dyspnea, and vomitting. LD_{50} for rats and mice - 210 and 4.4 mg/kg respectively [54]; MPSDC and MPADC - 0.005 and 0.001 mg/m^3 respectively [24].

Aniline (aminobenzene); C_6H_7N (93.14)

Oily liquid; m.p. -5.89°C, b.p. 184.4°C, f.p. 79°C, ρ 1.022; soluble in ethanol, diethyl ether, and water (34g/ℓ).

Present in wastes from the production of some dyes, textiles, pharmaceuticals, articles made of rubber, sulfuric acid, nylon, plastics, and ammonia [221].

Toxicity: has an overall toxic effect [5]. TC_{odor}, SC_{odor}, and $TC_{irritation}$ - 0.37, 0.34, and 0.05 mg/m^3 respectively [222]; MPSDC and MPADC - 0.05 and 0.03 mg/m^3 respectively; toxicity classification - II [4,20,21]; MPSDC and MPADC established in Bulgaria, GDR, Czechoslovakia, and Yugoslavia - 0.05 and 0.03 mg/m^3 respectively, and in Romania - 0.05 and 0.02 mg/m^3 respectively [44,51]; MPC established in FRG - 20 mg/m^3 [51].

Removal: with the aid of filters made of nylon, polyethylene, poly(vinyl chloride), or polypropylene [34]; by adsorption on activated carbon (efficiency 20-50%) [46].

Determination: colorimetric analysis (sensitivity 1 μg) [29]; photometric analysis (sensitivity 1.3 μg) [58].

o-Anisidine (2-aminoanisole); C_7H_9NO (123.0)

Yellowish liquid; m.p. 6.2°C, b.p. 225°C, f.p. 107°C, s.-i.p. 430°C, CLE 0.4-5%, ρ 1.092; soluble in ethanol and diethyl ether; slightly soluble in water.

Present in wastes from the production of some dyes and pharmaceuticals.

Toxicity: adsorbed by undamaged skin; has an overall toxic, irritating, and allergenic effect [5]; ASL - 0.003 mg/m^3 [18].

Determination: polarographic analysis [5].

Anisole (methylphenyl ether); C_7H_8O (108.13)

Liquid with a pleasant odor; m.p. -37.8°C, b.p. 155°C, f.p. 41°C, s.-i.p. 485°C, ρ 0.996; soluble in ethanol and diethyl ether; insoluble in water.

Present in wastes from the production of some basic organic chemicals, pesticides, dyes, and perfumes.

Toxicity: adsorbed by undamaged skin; causes irritation of the respiratory tracts [220]; has an overall toxic effect [55].

Anthracene; $C_{14}H_{10}$ (178.22)

White crystals with a purple fluorescence; m.p. 218°C, b.p. 342°C, ρ 1.25; soluble in organic solvents; insoluble in water.

Present in wastes of solid fuels [223], coal-tar chemicals, dyes; also in waste gases (0.008-0.96 g/kg) ejected during the laying of the bed of water reservoirs for hydropower stations; present in the air over cities and villages (0.5-7 and 1 g/m^3 respectively) [45].

Toxicity: carcinogenic and allergenic [46,47].

Anthraquinone; $C_{14}H_8O_2$ (208.22)

Yellowish crystals; m.p. 286°C, b.p. 379.8°C, CLE 34 g/m^3, ρ 1.419; soluble in nitrobenzene; insoluble in water.

Present in wastes from the production of dyes.

Toxicity: allergenic; has an overall toxic and irritating effect [5,47,55,70].

Aramite (2-(4-t-butylphenoxy)isopropyl(2-chloroethyl) sulfate); $C_{15}H_{23}ClO_4S$ (334.87)

Liquid (technical grade - dark, oily liquid); m.p. 31.7°C, b.p. 200-210°C, ρ 1.145-1.165; soluble in most organic solvents; insoluble in water.

Present in wastes from the production of some pesticides.

Toxicity: carcinogenic [5] and mutagenic [37]; has an overall toxic and irritating effect [5]. LD_{50} for rats, mice, and guinea pigs - 3.9-6.3, 2.0, and 3.9 g/kg respectively; recommended MPSDC - 2 mg/m^3 [46].

Atrazine (6-isopropylamino-2-chloro-4-ethylamino-sym-triazene); $C_8H_{14}ClN_5$ (215.70)

White crystalline powder; m.p. 173-175°C; soluble in water (33 mg/ℓ).

Present in wastes from the production of pesticides.

Toxicity: TC_{odor} - 5.4 mg/ℓ [5]; $TC_{irritation}$ - 70-120 mg/m^3 [55]; LD_{50} for rats and mice - 1.41-3.30 and 0.85-1.57 g/kg respectively; MPSDC and MPADC - 0.002 mg/m^3 [24].

Auramine (bis(p-dimethylaminophenyl)methylenimine hydrochloride; (318.88)

Yellow flakes; m.p. 267°C; soluble in water.

Present in wastes from the production of some dyes and pesticides.

Toxicity: adsorbed by undamaged skin; causes irritation of the skin and mucous membranes; has an overall toxic effect [5]; is carcinogenic [41,82].

Aziridine; C_2H_5N (43.07)

Oily liquid with an odor of ammonia; m.p. -78°C, b.p. 55-56°C, f.p. 11.1°C, CLE 3.6-46%, ρ 0.837; soluble in water and in organic solvents.

Present in wastes from the production of pesticides, some pharmaceuticals, textiles, and triethylamine.

Toxicity: has an overall toxic effect (causing excitment, acute dyspnea, impaired coordination of movements, cramps, necrosis of the liver and kidneys); causes irritation of mucous membranes, the respiratory tracts, and the skin; is carcinogenic, mutagenic, and teratogenic [5,37,100]. TC_{odor} - 1.5-3.0 mg/m^3; $TC_{sensitivity\ to\ light}$ - 0.15 mg/m^3; $TC_{c\ onic}$ - 0.4-0.7 mg/m^3 [99]; MPSDC and MPADC - 0.0001 mg/m^3; toxicity classification - I [20,21].

Determination: see [97,204,205].

Azobenzene (diphenyldiimide); $C_{12}H_{10}N_2$ (182.21)

Orange-red crystals; m.p. 68°C, b.p. 293°C, ρ 1.203.

Present in wastes from the production of pesticides.

Toxicity: has an irritating and overall toxic effect. LD_{50} for rats - 1000 mg per kg [47,55,70].

Azoles (a mixture of sodium salts of alkyl-, dialkyl-, and trialkylphenylsulfoxylates)

Present in wastes from the production of surface active agents; synthetic rubber, textiles; also in wastes of fur-processing plants and laundries.

Toxicity: slightly toxic. LD_{50} for rats - 2 g/kg [5].

Barban (O-(4-chloro-2-butyn-1-yl)-N-(3-chlorophenyl) carbamate); $C_{11}H_9Cl_2NO_2$ (258.2)

White crystals; m.p. 75-76°C(d.); soluble in organic solvents; slightly soluble in water (0.001%).

Present in wastes from the production of some pesticides.

Toxicity: affects the blood, the kidneys, and the functioning of enzymes; causes allergy and irritation of the skin and mucous membranes. $TC_{chronic}$ - 5 mg per m^3; LD_{50} for guinea pigs, rats, mice, and rabbits - 240, 1016, 820, and 600 mg/kg respectively [5,47,70]; MPADC and MPSDC - 0.01 and 0.006 mg/m^3 respectively [24].

Baytex (O,O-dimethyl-O-(3-methyl-4-methylthiophenyl) thiophosphate); $C_{10}H_{15}O_3$-PS_2 (278.3)

Oily liquid with an odor of garlic; b.p. 100°C(≅2.9 Pa); soluble in organic solvents; slightly soluble in water.

Present in wastes from the production of some pesticides.

Toxicity: has an overall toxic effect; causes nausea, vomiting, ptyalism, respiratory depression, and cramps; inhibits the functioning of enzymes. LD_{50} for rats - 190-245 mg/kg [47,70].

Benzal chloride (dichloromethylbenzene); $C_7H_6Cl_2$ (261.03)

Liquid with a sharp odor; m.p. -16.4°C, b.p. 205.2°C, ρ 1.26; soluble in organic solvents; undergoes hydrolysis in water.

Present in wastes from the production of some basic organic chemicals.

Toxicity: has an overall toxic and irritating effect [55,70,86].

Benzaldehyde (art. almond oil); C_7H_6O (106.12)

Liquid with an odor of bitter almonds; m.p. -26°C, b.p. 179.2°C, f.p. 64°C, s.-i.p. 205°C, ρ 1.046; soluble in ethanol; slightly soluble in water (≅0.3%).

Present in wastes from the production of some plastics, tanning agents, dyes, foodstuffs, perfumes, organic acids, petrochemicals, and sulfate pulp.

Toxicity: is allergenic; has an irritating and, at high concentrations, a narcotic effect.

Removal: with the aid of filters made of nylon, poly(vinyl chloride), polyesters, polyethylene, or teflon [34].

Benzanthrone; $C_{17}H_{10}O$ (230.27)

Pale yellow crystals; m.p. 170°C, lower CLE 41 g/m^3; soluble in organic solvents; in soluble in water.

Present in wastes from the production of dyes.

Toxicity: causes skin irritation; has an overall toxic effect (causes damage to the liver and cardiac muscle) [5,55,70].

Determination: see [97].

Benzene (benzol); C_6H_6 (78.12)

Liquid with a peculiar odor; m.p. 5.53°C, b.p. 80.1°C, f.p. -12°C, s.-i.p. 634°C, CLE 1.4-7.1%, ρ 0.879; soluble in organic solvents; slightly soluble in water (0.82 g/ℓ).

Present in wastes from the production of some basic organic compounds and pharmaceuticals, petrochemicals, synthetic rubber, plastics, explosives, nylon, ion-exchange resins, varnishes and paints, and artificial leather.

Toxicity: causes skin irritation [239]; suspected of being carcinogenic and mutagenic [38,240,241]. SC_{odor}, $SC_{irritation}$, and $SC_{chronic}$ - 2.4 [238]-2.5[237], 1.5, and 13 mg/m^3 respectively; $TC_{chronic}$ and $TC_{irritation}$ - 64 [238] and 2 mg/m^3 respectively [237]; MPSDC and MPADC - 1.5 and 0.8 mg/m^3 respectively; toxicity classification - II [20,21]; the maximum permissible concentration established in some countries is as follows (mg/m^3):

Country	MPSDC	MPADC
Hungary	1.5	0.8
GDR	1.5	0.8
Poland	1.0	0.3
Romania	2.4	0.8
Czechoslovakia	2.4	0.8
Yugoslavia	1.5	0.8

MPC established in Czechoslovakia - 24 kg/hr [49], and in FRG - 20 mg/m^3 [47]; recommended MPSDC for plants - 0.1 mg/m^3.

Removal: by adsorption from gas stream [242-244]; with the aid of scrubbers (efficiency 99%) [46]; with the aid of filters made of polypropylene, polyethylene, nylon, polyesters, or teflon [34]; by adsorption on activated carbon (efficiency 6%) [34] (this results in a decrease of the odor by 20-50%) [46]; by thermal catalytic oxidation [89].

Determination: colorimetric analysis (sensitivity 1 μg); gas chromatographic analysis (sensitivity 0.0005 mg) [29], polarographic analysis (sensitivity 2.5 μg) [58]; nitration (sensitivity 1 μg) [27]; automatic methods [245,246].

Benzene sulfonic acid; $C_6H_6O_3S \cdot 1.5H_2O$ (185.20)

White powder; m.p. 43-44°C; soluble in water.

Present in wastes from the production of some basic organic compounds.

Toxicity: has an overall toxic and irritating effect. LD_{50} for mice - 0.4-3.2 g/kg [70,72,86].

Benzene sulfonyl chloride; $C_6H_5ClO_2S$ (176.62)

Oily liquid; m.p. 14.5°C, b.p. 251.5°C(d.), ρ 1.384; soluble in organic solvents; insoluble in water.

Present in wastes from the production of some basic organic compounds.

Toxicity: causes irritation of mucous membranes and the respiratory tracts; is allergenic [5]. $TC_{irritation}$ - 0.003 mg/ℓ; ASL - 0.0005 mg/m^3 [18].

Determination: see [247].

Benzidine (4,4'-diaminobiphenyl); $C_{12}H_{12}N_2$ (184.23)

White crystals which darken in air; m.p. 127.5-128°C, b.p. 401.7°C, ρ 1.250; soluble in ethanol, diethyl ether; slightly soluble in water (0.4%).

Present in wastes from the production of some organic compounds and dyes.

Toxicity: very toxic; adsorbed by undamaged skin; affects the composition of the blood; causes damage to the liver and kidneys; is mutagenic and carcinogenic [5,37,52,55,56]. LD_{50} for dogs - 400 mg/kg. Measures for protecting the air from benzidine have been described in literature [5].

Benzoic acid; $C_4H_6O_2$ (122.12)

White crystals or leaves; m.p. 122.3°C, b.p. 249.2°C, s.-i.p. 574°C, ρ 1.316; soluble in ethanol; slightly soluble in water (0.21%).

Present in wastes from the production of varnishes and paints, leather, wood chemicals, textiles, synthetic rubber, and some foodstuffs, pharmaceuticals, and basic organic compounds.

Toxicity: causes leukocytosis and hormonal disturbances; irritation of the skin and mucous membranes. $SC_{chronic}$ and TC_{odor} - 50 and 1.49 mg/m^3 respectively; LD_{50} for rats - 3.3 g/kg [236].

1,2-Benzopyrene (benzopyrene); $C_{20}H_{12}$ (252.32)

Pale yellow crystals; m.p. 179°C, b.p. 310°C, ρ 1.351; soluble in ethanol, diethyl ether; insoluble in water.

Present in wastes from the production of ethylene oxide, petrochemicals, coal-tar and shale chemicals, and aluminum; also in wastes of plants making asphalt concrete and from the combustion of wood [7,84].

Toxicity: very carcinogenic [38,41,248,249]; mutagenic and teratogenic [41,248, 249]. $SC_{irritation}$ - 0.02 μg/100 m^3; MPADC - 0.1 μg/100 m^3; toxicity classification - I [20,21].

Removal: by combustion and adsorption [7,31,41,50].

Determination: spectrophotometric, fluorescence-spectral analysis (sensitivity 0.031 mg/100 m^3) [27,29,32,63,250,251].

p-Benzoquinone (quinone); $C_6H_4O_2$ (108.10)

Yellow crystals with a sharp odor; m.p. 115.7°C; soluble in ethanol, diethyl ether; slightly soluble in water (1%).

Present in wastes from the production of some dyes, hydroquinone, tanning agents, aromatic substances; also in wastes of fur-processing plants.

Toxicity: has an overall toxic and irritating effect; is allergenic; causes severe irritation of mucous membranes and the respiratory organs, conjunctivitis, and coloration of the eyeball and the skin around the eyes [5,55].

Benzoyl chloride; C_7H_5ClO (140.57)

Liquid with a sharp odor; m.p. -1°C, b.p. 197.2°C, f.p. 81°C, s.-i.p. 519°C, ρ 1.212; soluble in most organic solvents; undergoes hydrolysis in water.

Present in wastes from the production of some dyes, chemical reagents, and pharmaceuticals, and synthetic rubber.

Toxicity: has an overall toxic effect (causes damage to the cardiac muscle, the liver, and the kidneys, and lowering of blood pressure); causes lacrimation and coughing, $TC_{irritation}$ - 0.001 mg/ℓ [5].

Removal: with the aid of filters made of polyesters, polyethylene, or polypropylene [34].

Determination: colorimetric analysis (sensitivity 5 mg) [5].

Benzyl acetate; $C_9H_{10}O_2$ (150.17)

Liquid with an odor of pears; m.p. -51.5°C, b.p. 214.9°C, f.p. 102°C, CLE 0.6-7.3%, ρ 1.056-1.059; soluble in ethanol; insoluble in water.

Present in wastes from the production of varnishes and paints, some perfumes and plastics, and cooking fats.

Toxicity: causes irritation of mucous membranes and the respiratory tracts, vomiting, and diarrhea [5,55,71].

Removal: by filtering [34].

Benzyl alcohol (phenyl carbinol; α-hydroxytoluene); C_7H_8O (108.13)

Liquid with a slight odor; m.p. -15.3°C, b.p. 180°C(d.); soluble in ethanol and water (40 g/ℓ).

Present in wastes from the production of the corresponding derivatives, cellulose acetate, gelatin, shellac, and some perfumes.

Toxicity: adsorbed by undamaged skin; has an overall toxic and irritating effect; is allergenic. LD_{50} for rats - 3.1 g/kg [55,71].

Benzylamine (ω-aminotoluene); C_7H_9N (107.15)

Liquid with an ammonia odor; b.p. 184.5°C, ρ 0.981; soluble in organic solvents; insoluble in water.

Present in wastes from the production of some dyes, pharmaceuticals, and chemical reagents.

Toxicity: causes irritation of the skin and mucous membranes [70].

Benzyl chloride (chloromethylbenzene; ω-chlorotoluene); C_7H_7Cl (126.53)

Liquid with a mild odor; m.p. -39°C, b.p. 179.3°C, f.p. 60°C, s.-i.p. 585°C, lower CLE 1.1%, ρ 1.10; soluble in ethanol, diethyl ether; insoluble in water.

Present in wastes from the production of benzyl chloride derivatives, some dyes and pharmaceuticals, plastics, and tanning agents.

Toxicity: adsorbed by undamaged skin, has an overall toxic effect; causes irritation of mucous membranes, the eyes, the skin, and the respiratory organs [47, 55,70]; suspected of being carcinogenic [52].

Benzyl cyanide (phenylacetonitrile); C_8H_7N (117.14)

Liquid with a mild odor; m.p. -23.8°C, b.p. 234°C(d.), f.p. 106°C, s.-i.p. 55-139°C, ρ 1.018; soluble in ethanol; insoluble in water.

Present in wastes from the production of some basic organic compounds.

Toxicity: causes respiratory depression, cramps, paralysis [5,71].

Benzylmercaptan; C_7H_8S (124.19)

Liquid with a sharp odor; b.p. 195°C, ρ 1.06; soluble in organic solvents; insoluble in water.

Present in wastes from the production of paper.

Toxicity: causes headache, nausea, cyanosis; has an irritating effect [55,71].

Binapacryl (acricid, morocide, endosan; 2-sec-butyl-4,6-dinitrophenyl-3,3-dimethylacrylate); $C_{15}H_{18}N_2O_6$ (322.4)

Cream-colored crystals; m.p. 68-69°C; soluble in organic solvents; insoluble in water.

Present in wastes from the production of some pesticides.

Toxicity: adsorbed by undamaged skin; has an overall toxic effect. LD_{50} for rats, mice, dogs, and guinea pigs - 120-165, 1600, 50, and 300 mg/kg respectively [47,70].

Biphenyl-diphenyl ether mixture (26.5:73.5)

Light brown liquid with a sharp odor; b.p. 258°C, f.p. 115°C, s.-i. p. 695°C, CLE 1.35-2.5%.

Present in wastes from the production of some basic organic chemicals, synthetic fibers, synthetic rubber, plastics, varnishes and paints, and wood chemicals; also in wastes of metal-processing plants.

Toxicity: $SC_{irritation}$, $SC_{chronic}$, TC_{odor}, $TC_{sensitivity\ to\ light}$, $TC_{irritation}$ and $TC_{chronic}$ - 0.01, 0.01, 0.06, 0.04, 0.03, and 0.2 mg/m^3 respectively [300];

MPSDC and MPADC established in USSR, as well as in Bulgaria, Romania, and Yugoslavia - 0.01 mg/m^3; toxicity classification - III [20,21]; MPADC established in GDR - 0.003 mg/m^3 [44,51].

Determination: nitration (sensitivity 0.25 μg) [27]; gas chromatographic analysis (sensitivity 0.02 mg/m^3 in 10 ℓ; 0.002 mg/m^3 in 100 ℓ [301]) [20].

Bitumen (asphalt)

Dark mass; b.p. >470°C, ρ 0.95-1.1.

Present in wastes from the production of asphalt coatings, varnishes and paints, and plastic binders.

Toxicity: can cause cancer if it gets into lungs [248].

Borneol (1,7,7-trimethylbicyclo[2,2,1]-2-endo-heptanol); $C_{10}H_{18}O$ (154.24)

Powder with an odor of pepper; m.p. 204-208°C, b.p. 212°C, ρ 1.011; soluble in ethanol; insoluble in water.

Present in wastes from the production of some basic organic compounds, soap, pharmaceuticals, and perfumes.

Toxicity: has an overall toxic and irritating effect; is allergenic [70].

Bromoacetic acid (bromoethanoic acid); C_2H_3 BrO_2 (138.96)

White powder; m.p. 50°C, b.p. 208°C, ρ 1.94; soluble in water and organic solvents.

Present in wastes from the production of some basic organic chemicals.

Toxicity: has an overall toxic and irritating effect; causes chemical burns [5, 70].

Removal: with the aid of filters made of polyethylene or polypropylene [34].

Bromoacetone; C_3H_5BrO (136.9)

Liquid with a sharp odor; m.p. -54°C, b.p. 136.5°C(d.), ρ 1.634; soluble in organic solvents; slightly soluble in water.

Present in wastes from the production of bromoacetone.

Toxicity: causes irritation of mucous membranes and lacrimation. $TC_{irritation}$ - 18 mg/m^3 [55,70].

ω-Bromoacetophenone, (phenacyl bromide); C_8H_7BrO (199.06)

White crystals; m.p. 50°C, b.p. 140°C, ρ 1.647; soluble in organic solvents; insoluble in water.

Present in wastes from the production of bromide derivatives and ω-bromoacetophenone.

Toxicity: causes severe irritation of mucous membranes, the eyes and the skin [55,70].

Bromobenzene (phenyl bromide); C_6H_5Br (157.02)

Liquid with a peculiar odor; m.p. -30.6°C, b.p. 156.1°C, f.p. ≅30°C, s.-i.p. 545°C, ρ 1.495; soluble in organic solvents; insoluble in water.

Present in wastes from the production of motor fuels and some basic organic compounds.

Toxicity: causes irritation of the respiratory tracts and the skin; at high concentrations has a narcotic effect [55,70]. MPADC - 0.03 mg/m^3; toxicity classification - II [23].

Removal: with the aid of filters made of polyethylene, polypropylene, or polyvinyldene chloride [34].

Bromobenzyl cyanide (α-bromocyanotoluene); C_8H_6BrN (196.06)

White crystals with an odor of decaying fruits; m.p. 25.4°C, ρ 1.516; soluble in organic solvents; slightly soluble in water.

Present in wastes from the production of some basic organic chemicals and pharmaceuticals.

Toxicity: has an overall toxic and irritating effect. TC_{odor} and $TC_{irritation}$ - 0.00009 and 0.00015 mg/ℓ respectively; $TC_{irritation}$ for rats, rabbits, and guinea pigs - 131, 134, and 9 mg/m^3 respectively [5,70].

Bromoform (tribromomethane); $CHBr_3$ (252.77)

Liquid with an odor of chloroform; m.p. 7.7°C, b.p. 149.6°C, ρ 2.892; soluble in organic solvents; slightly soluble in water (0.3%).

Present in wastes from the production of some solvents and pharmaceuticals.

Toxicity: has an overall toxic effect (causes damage to the liver) and an irritating effect (causes lacrimation) [55,70]. MPADC - 0.05 mg/m^3; toxicity classification - III [23].

Bromomethane (methyl bromide); CH_3Br (95.0)

Liquid; m.p. -93°C, b.p. 3.6°C, ρ 1.73; slightly soluble in water (1.75%).

Present in wastes from the production of some pesticides.

Toxicity: adsorbed by undamaged skin, has an overall toxic effect; causes vomiting, impaired functioning of the central nervous system; at high concentrations causes cramps and pulmonary edema; has a narcotic effect [47,55,70].

Removal: by adsorption on activated carbon [46].

Bromoxynil (brominil; 3,5-dibromo-4-hydroxybenzonitrile); $C_7H_3Br_3NO$ (276.9)

White crystals; m.p. 194-195°C; soluble in water and organic solvents.

Present in wastes from the production of some pesticides.

Toxicity: adsorbed by undamaged skin, has an overall toxic effect. LD_{50} for rats and dogs - 190-1395 and 510 mg/m^3 respectively [47].

1,3-Butadiene (erythrene); C_4H_6 (54.09)

Gas with an odor of garlic; m.p. -108.92°C, b.p. -4.47°C, f.p. -40°C, s.-i.p. 420°C, CLE 2.0-11.5%, ρ 0.621; soluble in organic solvents; insoluble in water.

Present in wastes from the production of synthetic rubber and some intermediate petrochemicals.

Toxicity: causes irritation of mucous membranes, the skin, and the respiratory tracts; at high concentrations has a narcotic effect [56]; is carcinogenic [45]. SC_{odor}, $SC_{sensitivity\ to\ light}$, TC_{odor}, $TC_{sensitivity\ to\ light}$, and $TC_{irritation}$ - 1,1,4, 3.8, and 3.6 mg/m^3 respectively [250]; MPSDC and MPADC - 3 and 1 mg/m^3 respectively; toxicity classification - IV [20,21]; MPSDC and MPADC established in Bulgaria, GDR, and yugoslavia - 3 and 1 mg/m^3 respectively [46,51].

Removal: by adsorption on activated carbon [46].

Determination: spectrophotometric analysis (sensitivity 1 μg) [27]; photometric analysis (sensitivity 0.5 μg) [63].

1-Butanal (n-butyraldehyde); C_4H_8O (72.10)

Liquid with a sharp odor; m.p. -99°C, b.p. 75.15°C, f.p. -10-(-12)°C, CLE 2.42-10.5%, ρ 0.802; soluble in organic solvents; forms an azeotropic mixture with water.

Present in wastes from the production of articles made of rubber, and some plastics, solvents, and plasticizers.

Toxicity: has an overall toxic, irritating, and narcotic effect. LD_{50} for rats - 5.89 g/kg [58,70,86].

Butane (diethyl); C_4H_{10} (58.12)

Gas; m.p. -138.35°C, b.p. -0.5°C, f.p. -60°C, s.-i.p. -40.5°C, CLE 1.5-8.5%, ρ 0.579; slightly soluble in water (0.37 g/ℓ); soluble in organic solvents.

Present in wastes from the production of motor fuels, synthetic rubber, and petrochemicals; also in wastes of oil refineries.

Toxicity: causes irritation of the respiratory organs, asphyxia; at high concentrations causes narcosis. TC_{odor} - 305 mg/m^3 [251]; MPSDC - 200 mg/m^3; toxicity classification - IV [20,22]; MPSDC established in Bulgaria, GDR, and Yugoslavia - 200 mg/m^3; MPADC established in GDR - 50 mg/m^3 [44,51].

Removal: by thermal catalytic oxidation (efficiency 100%) [89].

Butanethiol-1 (n-butyl mercaptan); $C_4H_{10}S$ (90.18)

Liquid with a sharp odor; m.p. -115.9°C, b.p. 98.46°C, f.p.≅ 2°C, ρ 0.842; soluble in organic solvents; slightly soluble in water.

Present in wastes of natural gas processing plants; also in wastes from the production of cellulose and paper.

Toxicity: has an overall toxic effect: causes nausea, dizziness, weakness, blurring of vision, and impaired coordination of movements; and irritation of the respiratory tracts and mucous membranes; at high concentrations has a narcotic effect [5,55,70]. TC_{odor} - $0.7 \cdot 10^{-5}$ mg/ℓ.

Butanethiol-2 (sec-butyl mercaptan); 2-methylpropanthiol-1); $C_4H_{10}S$ (90.18)

Liquid with a sharp odor; m.p. -165°C, b.p. 84-85°C, ρ 0.830; soluble in organic solvents; slightly soluble in water.

Present in wastes of natural gas processing plants; also in wastes from the production of cellulose and paper.

Toxicity: has an overall toxic and irritating effect [55,70].

Butanol-1 (n-butyl alcohol); $C_4H_{10}O$ (74.12)

Liquid; m.p. -89.5°C, b.p. 117.5°C, f.p. 34°C, s.-i.p. 345°C, CLE 1.8-12.0%, ρ 0.810; soluble in organic solvents; forms an azeotropic mixture with water.

Present in wastes from the production of synthetic rubber, varnishes and paints, plastics; some esters, solvents, plasticizers, and petrochemicals.

Toxicity: adsorbed by undamaged skin, has an overall toxic and irritating effect [56]. $SC_{irritation}$ and $TC_{irritation}$ - 0.1 and 0.8 mg/m^3 respectively [257]; MPSDC - 0.1 mg/m^3; toxicity classification - III [20,21]; MPSDC and MPADC established in Bulgaria and Yugoslavia - 0.3 and 0.1 mg/m^3 respectively; MPSDC established in GDR - 0.3 mg/m^3 [44, 51]; MPC established in FRG - 300 mg/m^3 [51].

Removal: by scrubbing (using horizontal scrubbers), followed by adsorption on activated carbon [258]: by thermal catalytic oxidation over a double layer of the catalysts [259]; by adsorption on activated carbon [46,87].

2-Butene-1-al (crotonic acid α-aldehyde); C_4H_6O (70.09)

Colorless liquid with a sharp odor; m.p. -69°C, b.p. 102.2°C, f.p. 8°C, ρ 0.848; soluble in organic solvents and in water (18.1%).

Present in wastes from the production of butanol, butyric aldehyde, maleic acid; articles made of rubber; crotonyl alcohol (2-butene-1-ol); and some basic organic chemicals and pesticides.

Toxicity: has an overall toxic effect; causes irritation of the eyes, the skin, and the respiratory tracts; is allergenic [47,55,70].

Removal: by adsorption on activated carbon [46].

Determination: polarographic analysis (sensitivity 0.013 mg/m^3) [63].

1-Butene (butylene); C_4H_8 (56.1)

Gas; m.p. -185.8°C, b.p. -6.25°C, f.p. -40°C, s.-i.p. 384°C, CLE 1.6-9.4%.

Present in wastes of natural gas processing plants; also in wastes from the production of synthetic rubber and petrochemicals.

Toxicity: causes irritation of the upper respiratory tracts [56]. $SC_{irritation}$, TC_{odor}, $TC_{sensitivity\ to\ light}$, and $TC_{irritation}$ - 8, 15.4, 11, and 13.9-14.3 mg/m^3 respectively; MPSDC and MPADC - 3 mg/m^3; toxicity classification - IV [20,21]; MPSDC and MPADC established in Bulgaria and Yugoslavia - 3 mg/m^3, and in GDR - 3 and 2 mg/m^3 respectively [256].

Determination: gas chromatography [63].

Butonate (O,O-dimethyl(2,2,2-trichloro-1-butyrylhydroxyethyl)phosphonate); C_8H_{14}-Cl_3O_5P (327.55)

Liquid; b.p. 129°C, ρ 1.399; soluble in ethanol; insoluble in water.

Present in wastes from the production of pesticides.

Toxicity: very toxic; causes nausea, vomiting, ptyalism, cramps, coma [70].

Determination: chromatography [262].

2-Butoxyethanol-1 (n-butylcellosolve); $C_6H_{14}O_2$ (118.18)

Liquid with a sharp odor; b.p. 170.6°C, ρ 0.901; soluble in water and organic solvents.

Present in wastes from the production of nitrocellulose, some plastics, fats, and lubricants.

Toxicity: has an overall toxic effect: causes depression and vomiting. LD_{50} for rats - 1.48 g/kg [70]; MPSDC and MPADC - 1.00 and 0.3 mg/m^3 respectively; toxicity classification - III [23].

Removal: by adsorption on activated carbon [46].

n-Butyl acetate (butyl acetate); $C_6H_{12}O_2$ (116.16)

Liquid with an odor of ether; m.p. -73.3°C, b.p. 126.3°C, f.p. 25-29°C, s.-i.p. 370°C, CLE 1.35-8.35°C, ρ 0.881; soluble in organic solvents; insoluble in water.

Present in wastes from the production of wood chemicals, some perfumes, plastics, varnishes and paints, artificial leather, photographic films, antibiotics, and some foodstuffs.

Toxicity: SC_{odor} and TC_{odor} - 0.5 and 0.6[255]-3.1 mg/m^3 [32] respectively; $SC_{sensitivity\ to\ light}$ and $TC_{sensitivity\ to\ light}$ - 0.18 and 0.3 mg/m^3 respectively

[255]; $SC_{irritation}$ and $TC_{irritation}$ - 0.10 and 0.13 mg/m^3 respectively; MPSDC and MPADC - 0.10 mg/m^3; toxicity classification - IV [4,20,21]; MPSDC and MPADC established in Bulgaria and Yugoslavia - 0.1 mg/m^3, and in GDR - 0.3 and 0.1 mg/m^3 respectively [44,51]; MPC established in FRG - 300 mg/m^3 [51].

Removal: with the aid of filters made of polyvinylidine chloride, nylon, polyesters, polyethylene, polypropylene, or teflon [34].

Butyl acrylate; $C_7H_{13}O_2$ (128.17)

Liquid; b.p. 120°C, f.p. 18.9°C, ρ 0.879; soluble in organic solvents; slightly soluble in water.

Present in wastes from the production of synthetic rubber, some polymers, poly(methyl acrylate), acrylic acid, methyl- and butyl acrylate, leather, textiles, and dyes.

Toxicity: has an overall toxic effect: causes lacrimation, sneezing, ptyalism, and coughing; at high concentrations has a narcotic effect. $TC_{irritation}$ for mice - 0.02 mg/ℓ; LC_{50} - 7.8 mg/ℓ [5,55]; MPSDC -0.0075 mg/m^3; toxicity classification - II [96].

Removal: by adsorption on activated carbon (efficiency 90%) [87].

Determination: thin-film chromatography (sensitivity 0.2 μg) [252]; gas chromatography (sensitivity 0.002 mg/m^3 [253,254].

Butylcarbitol (monobutyl ether of diethylene glycol; diethylene glycol n-butyl ether; 2-(ß-butoxyethoxy) ethanol); $C_8H_{18}O_3$ (162.23)

Liquid with an odor of ether; m.p. -68°C, b.p. 231.2°C, f.p. 77.8°C, s.-i.p. 227.8°C, ρ 0.955; soluble in organic solvents and water.

Present in wastes from the production of some basic organic chemicals, plasticizers, solvents, natural and synthetic resins, vegetable oils, cellulose nitrate, alcohols, lacquers, leather, cosmetics, and household chemicals.

Toxicity: has an overall toxic effect; causes respiratory depression, sometimes cramps, and changes in the liver tissues; has a cumallative effect. $TC_{chronic}$ and TC_{odor} - 1.0 and 4 mg/m^3 respectively; LC_{50} for rats and mice - 4500 and 6050 mg/m^3 respectively; LD_{50} for rats and guinea pigs - 5.5-5.6 and 2.0 g/kg respectively [70].

Butyl-2,4-dichlorophenoxy acetate (2,4-butyl ether); $C_{12}H_{14}Cl_2O$ (277.2)

Liquid; m.p. 9°C, b.p. 146-147°C (≅130Pa), ρ 1.24; soluble in organic solvents; insoluble in water.

Present in wastes from the production of some pesticides.

Toxicity: has an overall toxic effect (impaires the central nervous system); causes dermititis; does damage to embryos, and is teratogenic. $TC_{irritation}$ - 0.004-

0.007 mg/ℓ [5,67]; MPSDC and MPADC - 0.009 and 0.004 mg/m^3 respectively [24].

iso-Butylene oxide (1,2-butylene oxide; α,α-dimethylethylene oxide); C_4H_8O (72.10)

Liquid with an odor of ethanol; m.p. -150°C, b.p. 58.5-59°C, ρ 0.831; soluble in ethanol, diethyl ether, and water (82.4 g/ℓ).

Present in wastes from the production of some organic chemicals.

Toxicity: is carcinogenic and mutagenic [93]. LD_{50} for rats - 0.5 g/kg [84].

n-Butyl formate; $C_5H_{10}O_2$ (101.12)

Liquid with a sharp odor; m.p. -90°C, b.p. 107°C, f.p. 15°C, s.-i.p. 285°C, CLE 1.73-8.15%, ρ 0.885; soluble in organic solvents; insoluble in water.

Present in wastes from the production of formamide, vitamin B_1, and some perfumes.

Toxicity: causes irritation of mucous membranes and the respiratory organs; at high concentrations has a narcotic effect [5]. TC_{odor} - 70 mg/m^3 [54].

n-Butyl furoate; $C_9H_{12}O_3$ (168.19)

Liquid with a sharp odor; b.p. 118-120°C(33kPa), ρ 1.055; soluble in organic solvents; insoluble in water.

Present in wastes from the production of plastics, polyester fibers, and some pesticides.

Toxicity: has an overall toxic effect: causes pneumonia, degeneration of the kidneys; causes irritation of mucous membranes and the respiratory tracts. $TC_{irritation}$ - 20 mg/m^3; LD_{50} for mice - 1.5 g/kg [5].

Butyl methacrylate; $C_8H_{14}O_2$ (142.20)

Liquid; m.p. -76°C, b.p. 163°C, ρ 0.894; soluble in organic solvents; insoluble in water.

Present in wastes from the production of polyacrylates, some plastics, synthetic rubber, varnishes and paints.

Toxicity: has an overall toxic, irritating, and narcotic effect. TC_{odor} - 0.006 mg/ℓ; $TC_{irritation}$ for rabbits - 0.5 mg/ℓ [5,55].

2-Butylthiobenzothiazole; $C_{11}H_{11}NS_2$ (223.36)

Viscous liquid; b.p. 185-188°C, ρ 1.164; soluble in ethanol and acetone; insoluble in water.

Present in wastes from the production of articles made of rubber and defoliants.

Toxicity: slightly toxic. $TCDC_{acute}$ and $TC_{chronic}$ - 0.042 and 0.024 mg/ℓ respectively; LD_{50} for rats - 1.3 g/kg [5].

p-t-Butyltoluene; $C_{11}H_{16}$ (149.39)

Liquid; m.p. -62.53°C, b.p. 192.8°C, ρ 0.857; soluble in organic solvents; slightly soluble in water.

Present in wastes from the production of some basic organic comounds.

Toxicity: has an overall toxic and irritating effect [55]. ASL - 0.023 mg/m^3 [18].

n-Butyric acid (butanoic acid); $C_4H_8O_2$ (88.10)

Liquid with an odor of rancid butter; m.p. -7.9°C, b.p. 163.5°C, f.p. 76.6°C, s.-i.p. 425°C, ρ 0.958; soluble in organic solvents and water.

Present in wastes from the production of some esters, liqueurs, and syrups.

Toxicity: has an overall toxic and irritating effect. LD_{50} for rats - 8.79 g/kg [55]; MPSDC and MPADC for rats - 0.015 and 0.01 mg/m^3 respectively [44]; MPSDC and MPADC established in Bulgaria and Yugoslavia - 0.015 and 0.01 mg/m^3 respectively; MPC established in FRG - 20 mg/m^3 [51].

Removal: with the aid of filters made of polyesters, polyethylene, polyacrylates, poly(vinylidine chloride), teflon, or polypropylene [34].

Determination: paper chromatography (sensitivity 5 μg) [29].

γ-Butyrolactone; $C_4H_6O_2$ (86.0)

Liquid with a slight odor; m.p. -44°C, b.p. 204°C, ρ 1.129; soluble in organic solvents; undergoes hydrolysis in water.

Present in wastes from the production of some basic organic compounds.

Toxicity: slightly toxic; causes irritation of mucous membranes; is mutagenic [37,55].

n-Butyronitrile (n-propyl-iso-cyanide; n-propyl carbylamine); C_4H_7N (69.10)

Liquid; m.p. -112.6°C, b.p. 118°C, ρ 0.80; soluble in diethyl ether; slightly soluble in water.

Present in wastes from the production of some basic organic chemicals.

Toxicity: very toxic; adsorbed by undamaged skin; has an irritating effect. LD_{50} for rats - 100 mg/kg [5,70].

Cacodylic acid (dimethylarsenic acid); $C_2H_7AsO_2$ (137.99)

White crystals; m.p. 200°C; soluble in organic solvents; insoluble in water.

Present in wastes from the production of some pesticides.

Toxicity: has an overall toxic and irritating effect; is allergenic. LD_{50} for rats - 850-1350 mg/kg [47,55].

Camphor (1,7,7-trimethylbicyclo[2.2.1]-heptan-2-one); $C_{10}H_{16}O$ (152.32)

White crystals; m.p. 178.5-179.5°C, b.p. 207.4-209.1°C, f.p. 50°C, lower CLE 10.1 g/m^3, ρ 0.992; soluble in organic solvents; slightly soluble in water (0.1%).

Present in wastes from the production of some pharmaceuticals, plastics, esters, and varnishes and paints.

Toxicity: camphor is used for medical purposes. At high concentrations has an overall toxic effect (causes vomiting, cramps); is allergenic [55,70]. Recommended MPSDC - 0.25 mg/m^3 [46].

Removal:: by adsorption on acivated carbon [46].

n-Caproic acid (n-hexanoic acid); $C_6H_{12}O_2$ (116.16)

Liquid with an odor of cheese; m.p. -3.4-(-3.9)°C, b.p. 205-208°C, ρ 0.976; soluble in ethanol and diethyl ether; slightly soluble in water (0.97%).

Present in wastes from the production of some esters and perfumes.

Toxicity: has an overall toxic and irritating effect. SC_{odor} and TC_{odor} - 0.01 and 0.07-0.1 mg/m^3 respectively [319]; MPSDC and MPADC - 0.01 and 0.005 mg/m^3 respectively; toxicity classification - III [20,21]; MPSDC and MPADC established in Yugoslavia - 0.01 and 0.005 mg/m^3 respectively; MPC established in FRG - 20 mg/m^3 [51].

ε-Caprolactam; $C_6H_{11}NO$ (113.16)

White crystals; m.p. 68-70°C, b.p. 262°C, f.p. 135°C, s.-i.p. 673°C, ρ 1.02; soluble in water and organic solvents.

Present in wastes from the production of some synthetic fibers and polyamide plastic materials.

Toxicity: SC_{odor} and TC_{odor} - 0.21 and 0.30 mg/m^3 respectively; $SC_{sensitiviity\ to\ light}$ and $TC_{sensitivity\ to\ light}$ - 0.11 and 0.20 respectively; $SC_{irritation}$ and $TC_{irritation}$ - 0.06 and 0.11 mg/m^3 respectively; $SC_{c.n.s.}$ and $TC_{c.n.s.}$ - 0.11 and 0.18 mg/m^3 respectively; toxicity classification - III [20,21]; MPSDC and MPADC established in Bulgaria and Yugoslavia - 0.06 mg/m^3, and in GDR - 0.1 and 0.06 mg/m^3 respectively [44,51].

Determination: see [20,318].

n-Caprylic acid (octanoic acid); $C_8H_{10}O_2$ (144.21)

Liquid; m.p. 16-16.7°C, b.p. 237-239.7°C, f.p. 245°C, s.-i.p. 286°C, ρ 0.91; soluble in organic solvents; very slightly soluble in water (0.068%).

Present in wastes from the production of fats and some plastics.

Toxicity: adsorbed by undamaged skin; causes damage to the kidneys, the liver, the cardiac muscle, and the brain; and irritation of the eyes, mucous membranes, and the respiratory tracts [5]. MPSDC and MPADC - 0.01 and 0.005 mg/m^3 respectively [24], MPSDC and MPADC established in Bulgaria and GDR - 0.01 and 0.005 mg/m^3 respectively [56].

Removal: with the aid of filters made of polyethylene, poly(vinylidene chloride) or polypropylene [34]; by adsorption on activated carbon [46].

Captan (1,2,5,6 -tetrahydro-N-(trichloromethylthio)phthalimide); $C_9H_8Cl_3NO_2S$ (300.61)

White crystals; m.p. 178°C; slightly soluble in organic solvents; insoluble in water.

Present in wastes from the production of some pesticides.

Toxicity: has an overall toxic and irritating effect; is mutagenic and teratogenic; does damage to embryos and gonads [5,37,47,67,70,73,320]. LD_{50} for rats, guinea pigs, and rabbits - 2.65, 0.925, and -.74 g/kg respectively [5,70].

Carbaryl (sevin; N-methyl-O-(1-naphthyl)carbamate); $C_{12}H_{11}NO_2$ (201.2)

White crystals; m.p. 142°C, f.p. 196°C, s.-i.p. 561°C, ρ 1.23; soluble in organic solvents; slightly soluble in water.

Present in wastes from the production of some pesticides.

Toxicity: causes ptyalism, lacrimation, anxiety, excitement followed by depression, impaired coordination of movements, suppression of the functioning of enzymes; dysfunctioning of the liver, kidneys, and heart; is mutagenic, teratogenic, and carcinogenic; does damage to embryos [5,67,70,321]. MPSDC and MPADC - 0.02 and 0.01 mg/m^3 respectively [24]; MPSDC recommended and established in USA - 0.05 [46] and 5 mg/m^3 [40] respectively.

Determination: colorimetric analysis (sensitivity 0.08 mg/m^3) [322].

Carbitol (diethylene glycol monoethyl ether); $C_6H_{14}O_3$ (134.17)

Liquid with a musty odor; m.p. 201.9°C, f.p. 94.4°C, ρ 0.990; soluble in water and organic solvents.

Present in wastes from the production of solvents for cellulose esters, lacquers, wax, enamels, antifreeze, and some pharmaceuticals.

Toxicity: has an overall toxic effect: at first stimulates and then suppresses the activity of the central nervous system; causes vomiting, drowsiness, respiratory depression, cramps, and damage to the liver. LD_{50} for rats and guinea pigs - 8.69-9.74 and 3.67-4.97 g/kg respectively [55,70,323].

Carbofuran (furadan; O-(2,3-dihydro-2,2-dimethylbenzofuran-7-yl)-N-methyl carbamate); $C_{12}H_{15}NO_3$ (221.3)

White crystals; m.p. 150-152°C; soluble in organic solvents; slightly soluble in water (0.7 g/kg).

Present in wastes from the production of some pesticides.

Toxicity: very toxic; suspected of being carcinogenic [47].

Determination: colorimetric analysis (sensitivity 0.5 μg) [324].

Carbon tetrachloride (tetrachloromethane); CCl_4 (153.82)

Liquid; m.p. -22.87°C, b.p. 76.75°C, ρ 1.631; soluble in organic solvents; slightly soluble in water (0.08%).

Present in wastes from the production of some pesticides, solvents, and photographic chemicals.

Toxicity: is carcinogenic and mutagenic [52.93], MPSDC and MPADC - 4 and 0.7 mg/m^3 respectively [20].

Removal: by catalysis [406]; by using filters made of polyethylene or polypropylene [34].

Determination: see [20,29].

Carbophenothion (O,O-diethyl-S-(4-chlorophenylthiomethyl) dithiophosphate); $C_{11}H_{16}ClO_2PS$ (342.9)

Yellow liquid; b.p. 130°C (≅1.3 Pa), ρ 1.26-1.28; soluble in organic solvents; slightly soluble in water.

Present in wastes from the production of some pesticides.

Toxicity: very toxic; causes vomiting, ptyalism, cramps, and disturbance of metabolism. LD_{50} for rats - 24 mg/kg [47,70].

Cellosolve acetate; $C_6H_{12}O_3$ (132.17)

Liquid with an odor of ether; m.p. -61.7°C, b.p. 156.3°C, f.p. 65.4°C, ρ 0.974; soluble in organic solvents and in water (23%).

Present in wastes from the production of some lacquers, solvents, alkyd resins, leather finishes; also in wastes of wood-processing plants.

Toxicity: has an overall toxic effect: causes suppression of the central nervous system, vomiting, and respiratory depression; has an irritating effect [55,70].

Removal: by adsorption on activated carbon [46].

Cetyl alcohol (hexadecanol); $C_{16}H_{34}O$ (242.43)

White crystals; m.p. 49.3°C, b.p. 324.08°C, f.p. 160°C, s.-i.p. 238°C, ρ 0.818; soluble in organic solvents; insoluble in water.

Present in wastes from the production of some surface-active substances, pharmaceuticals, plasticizers, antiseptics, and aromatic substances.

Toxicity: slightly toxic. LC_{100} for rats - 22.2 mg/ℓ (after a 2-hr exposure); LD_{50} for rats and mice - 6.4-12.8 and 3.2-6.4 g/kg respectively [55,70,86].

Chloramben (ambien; 3-amino-2,5-dichlorobenzoic acid); $C_7H_5Cl_2NO_2$ (206.0)

White (or cream-colored) crystals; m.p. 200-201°C; soluble in alkane solvents.

Present in wastes from the production of some pesticides.

Toxicity: has an irritating effect; in USA considered to be potentially carcinogenic [70]. LD_{50} for rats - 3.5 g/kg [47], 4.2 g/kg [54]; LD_{50} for mice - 2.3 g/kg [54]; MPSDC and MPADC - 0.01 and 0.006 mg/m^3 respectively [24].

Chloramine B (sodium chloramidbenzosulfate trihydrate); $C_6H_5ClNNaO_2S \cdot 3H_2O$ (213.64)

White powder with an odor of chlorine; m.p. 180-185°C(d.); soluble in water; insoluble in organic solvents.

Present in wastes from the production of some desinfectants and degassing and bleaching agents.

Toxicity: has an irritating effect. $TC_{irritation}$ - 0.004-8.3 mg/m^3; LD_{50} for rats and mice - 1.0 and 0.8 g/kg respectively [70].

Chloramine T (sodium chloramidtoluenesulfate trihydrate); $C_7H_8ClNNaO_2S \cdot 3H_2O$ (245.66)

White powder; m.p. 175-180°C(d.); soluble in water (14 and 50% in hot and cold water respectively).

Present in wastes from the production of some disinfectants, bleaching agents, and analytical reagents.

Toxicity: has an irritating effect; is allergenic: causes rhinitis and asthma [5,55,70].

Chlordane (octachlor; 2,3,3a,4,7,7a-hexahydro-4,7-methano-1,2,6,5,6,7,8,8a-octachlororindene); $C_{10}H_6Cl_8$ (409.8)

Light yellow liquid; b.p. 175(≅270 Pa); soluble in organic solvents; insoluble in water.

Present in wastes from the production of some pesticides.

Toxicity: has an overall toxic effect: causes cramps, supression of the activity of the central nervous system, degeneration of liver; has an irritating effect; is adsorbed by undamaged skin. LD_{50} for rats - 457-590 mg/kg [55,70].

Chloroacetic acid (chloroethanoic acid); $C_2H_3ClO_2$ (94.50)

White crystals; m.p. 63°C, b.p. 189.35°C, f.p. 132°C, s.-i.p. 446°C; soluble in water and organic solvents.

Present in wastes from the production of some organic compounds, dyes, pharmaceuticals, and pesticides.

Toxicity: has an overall toxic effect; causes irritation of mucous membranes and the skin. $TC_{irritation}$ - 23.7 mg/m^3; LD_{50} for rats - 55 mg/kg [5,55,70].

Removal: by using filters made of polyesters, poly(vinyledene chloride), polyethylene, teflon, or polypropylene [34].

p-Chloroaniline (4-chloroaniline); C_6H_6ClN (127.6)

White crystals; m.p. 70-71°C, b.p. 231-231°C, f.p. 123°C, s.-i.p. 688°C, ρ 1.42; soluble in organic solvents and hot water.

Present in wastes from the production of some pesticides and dyes.

Toxicity: has an overall toxic effect: causes headache, dizziness, nausea, retrosternal pain, dyspnea; causes irritation of the respiratory tracts; is allergenic; is adsorbed by undamaged skin. $TC_{chronic}$ - 0.0015 mg/ℓ; LD_{50} for mice and guinea pigs - 228 mg/kg [5,55,80]; MPSDC and MPADC - 0.04 and 0.01 mg/m^3 respectively; toxicity classification - II [20,21].

Chlorobenzene (phenyl chloride); C_6H_5Cl (112.56)

Liquid; m.p. -45.5°C, b.p. 132°C, f.p. 29°C, CLE 1.3-7.1%, ρ 1.06; soluble in organic solvents; slightly soluble in water (0.049%).

Present in wastes from the production of phenol, aniline, some pesticides and solvents.

Toxicity: has an overall toxic effect: causes depression, cyanosis; has an irritating and narcotic effect; is adsorbed by undamaged skin [47,70]. MPSDC and MPADC - 0.1 mg/m^3 ; toxicity classification - III [20,21]; MPC established in FRG - 150 mg/m^3 [51].

Removal: by adsorption on activated carbon [46].

2-Chlorobuta-1,3-diene (chloroprene); C_4H_5Cl (88.54)

Liquid with a sharp odor; m.p. -130°C, b.p. 59.4°C, f.p. 20°C, CLE 2.5-12%, ρ 0.95; soluble in organic solvents; slightly soluble in water.

Present in wastes from the production of synthetic rubber, acetylene, and acrylonitrile.

Toxicity: has an overall toxic effect: causes degenerative changes in the liver and kidneys; causes irritation of mucous membranes and the respiratory tracts; is carcinogenic, mutagenic, and teratogenic [55,70,79]. SC_{odor} and TC_{odor} - 0.25 and 0.4 mg/m^3 respectively; $TC_{sensitivity\ to\ light}$ - 0.4 mg/m^3; $SC_{chronic}$ and $TC_{chronic}$ - 0.08 and 0.22 mg/m^3 respectively [429]; MPSDC and MPADC - 0.1 mg/m^3 [20,21]; MPSDC and MPADC established in Bulgaria and Yugoslavia - 0.1 mg/m^3, and in GDR - 0.1 and 0.05 mg/m^3 respectively; MPC established in FRG - 150 mg per m^3 [44,51].

Determination: spectrophotometric analysis (sensitivity 0.5 μg) [27]; chromatographic analysis (sensitivity 0.005 mg/m^3) [20,430].

1-Chlorobutane (n-butyl chloride); C_4H_9Cl (92.57)

Liquid with a sharp odor; m.p. -123.1°C, b.p. 78°C, CLE 1.85-10.10%), ρ 0.892; insoluble in water; soluble in ethanol and diethyl ether.

Present in wastes from the production of some pesticides and basic organic compounds.

Toxicity: LD_{50} for rats - 2.67 g/kg [61].

Removal: by adsorption on activated carbon [46].

1-Chloro-2-(ß-chloroethoxy)ethane (sym-dichloroethyl ether; ß,ß'-sym-dichloro ether) C_4H_8ClO (143.0)

Liquid with a sharp odor; m.p. -50°C, b.p. 178°C, ρ 1.122; soluble in organic solvents; slightly soluble in water (1.02%).

Present in wastes from the production of some pesticides; textiles, and fats and oils.

Toxicity: has an overall toxic effect: causes changes in the kidneys and liver, and pulmonary edema; causes irritation of mucous membranes, the eyes, and the respiratory tracts, has a narcotic effect; suspected of being carcinogenic. LD_{50} for rats - 105 mg/kg [47,55,70].

Removal: by adsorption on activated carbon.

Chloroform (trichloromethane); $CHCL_3$ (119.39)

Liquid with a specific odor; m.p. -63.5°C, b.p. 61.2°C, ρ 1.483; soluble in organic solvents; slightly soluble in water (0.32%).

Present in wastes from the production of some plastics, pharmaceuticals, articles made of rubber, solvents, alkaloids, waxes, and refrigerants.

Toxicity: has an overall toxic effect: does damage to the heart, liver, and kidneys; causes irritation of mucous membranes; is carcinogenic and mutagenic; has a narcotic effect [55,70,79].

Removal: by adsorption on activated carbon [46]; with the aid of filters made of poly(vinylidene chloride), nylon, polyesters, or teflon [34].

Determination: gas chromatographic analysis [63].

bis-ß-Chloroisopropyl ether (dichloroisopropyl ether); $C_6H_{12}Cl_2O$ (171.07)

Liquid; m.p. -97-(-102)°C, b.p. 189°C, ρ 1.112; soluble in water (1.7 g/ℓ).

Present in wastes from the production of some organic chemicals, propylene glycol, and dyes; textiles, fats, waxes, and oils.

Toxicity: LD_{50} for rats - 0.24 g/kg [55].

α-Chloronaphthalene (α-naphthyl chloride; 1-chloronaphthalene); $C_{10}H_{17}Cl$ (162.61)

Oily liquid; m.p. -17°C, b.p. 259.3°C, ρ 1.19; soluble in organic solvents; insoluble in water.

Present in wastes from the production of solvents for oils, mixtures of lead tetraethyl with organic chlorides and bromides, and chemical preparations used in microscopic research.

Toxicity: has an overall toxic effect (causes dysfunctioning of the liver; causes dermatitis; is adsorbed by undamaged skin [55,70,86,98].

Chloroneb (2,5-dimethoxy-1,4-dichlorobenzene); $C_8H_8Cl_2O_2$ (207.1)

White crystals; m.p. 133-135°C, b.p. 268°C, ρ 1.66; soluble in organic solvents and in water (8 mg/ℓ).

Present in wastes from the production of some pesticides.

Toxicity: has an irritating effect; is adsorbed by undamaged skin. LD_{50} for rats - 5 g/kg [47].

m-Chloronitrobenzene (nitrochlorobenzene); $C_6H_4ClNO_2$ (157.56)

Yellow crystals; m.p. 44.4°C, b.p. 235.6°C, ρ 1.534; soluble in dimethyl ether, chloroform, and hot ethanol; insoluble in water.

Present in wastes from the production of some dyes and medicinal preparations.

Toxicity: has an overall toxic and irritating effect; is allergenic [5,55,70]. SC_{odor}, $SC_{sensitivity\ to\ light}$, TC_{odor}, and $TC_{sensitivity\ to\ light}$ - 0.015, 0.008, 0.02, and 0.012 mg/m^3 respectively [365]; MPSDC and MPADC - 0.004 mg/m^3; toxicity classification - II [20,21].

o-Chloronitrobenzene (nitrochlorobenzene); $C_6H_4ClNO_2$ (157.56)

White crystals; m.p. 32.5°C, b.p. 245.5(753 mm Hg), ρ 1.305; soluble in diethyl ether, hot ethanol; slightly soluble in benzene; insoluble in water.

Present in wastes from the production of some dyes and medicinal preparations.

Toxicity: has an overall toxic and irritating effect; is allergenic [5,55,70]. SC_{odor}, $SC_{sensitivity\ to\ light}$, TC_{odor}, $TC_{sensitivty\ to\ light}$ - 0.015, 0.008, 0.02, and 0.012 mg/m^3 respectively [365]; MPSDC and MPADC - 0.004 mg/m^3; toxicity classification - II [20,21].

p-Chlronitrobenzene (nitrochlorobenzene); $C_6H_4ClNO_2$ (157.56)

White crystals; m.p. 83.4°C, b.p. 242°C(761 mm Hg), ρ 1.520; soluble in diethyl ether and hot ethanol; insoluble in water.

Present in wastes from the production of some dyes and medicinal preparations.

Toxicity: has an overall toxic and irritating effect; is allergenic [5,55,70]. SC_{odor}, $SC_{sensitivity\ to\ light}$, TC_{odor}, $TC_{sensitivity\ to\ light}$ - 0.015, 0.008, 0.002, and 0.012 mg/m^3 respectively [365]; MPSDC and MPADC - 0.004 mg/m^3; toxicity classification - II [20,21].

m-Chlorophenyl isocyanate; C_7H_4ClNCO (153.57)

Liquid; m.p. -4°C, b.p. 7.2°C (≅1.3 kPa), ρ 1.270; soluble in organic solvents; reacts with water.

Present in wastes from the production of some pesticides.

Toxicity: has an overall toxic effect: causes dyspnea, dystrophic changes in the cardiac muscle, the liver and the kidneys; causes irritation of mucous membranes and the respiratory tracts; is allergenic [5], SC_{odor}, $SC_{c.n.s.}$, $SC_{irritation}$, and $TC_{irritation}$ - 0.008, 0.005, 0.005, and 0.008 mg/m^3 respectively; MPSDC and MPADC - 0.005 mg/m^3; toxicity classification - II [20,21]; MPSDC and MPADC established in Bulgaria and Yugoslavia - 0.005 mg/m^3; MPSDC and MPADC established in GDR - 0.005 and 0.003 mg/m^3 respectively [44,51].

p-Chlorophenyl isocyanate; C_7H_6ClNCO (153.57)

Crystals with a sharp odor; m.p. 31-32°C, b.p. 78°C(≅1.3 kPa), ρ 1.249; soluble in organic solvents; reacts with water.

Present in wastes from the production of some pesticides.

Toxicity: has an overall toxic effect: causes dyspnea, dystrophic changes in the cardiac muscle, the liver, and the kidneys; causes irritation of mucous membranes and the respiratory tracts; is allergenic [5]. SC_{odor} and $SC_{c.n.s.}$ - 0.0064 and 0.0015 mg/m^3 respectively; MPSDC and MPADC - 0.0015 mg/m^3; toxicity classification - II [20,21]; MPSDC and MPADC established in Bulgaria, Yugoslavia, and GDR - 0.0015 mg/m^3 [44,51].

Determination: see [20,27,431].

1,1-bis-(4-Chlorophenyl)-2,2,2-trichloroethane (1,1,1-trichloro-2,2-bis-(p-chlorophenyl)ethane; DDT); $C_{14}H_9Cl_5$ (345.5)

White crystals; m.p. 108.5-109°C, b.p. 185°C; soluble in benzene, xylene; insoluble in water.

Present in wastes from the production of some pesticides (the use of DDT in the USSR has been banned since 1970).

Toxicity: very toxic; is adsorbed by undamaged skin; does damage to the gonads and embryos; is mutagenic and carcinogenic. LD_{50} for rats - 113 mg/kg [24,

45,55,67,70,73]; MPSDC established in USA - 1 mg/m^3 [40]; the use of DDT is banned in Bulgaria, Hungary, and Poland; in Yugoslavia and Czechoslovakia its use is allowed in limited amounts [85].

Chlorophos (dipterex; 1-hydroxy-2,2,2-trichloroethyl)- O,O-dimethylphosphonate); $C_4H_2Cl_3O_4P$ (257.5)

White crystals; m.p. 83-84°C; soluble in water and organic solvents.

Present in wastes from the production of pesticides.

Toxicity: has an overall toxic effect: suppresses the functioning of enzymes, and causes nausea, vomiting, ptyalism, respiratory depression and cramps; has a cumulative effect. LD_{50} for rats - 400 mg/kg [70].

Chlorothion (O,O-dimethyl-O-(3-chloro-4-nitrophenyl) thiophosphate); $C_8H_8ClNO_5PS$ (297.68)

Yellow oily liquid; m.p. 21°C, b.p. 125°C(≅13 Pa), ρ 1.56; soluble in organic solvents; very slightly soluble in water (1:25000).

Present in wastes from the production of some pesticides.

Toxicity: very toxic; causes vomiting, ptyalism, respiratory depression, cramps, coma; suppresses the functioning of enzymes; has a cumulative effect. LD_{50} for rats - 880 mg/kg [70].

Chrysoidine (2,4-diaminoazobenzene hydrochloride); $C_{12}H_{12}N_4 \cdot HCl$ (349.73)

Reddish brown crystals with a greenish hue; m.p. 117°C; soluble in water; insoluble in diethyl ether.

Present in wastes from the production of dyes for cotton, silk, leather, and lacquers.

Toxicity: has an overall toxic and irritating effect; causes excitement, vomiting, ptyalism; does damage to the cornea; is adsorbed by undamaged skin; is carcinogenic [5].

Cidial (O,O-dimethyl-S[α-(ethoxycarbonyl)benzyl] dithiphospate); $C_{12}H_{17}O_4PS_2$ (320.4)

Oliy liquid with a specific odor; b.p. 70-80°C (≅3.2.10^{-3} Pa); soluble in organic solvents; slightly soluble in water (0.02%).

Present in wastes from the production of some pesticides.

Toxicity: very toxic; is adsorbed by undamaged skin; causes dizziness, nausea, vomiting, ptyalism, and irritation of mucous membranes. LD_{50} for laboratory animals - 138-172 mg/kg [67]; MPSDC and MPADC - 0.006 and 0.002 mg/m^3 respectively [20, 21,24].

Complexone (Na-EDTA; disodium ethylenediaminotetracetate dihydrate); $C_{10}H_{14}N_2$-$Na_2O_8 \cdot 2H_2O$ (372.25)

White crystalline powder; soluble in water (10.8%); insoluble in organic solvents.

Present in wastes from the production of synthetic rubber, color-photography chemicals, and some analytical reagents and medicinal preparations.

Toxicity: has an overall toxic effect: does damage to the kidneys; is mutagenic [5,37,93].

Crotoxyphos (ciodrin; O,O-dimethyl-O[1-(α-phenylethyl)hydroxycarbonyl)-1-propen-2-yl] phosphate); $C_{14}H_{19}O_5P$ (298.18)

Yellowish liquid with a slight odor; b.p. 135°C(≅4 Pa); soluble in organic solvents; slightly soluble in water (1 g/ℓ).

Present in wastes from the production of some pesticides.

Toxicity: very toxic; causes vomiting and spasms. LD_{50} for rats and rabbits - 125 and 384 mg/kg [47].

Cumene (isopropylbenzene); C_9H_{12} (120.19)

Liquid; m.p. -96°C, b.p. 152.39°C, f.p. 34°C, s.-i.p. 424°C, CLE 0.88-6.5%, ρ 0.861; soluble in organic solvents; slightly soluble in water (0.01%).

Present in wastes from the production of phenol, acetone, some monomers, solvents, varnishes and paints.

Toxicity: is adsorbed by undamaged skin; has an overall toxic and irritating effect; and at high concentrations also a narcotic effect. LD_{50} for rats - 2.91 g/kg [55,70]; recommended MPSDC - 0.014 mg/m^3 [328]; MPSDC and MPADC - 0.014 mg per m^3; toxicity classification - IV [20,21]; MPSDC and MPADC established in Bulgaria - 0.014 mg/m^3, and in GDR - 0.014 and 0.05 mg/m^3 respectively [44,51].

Determination: gas chromatography (sensitivity 0.1 μg) [63]; the nitration method (sensitivity 1 μg) [27].

Cupric 2,4,5-trichlorophenolate; $C_{12}H_4Cl_6CuO_2$ (456.4)

Dark red powder with an odor of phenol; soluble in ethanol; insoluble in water.

Present in wastes from the production of some pesticides.

Toxicity: has an overall toxic effect: causes headache, throat burns, irritation of the respiratory tracts, heart pain, and dysfunctioning of internal organs; does damage to vascular walls [67]. MPSDC and MPADC - 0.005 and 0.001 mg/m^3 respectively [24].

Cuproson (copper oxychloride mixture with zineb, $C_4H_6N_2S_4Zn$)

Greenish and bluish gray powder; insoluble in water.

Present in wastes from the production of some pesticides.

Toxicity: has an overall toxic effect: does damage to the liver and endocrine glands; does damage to embryos and gonads [5]. MPSDC and MPADC - 0.003 and 0.0003 mg/m^3 respectively [24].

Cyanamide (carbamic nitrile); CH_2N_2 (42.04)

White crystals; m.p. 43°C, s.-i.p. 126°C, ρ 1.072; soluble in organic solvents and water.

Present in wastes from the production of some polymers.

Toxicity: has an overall toxic effect: causes headache, palpitation, and rapid breathing [55,70]. MPSDC and MPADC - 0.01 and 0.006 mg/m^3 respectively [24].

Cyanogen (oxalic nitrile); C_2N_2 (52.03)

Gas with an odor of bitter almond; m.p. -34.4°C, b.p. -20.7°C, ρ 2.335 g/ℓ; soluble in ethanol (2300 cc per 100 parts at 20°C), diethyl ether (500 cc per 100 parts at 20°C), and water (450 cc per 100 parts at 20°C).

Present in wastes from the production of cyanogen derivatives.

Toxicity: very toxic; causes vomiting, respiratory depression, cramps, paralysis; has an irritating effect [55,70].

Cyanogen chloride; CClN (61.48)

Gas; m.p. -6.9°C, b.p. 12.6°C; soluble in water and organic solvents.

Present in wastes from the production of some organic chemicals.

Toxicity: very toxic; causes asthenia, fatigability, headache, nausea, vomiting, respiratory depression, cramps, paralysis; has a precarcinogenic and premutagenic effect [55,70,79].

Cyanuric acid; $C_3H_3N_3O_3$ (129.08)

White crystalline powder; m.p. 320°C(d.); soluble in water (0.25%).

Present in wastes from the production of some organic chemicals and pesticides.

Toxicity: very toxic; has an overall toxic and irritating effect; is carcinogenic [5,55,70,86].

Cyanuric chloride (2,4,6-trichloro-1,3,5-triazene); $C_3Cl_3N_3$ (184.43)

White crystals, m.p. 146°C, b.p. 194°C, ρ 1.32; soluble in organic solvents; slightly soluble in water.

Present in wastes from the production of some dyes, bleaching agents, pesticides, and pharmaceuticals.

Toxicity: has an overall toxic effect: causes dystrophic changes in the liver, kidneys, and cardiac muscle, anemia, dysfunctioning of the central nervous system;

causes irritation of mucous membranes and the respiratory tracts; is carcinogenic and allergenic. $SSDC_{acute}$ and $TCDC_{acute}$ - 0.13 and 0.3 mg/m^3 respectively after exposure for 1 minute); LC_{50} for mice - 0.01 mg/ℓ (after exposure for 2 hours); LD_{50} for mice - 485 mg/kg [5,55].

Cyclohexane (benzene hexahydride); C_6H_{12} (84.16)

Liquid with a sharp odor; m.p. 6.55°C, b.p. 80.74°C, f.p. -18°C, s.-i.p. 260°C, CLE 1.3-8.4%, ρ 0.778; soluble in organic solvents; insoluble in water.

Present in wastes from the production of some solvents, pesticides, and steroids, adipic acid, benzene, chlorocyclohexane, nitrocyclohexanen cyclohexanol, and cyclohexanone.

Toxicity: has an overall toxic effect; causes irritation of mucous membranes, respiratory tracts, and the skin [55,70]. MPSDC and MPADC - 1.4 mg/m^3; toxicity classification - IV [20,21].

Removal: by adsorption on activated carbon [46]; by thermal catalytic oxidation (efficiency 100%) [89].

Determination: gas chromatographic analysis (sensitivity 0.0005 mg/m^3 [29].

Cyclohexanol (hexahydrophenol); $C_6H_{12}O$ (100.16)

Viscous liquid with an odor of camphor; m.p. 25.15°C, b.p. 161.1°C, f.p. 61°C, s.-i.p. 440°C, CLE 1.5-11.1%, ρ 0.941; soluble in organic solvents and in water (≅4%).

Present in wastes from the production of some organic chemicals, emulsion stabilizers, and polymers.

Toxicity: has an overall toxic effect: causes pathological changes in the liver, kidneys, and blood vessels; causes irritation of mucous membranes and the respiratory tracts; at high concentrations has a narcotic effect [55,70]; is adsorbed by undamaged skin. SC_{odor}, TC_{odor}, $SC_{c.n.s.}$, and $TC_{c.n.s.}$ - 0.20, 0.24, 0.06, and 0.11 mg/m^3 respectively [432]; MPSDC and MPADC - 0.06 mg/m^3; toxicity classification - III [20,21]; MPSDC and MPADC established in Bulgaria and Yugoslavia - 0.06 mg/m^3; MPSDC and MPADC established in GDR - 0.15 and 0.06 mg/m^3 respectively [44,51]; MPC established in FRG - 300 mg/m^3 [44,51].

Removal: by adsorption on activated carbon [46].

Determination: see [20].

Cyclohexanone (pimelin ketone); $C_6H_{10}O$ (98.14)

Oily liquid; m.p. -40.2°C, b.p. 155.6°C, f.p. -18°C, s.-.i.p. 495°C, CLE 0.92-3.5%, ρ 0.946; soluble in organic solvents and in water (7%).

Present in wastes from the production of some organic chemicals, dyes, and solvents.

Toxicity: has an overall toxic effect; causes irritation of mucous membranes, the respiratory tracts, and the skin; has a narcotic effect [55,70]. SC_{odor}, $SC_{sensitivity\ to\ light}$, TC_{odor}, $TC_{sensitivity\ to\ light}$, and $TC_{c.n.s.}$ - 0.14, 0.06, 0.21, 0.11, and 0.06 mg/m^3 respectively [432]; MPSDC - 0.004 mg/m^3; toxicity classification - III [20,21]; MPSDC and MPADC establishrd in Bulgaria, Hungary, and Yugoslavia - 0.004 mg/m^3; MPSDC and MPADC established in GDR - 0.1 and 0.004 mg/m^3 respectively; MPC established in FRG - 150 mg/m^3 [44,51].

Removal: by adsorption on activated carbon [46]; by using filters made of nylon, polyesters, or teflon [34].

Determination: see [20,27].

Cyclohexyl acetate; $C_8H_4O_2$ (142.20)

Liquid; b.p. 177°C, ρ 0.93; soluble in organic solvents; insoluble in water.

Present in wastes from the production of solvents for nitrocellulose, plastics, fats, and aromatic substances.

Toxicity: has an overall toxic effect; causes irritation of mucous membranes and the respiratory tracts [5].

Cyclohexylamide (actidione; ß-[2-(3,5-dimethyl-2-hydroxyhexyl)-2-hydroxyethyl] glutamic amide); $C_{15}H_{23}NO_4$ (281.4)

White crystalline powder; m.p. 115.5-117°C; soluble in organic solvents; slightly soluble in water.

Present in wastes from the production of pesticides and antibiotics.

Toxicity: very toxic. LD_{50} for rats - 1.8-2.5 mg/kg [52].

Cycloxehylamine (aminohexahydrobenzene); $C_6H_{13}N$ (99.17)

Liquid with a fishy odor; m.p. -17°C, b.p. 134°C, f.p. -18°C, ρ 0.89; soluble in organic solvents and in water.

Present in wastes from the production of some organic chemicals, caprolactam, pesticides, plasticizers, corrosion inhibitors, articles made of rubber, dyes, emulsifiers, soap, special absorbers.

Toxicity: has an overall toxic effect: causes vomiting and flaccidity, and impairs the functioning of the central nervous system; causes irritation of mucous membranes and the respiratory tract; is narcotic, mutagenic, and teratogenic; is adsorbed by undamaged skin [5,37,55],70].

Cyclopentadiene-1,3; C_5H_6 (66.10)

Liquid; m.p. -85°C, b.p. 42°C, f.p. -50°C, s.-.i.p. 640°C, ρ 0.802; soluble in organic solvents; slightly soluble in water.

Present in wastes from the production of some pesticides, organic chemicals, synthetic rubber, plastics, alkaloids, and camphor.

Toxicity: has an overall toxic effect: causes nausea and headache; is adsorbed by undamaged skin. TC_{odor} and $TC_{chronic}$ - 0.0001-0.0003 mg/m^3 and 0.35 mg/ℓ respectively [55,70,433]; ASL - 0.05 mg/m^3 [90].

Cyclopentane (pentamethylene); C_5H_{10} (70.13)

Liquid; m.p. -93.77°C, b.p. 49.26°C, f.p. -37°C, ρ 0.745; soluble in organic solvents; insoluble in water.

Present in wastes from the production of petrochemicals; also in wastes of oil refineries.

Toxicity: causes irritation of the respiratory tracts; at high concentrations has a narcotic effect [55,70]; ASL - 0.1 mg/m^3 [90].

Cyclopentanone (ketopentamethylene); C_5H_8O (84.11)

Oily liquid with a sharp odor; m.p. -52.8°C, b.p. 130°C, f.p. 30.5°C, ρ 0.948; soluble in organic solvents and in water.

Present in wastes from the production of adipic acid and some solvents.

Toxicity: has an irritating effect, and at high concentrations a narcotic effect [55,70].

Removal: by using filters made of nylon, polyacrylates, or teflon [34].

Cyclophosphamide (N'-bis(ß-chloroethyl)-N,O-trimethylene ester of phosphoric acid diamide); $C_7H_{15}Cl_2N_2O_2P$ (261.10)

Crystalline powder; m.p. 41-45°C; soluble in water (40 g/ℓ).

Present in wastes from the production of some organic chemicals and pharmaceuticals.

Toxicity: has an overall toxic effect: causes vomiting, dysfunctioning of the liver, leukopenia, and alopecia; is carcinogenic [47,70,81].

Cyclotrimethylenetrinitramine (hegogen); $C_3H_6N_6O_6$ (222.26)

Crystalline powder; m.p. 204-205°C, ρ 1.816; soluble in acetone; insoluble in water.

Present in wastes from the productiuon of some explosives.

Toxicity: has an overall toxic effect: causes general depression, weakness, headache, vomiting, and cramps [55,70].

Decahydronaphthalene (decalin; bicyclo[4.4.0]decane); $C_{10}H_{18}$ (138.24)

Liquid with an odor of menthol; cis-isomer: m.p. -43.2°C, b.p. 194.6°C, ρ 0.927; trans-isomer: m.p. -31.5°C, b.p. 185.2°C, ρ 0.8799, f.p. 58°C, s.-i.p.

250°C; soluble in organic solvents; insoluble in water.

Present in wastes from the production of naphthalene, fats, some plastics, oils, wax, and lacquers.

Toxicity: has an overall toxic effect; causes irritation of mucous membranes and the skin; is allergenic. LD_{50} for rats - 4.2 g/kg [55,79].

n-Decane (decane); $C_{10}H_{12}$ (142.3)

Liquid; m.p. -29.67°C, b.p. 174.12°C, f.p. 47°C, s.-i.p 208°C, CLE 0.6-5.5%, ρ 0.730; soluble in ethanol; insoluble in water.

Present in wastes of petrochemical plants.

Determination: chromatography [14].

Demeton (mercaptophos; 7:3 mixture of O,O-diethyl-O-(2-ethylthioethyl) thiophosphate and O,O-diethyl-S-(2-ethylthioethyl) thiophosphate); $C_8H_{19}O_3PS_2$ (258.3)

Oily liquid; b.p. 106-110°C(≅53 Pa); soluble in organic solvents; slightly soluble in water.

Present in wastes from the production of some pesticides.

Toxicity: is adsorbed by undamaged skin; has an overall toxic effect; is especially harmful to warm-blooded animals. LD_{50} for rats - 7.5 mg/kg [47,70]; the use of Demeton is banned in the USSR.

Diacetone (diacetone alcohol; 4-hydroxy-4-methylpentanone-2); $C_6H_{12}O_2$ (116.16)

Liquid with a peculiar odor; m.p. -42.8°C, b.p. 169.1°C, CLE 1.8-6.9%, ρ 0.938; soluble in water and organic solvents.

Present in wastes from the production of solvents for cellulose esters, celluloid, fats, oils, resins, and some pharmaceuticals.

Toxicity: causes irritation of mucous membranes and the respiratory tracts; at high concentrations has a narcotic effect. LD_{50} for rats - 4 g/kg [55,70, 72].

sym-Diacetylethane, (2,5-hexandione); $C_6H_{10}O_2$ (114.14)

Yellow liquid; m.p. -9°C, b.p. 192-194°C, f.p. 79°C, s.-i.p. 490°C, ρ 0.974; soluble in water, ethanol, and diethyl ether.

Present in wastes from the production of acetaldehyde, acetic acid; and printer's inks.

Toxicity: has an overall toxic and irritating effect , and at high concentrations - a narcotic effect [55,70].

Diallyl amine (di-2-propenyl amine); $C_6H_{11}N$ (97.16)

Colorless liquid with a sharp odor; m.p. -88.4°C, b.p. 110.4°C, f.p. 16°C, ρ 0.787; soluble in organic solvents and in water (8.6%).

Present in wastes from the production of some polymers.

Toxicity: is adsorbed by undamaged skin; has an overall toxic effect: is harmful to the central nervous system and the heart; causes irritation of mucous membranes and the skin. LD_{50} for rats - 578 mg.kg [5,70,86].

4,4-Diaminocyclohexylmethane (bis(4-aminocyclohexyl)methane); $C_{13}H_{26}N$ (210.37)

Brown wax-like substance; solidification p. 39-43°C, ρ 0.96; insoluble in water.

Present in wastes from the production of some plastics.

Toxicity: has an overall toxic effect: causes dystrophic changes in organs; causes irritation of the skin and mucous membranes. $TC_{irritation}$ - 0.02 mg/ℓ; LC_{50}(chronic) - 0.4 mg/ℓ [5].

4,4'-Diamino-3,3'-dichlorophenyl (dichlorobenzidine); $C_{12}H_{10}Cl_2N_2$ (253.13)

Brown crystals; m.p. 133°C; soluble in in benzene; insoluble in water.

Present in wastes from the production of some plastics and dyes, and chloronitrobenzene.

Toxicity: has an overall toxic and irritating effect; is allergenic, carcinogenic, and mutagenic [37,52,70,93].

4,4'-Diaminodiphenyl ether; $C_{12}H_{12}N_2O$ (200.24)

White crystals; m.p. 188-190°C, f.p. 217°C, s.-i.p. 570°C, lower CLE 47.5 g per m^3; soluble in organic solvents; insoluble in water.

Present in wastes from the production of some thermally stable polymers and dyes.

Toxicity: has an overall toxic effect: supresses the functioning of the central nervous system, and causes changes in the composition of the blood; is carcinogenic [5]. LD_{50} for rats and mice - 725 and 685 mg/kg respectively; $TC_{chronic}$ - 5.5 mg/m^3 [5].

4,4'-Diaminodiphenylmethane; $C_{13}H_{14}N_2$ (198.26)

Crystalline powder; m.p. 92.93°C, b.p. 398-399°C; soluble in organic solvents; slightly soluble in water.

Present in wastes from the production of diisocyanates, polyamides, synthetic rubber, azodyes, corrosion inhibitors, and reagents for determining tungsten and sulfates.

Toxicity: has an overall toxic effect: causes changes in the liver, the central nervous system, and the composition of the blood; is carcinogenic. LC_{50} for mice - 0.17 mg/ℓ [5,55,70,93].

4,4'-Diaminodiphenylsulfone (dapsone); $C_{12}H_{12}N_2O_2S$ (248.30)

White crystals; m.p. 175-181°C; soluble in organic solvents; slightly soluble in water.

Present in wastes from the production of some polymers, synthetic fibers, epoxides, and pharmaceuticals.

Toxicity: $SC_{chronic}$, $TC_{chronic}$, and $TC_{irritation}$ - 0.05, 0.13 [286], and 44.0 mg/m^3 [5] respectively; MPSDC- 0.05 mg/m^3; toxicity classification - III [23].

1,6-Diaminohexane (hexamethylenediamine); $C_6H_{16}N_2$ (116.2)

White leaves; m.p. 40-43°C, b.p. 205°C, f.p. 73°C, s.-i.p. 280°C, CLE 0.93-4.05%; soluble in water and organic solvents.

Present in wastes from the production of some plastics, synthetic rubber, and some synthetic fibers.

Toxicity: has an overall toxic effect: is harmful to cardiac vessels, the lungs, and the kidneys; causes irritation of the respiratory tracts, the skin, and mucous membranes [5]. SC_{odor} and TC_{odor} - 0.0027 and 0.0032 mg/m^3 respectively; $SC_{sensitivity\ to\ light}$ and $TC_{sensitivity\ to\ light}$ - 0.0017 and 0.0027 mg/m^3 respectively; $SC_{electric\ activity}$ and $TC_{electric\ activity}$ - 0.001 and 0.0017 mg/m^3 respectively [280], $TC_{chronic}$ (for rats) - 0.007 mg/ℓ [5]; $SC_{irritation}$ and $TC_{irritation}$ - 0.001 and 0.04 mg/m^3 respectively [49]; MPSDC and MPADC - 0.001 mg/m^3; toxicity classification - II [4,20,21]; MPADC established in Bulgaria, GDR, and Yugoslavia - 0.001 mg/m^3; MPSDC established in Bulgaria, GDR, and Yugoslavia - 0.001, 0.003, and 0.01 mg/m^3 respectively.

Removal: by combustion in cyclone furnaces [281].

Determination: with the aid of 2,4-dinitrochoro benzoate as a reagent (sensitivity 0.3 μg) [20,27].

Dianisidine, (3,3'-dimethoxybenzidine); $C_{14}H_{16}N_2O_3$ (244.28)

Purple crystals; m.p. 133-137°C; soluble in organic solvents; insoluble in water.

Toxicity: very toxic; has an irritating effect; is carcinogenic and mutagenic [37,70].

Diazinon (O,O-diethyl-O-(2-isopropyl-4-methylpyrimidine-6-yl) thiophosphate); $C_{12}H_{21}N_2O_3PS$ (304.36)

Oily liquid; b.p. 89°C(13 Pa), ρ 1.115; soluble in organic solvents and in water (40 mg/ℓ).

Present in wastes from the production of some pesticides.

Toxicity: is adsorbed by undamaged skin; has an overall toxic effect [55]. $SC_{irritation}$, SC_{odor}, and TC_{odor} - 0.009, 0.032, and 0.047 mg/m^3 respectively; LD_{50} for rats, mice, and rabbits - 150-250 [55], 240 [287], and 370 mg/kg respectively; MPSDC and MPADC - 0.01 mg/m^3; toxicity classification - II; MPSDC recommended in USA - 0.1 mg/m^3 [40].

Determination: gas chromatography [287,288].

Diazomethane; CH_2N_2 (42.04)

Yellow gas; m.p. -145°C, b.p. -23°C; soluble in ethanol and diethyl ether; decomposes in water.

Present in wastes of some chemical plants.

Toxicity: very toxic; affects the functioning of the heart, the kidneys, and the central nervous system; causes irritation of the skin and mucous membranes; is carcinogenic [5,55,70,79].

1,2,5,6-Dibenzanthracene; $C_{22}H_{14}$ (278.35)

White leaves; m.p. 267.5°C, b.p. 275°C; insoluble in water; slightly soluble in acetone; soluble in benzene.

Present in wastes of solid fuel processing plants and petrochemical plants.

Toxicity: is carcinogenic and mutagenic [38,84,88]. Permissible concentration in carbon black and soot ≤2 $\mu g/m^3$ [88].

Determination: chromatography [70]; spectroscopy [32].

Dibenzyl sulfide; $C_{14}H_{14}S$ (214.32)

White crystals with a sharp odor; m.p. 49°C, ρ 1.07 (at 50°C); soluble in organic solvents; insoluble in water.

Present in wastes from the production of some basic organic compounds.

Toxicity: has an overall toxic and irritating effect [71].

1,2-Dibromoethane; $C_2H_4Br_2$ (187.88)

Liquid with an odor of chloroform; m.p. 10°C, b.p. 131.6°C, ρ 2.178; soluble in organic solvents; slightly soluble in water (0.4%).

Present in wastes from the production of some pesticides.

Toxicity: is adsorbed by undamaged skin; has an overall toxic effect: causes functional disturbances of organs, necrosis of the liver; causes irritation of the eyes and the respiratory organs; is mutagenic [249]. LD_{50} for rats - 117 mg/kg [47,70]; MPC established in FRG - 150 mg/m^3 [51].

Removal: by adsorption on activated carbon [46].

n-Dibutylamine; $C_8H_{19}N$ (129.3)

Liquid with an ammonia odor; m.p. -62°C, b.p. 159.6°C, f.p. 45°C, s.-i.p. 580°C, ρ 0.76; soluble in water, ethanol, and diethyl ether.

Present in wastes from the production of some basic organic compounds, pharmaceuticals, flotation agents, pesticides, corrosion inhibitors, and ion-exchange resins.

Toxicity: causes irritation of the eyes and the skin. TC_{odor} - 2.5 mg/m^3; LD_{50} for rats - 0.50-0.55 g/kg [5,55,70,86].

n-Dibutyl ether (1-butoxybutane); $C_8H_{18}O$ (130.22)

Liquid with a fruity odor; m.p. -98°C, b.p. 142°C, f.p. 25°C, s.-i.p. 160°C, ρ 0.770; soluble in organic solvents; slightly soluble in water (0.3%).

Present in wastes from the production of some basic organic compounds.

Toxicity: has an overall toxic and irritating effect. LD_{50} for rats - 7.4 g/kg [55,70].

Di-n-butyl-o-phthalate; $C_{16}H_{22}O_4$ (278.35)

Liquid with a fruity odor; m.p. -40°C, b.p. 340°C, f.p. 148°C, s.-i.p. 404°C, ρ 1.049; soluble in organic solvents; slightly soluble in water.

Present in wastes from the production of synthetic rubber, polystyrene, poly(vinyl chloride), and some pesticides.

Toxicity: causes irritation of mucous membranes and the respiratory tracts. $TC_{irritation}$ - 0.2 mg/ℓ [5]; recommended MPSDC - 0.1 mg/m^3 [289]; ASC - 0.1 mg per m^3 [90].

Determination: polarographic analysis (sensitivity 0.3 μg/ml of solvent) [63].

Dichloroacetic acid; $C_2H_2Cl_2O_2$ (128.95)

Liquid with a sharp odor; m.p. 13.5°C, b.p. 194.4°C, f.p. 139°C, s.-.i.p. 660°C, ρ 1.563; soluble in water and organic solvents.

Present in wastes from the production of some organic chemicals and pharmaceuticals.

Toxicity: has an overall toxic effect; causes irritation of mucous membranes, the skin, and the respiratory tracts [55,70].

Removal: with the aid of filters made of polyesters, polyethylene, polypropylene, or polyacrylates [34].

3,4 dichloroaniline (dichloroaniline); $C_6H_5Cl_2N$ (162.03)

Grayish crystals; m.p. 71.5°C, b.p. 272°C, f.p. 164°C, s.-i.p. 677°C; soluble in ethanol, and benzene; slightly soluble in water.

Present in wastes from the production of some dyes, pharmaceuticals, and pesticides.

Toxicity: has an overall toxic effect. SC_{odor} and TC_{odor} - 0.032 and 0.047 mg/m^3 respectively; $SC_{sensitivity\ to\ light}$ and $TC_{sensitivty\ to\ light}$ - 0.020 and 0.025 mg/m^3 respectively [303]. LD_{50} for rats, mice, guinea pigs, and rabbits -

0.7, 1.0, 0.68, and 0.68 g/kg respectively; toxicity classification - III [23].

1,4-Dichlorobenzene (dichlorobenzene; p-dichlorobenzene); $C_6H_4Cl_2$ (147.0)

White crystals with a sharp odor; m.p. 53°C, b.p. 174°C, f.p. 66°C, ρ 1.458; soluble in benzene and diethyl ether; slightly soluble in water (0.0079%).

Present in wastes from the production of some solvents, pesticides, and dyes; leather, articles made of rubber, resins, pitch; also in wastes of metal-working plants.

Toxicity: has an overall toxic effect: causes functional disturbances of the liver, kidneys, lungs, and the central nervous system, anemia, cataract; causes irritation of mucous membranes and the respiratory tracts; is allergenic and carcinogenic [47,52,55,70,93]. ASL - 0.35 mg/m^3 [18]; in USA recommended MPSDC - 3 mg/m^3 [46]; MPC established in FRG - 150 mg/m^3 [51].

Removal: with the aid of filters made of poly(vinylidene chloride), polyethylene, or polypropylene [34].

1,1-Dichloro-2,2-bis(4-chlorophenyl)ethane; $C_{14}H_{10}Cl_4$ (320.1)

White crystals; m.p. 112°C; soluble in organic solvents; insoluble in water.

Present in wastes from the production of some pesticides.

Toxicity: LD_{50} for rats - 3.4 g/kg [47,55,70].

2,2-Dichlorodiethyl ether (ß,ß'-sym-dichloroethyl ether); $C_4H_8Cl_2O$ (143.0)

Liquid with a sharp odor; m.p. -51.8°C, b.p. 178.5°C, f.p. 55°C, s.-.i.p. 368.9°C, ρ 1.122; soluble in organic solvents; slightly soluble in water (0.03%).

Present in wastes from the production of some pesticides.

Toxicity: has an overall toxic effect: causes functional disturbances of the lungs and kidneys; has an irritating and narcotic effect; is carcinogenic. LD_{50} for rats - 105 mg/kg [47].

sym-Dichlorodimethyl ether; $C_2H_4Cl_2O$ (114.92)

Liquid; m.p. -41.3°C, b.p. 102°C, ρ 0.339; soluble in organic solvents; insoluble in water.

Present in wastes from the production of some plastics and basic organic chemicals.

Toxicity: is carcinogenic [86,98].

1,1-Dichloroethane (ethylidene dichloride); $C_2H_4Cl_2$ (98.97)

Oily liquid with an odor of ether; m.p. -96.98°C, b.p. 57.28°C, f.p. 8°C, s.-i.p. 413°C, CLE 4.8-15.9%, ρ 1.175; soluble in ethanol and diethyl ether; slightly soluble in water.

Present in wastes from the production of some pesticides.

Toxicity: is adsorbed by undamaged skin; has an overall toxic effect: causes functional disturbances of organs, headache, dizziness, vomiting, loss of consciousness, and is harmful to the cornea; causes irritation of mucous membranes, the skin, and the respiratory tracts; is mutagenic and at high concentrations - narcotic [304]. TC_{odor} and $TC_{irritation}$ - 17.5 and 6 mg/m^3 respectively [305]; MPSDC and MPADC - 3 and 1 mg/m^3 respectively; toxicity classification - II [20,21]; MPSDC and MPADC established in Bulgaria, Romania, and Yugoslavia - 3 and 1 mg/m^3 respectively; MPC established in FRG - 20 mg/m^3 [51].

Removal: by adsorption on activated carbon [46].

Determination: some methods (sensitivity 1 μg) have been described in literature [27].

1,2-Dichloroethane (ethylene chloride); $C_2H_4Cl_2$ (98.97)

Liquid with an odor of ether; m.p. -35.36°C, b.p. 83.47°C, f.p. 13°C, s.-i.p. 413°C, CLE 6.2-16.9%, ρ 1.175; soluble in organic solvents; slightly soluble in water (0.87%).

Present in wastes from the production of acetylcellulose, phenols, camphor, articles made of rubber, varnishes and paints, some perfumes, pharmaceuticals, and organic chemicals, thermoplastics, and solvents for fats.

Toxicity: has an overall toxic effect: causes functional disturbances of the liver and kidneys; causes irritation of the eyes and the respiratory tracts. MPSDC and MPADC - 3 and 1 mg/m^3 respectively; toxicity classification - II [20,21].

Removal: see [306].

Determination: see [307].

1,2-*trans*-Dichloroethylene; $C_2H_2Cl_2$ (96.95)

Liquid with an odor of ether; m.p. -49.44°C, b.p. 47.5°C, f.p. 6.1°C, CLE 3.8-17.9%, ρ 1.257; soluble in organic solvents; slightly soluble in water (0.63%)

Present in wastes from the production of solvents for fats; phenols, camphor, and some pharmaceuticals.

Toxicity: is adsorbed by undamaged skin; has an overall toxic, irritating, and narcotic effect [55,70].

2,4-Dichloro-1-hydroxybenzene (dichlorophenol); $C_6H_4Cl_2O$ (163.0)

White crystals; m.p. 45°C, b.p. 210°C, s.-i.p. 761°C, ρ 1.38; soluble in organic solvents; slightly soluble in water (0.46%).

Present in wastes from the production of some organic chemicals.

Toxicity: has an overall toxic and irritating effect. LD_{50} for rats - 0.58 mg per m^3 [18]; MPC established in FRG - 20 mg/m^3 at a rate of ejection of >0.1 kg per hour [55,72,86].

Dichloromethane (methylene chloride); CH_2Cl_2 (84.94)

Liquid; m.p. -96.7°C, b.p. 40.1°C f.p. 14°C, s.-i.p. 556°C, CLE 12-22%, ρ 1.336; soluble in organic solvents and in water (2%).

Present in wastes from the production of some dyes, textiles, leather, pharmaceuticals, basic organic chemicals, photographic films, and synthetic fibers.

Toxicity: has an irritating effect and at high concentrations - a narcotic effect [70]. MPSDC - 8.8 mg/m^3; toxicity classification - IV [23].

sym-Dichloromethyl ether; $C_2H_4CL_2O$ (115.0)

Liquid with a very strong odor; m.p. -41.5°C, b.p. 105°C, ρ 1.315; soluble in organic solvents; undergoes decomposition in water.

Present in wastes from the production of some organic chemicals.

Toxicity: has an overall toxic effect; causes irritation of the eyes, and the respiratory tracts; is carcinogenic [55,70].

2,3-Dichloro-1,4-naphthoquinone (dichlone); $C_{10}H_4Cl_2O_2$ (227.1)

Yellow crystals; m.p. 193°C; slightly soluble in organic solvents; insoluble in water.

Present in wastes from the production of some pesticides.

Toxicity: has an overall toxic and irritating effect. $TC_{irritation}$ - 1 mg/m^3; LD_{50} for rats and mice - 1300 and 440-750 mg/kg respectively [5,47,54,70]; MPSDC and MPADC - 0.01 and 0.006 mg/m^3 respectively [24,54,70]; MPSDC established in USA - 1 mg/m^3 [40].

Determination: spectrophotometric analysis (sensitivity 1 μg) [27].

1,1-Dichloro-1-nitroethane; $C_2H_3ClNO_2$ (143.97)

Liquid with an odor of chloroform; b.p. 124°C, ρ 1.427; soluble in organic solvents; slightly soluble in water (0.25%).

Present in wastes from the production of some solvents and pesticides.

Toxicity: is adsorbed by undamaged skin; causes functional disturbances of organs; irritation of the respiratory tracts; at high concentrations has a narcotic effect. LD_{50} for mice and rabbits - 0.62 and 0.15 g/kg respectively [5,47,55].

Removal: by adsorption on activated carbon (efficiency 20-50%) [46].

2,4-Dichlorophenoxyacetic acid (2,4-D); $C_8H_6Cl_2O_3$ (221.1)

White crystals with a slight odor of phenol; m.p. 141°C, b.p. 160°C; soluble in water and organic solvents.

Present in wastes from the production of some pesticides.

Toxicity: causes vomiting; functional disturbances of the kidneys and the liver; suspected of being mutagenic [5,47,55,67,70]. $TC_{irritation}$ - 10 mg/m^3 [40]; maximum permissible concentration established in USA - 10 mg/m^3 [40].

Determination: gas chromatography [285].

2,4-Dichlorophenoxyaceticdimethylammonium (2,4-D amine salt); $C_{10}H_{13}ClNO_3$ (266.1)

White crystals; m.p. 85-87°C; soluble in methanol and in water (4200 g/ℓ).

Present in wastes from the production of some pesticides.

Toxicity: LD_{50} for rats and mice - 980-1200 mg/ℓ; MPSDC - 0.02 mg/m^3.

1,2-Dichloropropane (propylene chloride); C_3H_6Cl (112.99)

Liquid; m.p. -100.44°C, b.p. 96.37°C, f.p. 15°C, s.-i.p. 557°C, CLE 3.4-14.5°C, ρ 1.156; soluble in organic solvents; insoluble in water.

Present in wastes of plants where 1,2-chloropropane is used as a solvent for oils and fats.

Toxicity: has an overall toxic effect (at high concentrations causes necrosis of the liver and the kidneys); causes irritation of mucous membranes; has a narcotic effect. LD_{50} for mice - 860 mg/kg [55,70].

1,2-Dichloropropane (propylene chloride); C_3H_6Cl (112.99)

Liquid; m.p. -100.44°C, b.p. 96.37°C, f.p. 15°C, s.-i.p. 557°C; CLE 3.4-14.5°C, ρ 1.156; soluble in organic solvents; insoluble in water.

Present in wastes of plants where 1,2-chloropropane is used as a solvent for oils and fats.

Toxicity: has an overall toxic effect (at high concentrations causes necrosis of the liver and the kidneys); causes irritation of mucous membranes; has a narcotic effect. LD_{50} of mice - 860 mg/kg [55,70].

1,3-Dichloropropane (trimethylene dichloride) $C_8H_4Cl_2$ (110.98)

Yellowish liquid; mixture of cis-, and trans-isomers m.p. 50°C, b.p. 108°C, f.p. 34°C(trans-), ρ 1.218 (trans-); soluble in organic solvents; insoluble in water.

Present in wastes from the production of some pesticides.

Toxicity: causes functional disturbances of the kidneys and the liver; causes irritation of mucous membranes, the skin, and the respiratory tracts. LD_{50} for rats - 250-500 mg/kg [47,55,70]. MPSDC and MPADC - 0.1 and 0.01 mg/m^3 respectively; toxicity classification - II [23].

1,3-Dichloro-2-propanol (α-dichlorohydrin glycerol); $C_3H_6Cl_2O$ (128.99)

Liquid with an odor of chloroform; m.p. -4°C, b.p. 174.3°C, ρ 1.367; soluble in ethanol, diethyl ether; soluble in water (11%).

Present in water from the production of articles made of rubber, nitrocellulose, celluloid, and some plastics.

Toxicity: has an overall toxic effect; causes irritation of mucous membranes and the respiratory tracts [55,70,98].

sym-Dichlorotetrafluoroethane (freon 114); $C_2Cl_2F_4$ (170.93)

Gas with an odor of ether; m.p. -94°C, b.p. 3.8°C, ρ 1.473; insoluble in water.

Present in wastes from the production of refrigerants and aerosols.

Toxicity: causes irritation of the respiratory tracts; at high concentrations has a narcotic effect [55,70,86].

N,N-Dichloro-p-toluenesulfamide; $C_7H_7Cl_2NO_2S$ (240.11)

Yellowish crystals; m.p. 83°C; soluble in benzene; slightly soluble in water.

Present in wastes from the production of bleaching agents for textiles; degasifying agents, and some pesticides.

Toxicity: causes irritation of mucous membranes, the respiratory tracts, and the skin [98].

Dicrotophos (bidrin; carbicron; O,O-dimethyl-O-[1-(N,N-dimethylcarbamoyl)-1-propene-2-yl] phosphate); $C_8H_{16}NO_5P$ (237.2)

Liquid; b.p. 400°C, ρ 1.216; soluble in water and most of the organic solvents.

Present in wastes from the production of some pesticides.

Toxicity: is adsorbed by undamaged skin; has an overall toxic effect. LD_{50} for rats, mice, and rabbits - 22.0, 11.3, and 108.0 mg/kg respectively [47,70].

Dicyclohexylamine (dodecahydrophenylamine); $C_{12}H_{23}N$ (181.32)

Liquid; m.p. -0.1°C, b.p. 256°C, f.p. 99°C, s.-i.p. 240°C, ρ 0.910; soluble in ethanol and diethyl ether; slightly soluble in water.

Present in wastes from the production of corrosion inhibitors and lacquers.

Toxicity: causes degenerative changes in the liver and the kidneys. LC_{min} - 0.15 mg/ℓ [5].

Dicyclopentadiene; $C_{10}H_{12}$ (132.2)

White crystals with an unpleasant odor; m.p. 33°C, b.p. 170°C(d.), f.p. 32°C, s.-i.p. 510°C, ρ 0.976; soluble in organic solvents; insoluble in water.

Present in wastes from the production of some pesticides and plastics and synthetic rubber.

Toxicity: is adsorbed by undamaged skin; has an overall toxic effect. TC_{odor} and $TC_{chronic}$ - 0.0004 and 0.02 mg/ℓ respectively [308]; ASL - 0.01 mg/m^3 [90].

Diethanolamine (iminoethyl alcohol; 2,2'-dihydroxydiethylamine); $C_4H_{11}NO_2$ (105.14)

Solid crystals with an odor of ammonia; m.p. 28°C, b.p. 270°C(748 mm Hg), ρ 1.097; soluble in water and ethanol; slightly soluble in diethyl ether; insoluble in benzene.

Present in wastes from the production of some petrochemicals, pharmaceuticals, and organic chemicals, surface-active agents, articles made of rubber, textiles, some perfumes and pesticides, and corrosion inhibitors.

Toxicity: causes degenerative changes in the liver and the kidneys; irritation of mucous membranes and the respiratory tracts. LD_{50} for rats and mice - 3.5 and 3.3 g/kg respectively; and for guinea pigs and rabbits - 2.2 g/kg [5,55,70].

Determination: thin-layer chromatography [309].

Diethylamine; $C_6H_{11}N$ (73.14)

Liquid with an odor of ammonia; m.p. -50°C, b.p. 55.5°C, f.p. -26°C, s.-i.p. 310°C, CLE 1.8-10.1%, ρ 0.705; soluble in water and organic solvents.

Present in wastes from the production of some organic chemicals, petrochemicals, and pharmaceuticals, articles made of rubber, dyes, some flotation agents and pesticides, solvents, and corrosion inhibitors.

Toxicity: causes corneal opacity, necrosis of the liver; irritation of mucous membranes, the skin, and the respiratory tracts [5,55,70]. TC_{odor}, TC_{odor}, $SC_{irritation}$ and $TC_{irritation}$ - 0.067, 0.084, 0.04, and 0.4 mg/m^3 respectively [310]; MPSDC and MPADC - 0.05 mg/m^3; toxicity classification - IV [20,21]; MPSDC and MPADC established in Bulgaria, Romania, and Yugoslavia - 0.05 mg/m^3 [44]; MPC established in FRG - 20 mg/m^3 [51].

Determination: coulometric analysis (sensitivity 0.1 mg/m^3) [311].

N,N-Diethylaminobenzene (diethylaniline); $C_{10}H_{15}N$ (149.23)

Yellowish viscous liquid; m.p. -38.8°C, b.p. 216.27°C, f.p. 83°C, ρ 0.935; soluble in water (1.4%).

Present in wastes from the production of some dyes and pharmaceuticals.

Toxicity: is adsorbed by undamaged skin; causes depression of the central nervous system and changes in the blood; has an irritating effect; is allergenic [55,70]; ASL - 0.01 mg/m^3 [91].

2-Diethylaminoethanethiol (ß-diethylaminoethyl mercaptan); $C_6H_{15}NS$ (133.26)

Liquid with a sharp odor; m.p. 160°C, ρ 0.875; soluble in water and diethyl ether.

Present in wastes from the production of some antioxidants, plastics, analytical reagents, some pharmaceuticals.

Toxicity: is adsorbed by undamaged skin; has an overall toxic effect: does damage to the central nervous system, and causes corneal opacity; causes irritation of mucous membranes and the skin. LD_{50} for mice - 231.4 mg/kg [5]; MPSDC and MPADC - 0.6 mg/m^3; toxicity classification - IV [20,21].

Determination: see [5].

1,4-Diethylbenzene (p-diethylbenzene); $C_{10}H_{14}$ (134.22)

Liquid; m.p. -42.85°C, b.p. 183.75°C, ρ 0.861; soluble in organic solvents; insoluble in water.

Present in wastes from the production of some organic chemicals.

Toxicity: is adsorbed by undamaged skin; has an overall toxic effect; causes irritation of mucous membranes and the skin [55,86]; ASL - 0.005 mg/m^3 [18].

Diethyl carbonate; $C_5H_{10}O_3$ (118.3)

Liquid; m.p. -43°C, b.p. 126.8°C, f.p. 26°C, ρ 0.975; soluble in organic solvents; insoluble in water.

Present in wastes from the production of some solvents, plastics, and pharmaceuticals.

Toxicity: has an irritating effect. $TC_{irritation}$ - 10 mg/ℓ; LD_{50} for rats - 1.57 g/kg [5].

Diethyl ether (ethenyl ether); $C_4H_{10}O$ (74.12)

Liquid with a peculiar odor; m.p. -116.2°C, b.p. 34.48°C, f.p. -43°C, s.-i.p. 180°C, CLE 1.9-48%, ρ 0.713; soluble in ethanol, benzene, and in water (6.5%).

Present in wastes from the production of some pharmaceuticals, organic chemicals, and solvents.

Toxicity: has an overall toxic and irritating effect, and at high concentrations - a narcotic effect [55,70].

Diethyl ketone (pentanone-3); $C_5H_{10}O$ (86.13)

Liquid with an odor of acetone; m.p. -42°C, b.p. 101.7°C, f.p. 13.0°C, ρ 0.815; soluble in organic solvents and in water (3.4%).

Present in wastes from the production of some solvents and varnishes and paints; also in wastes of petroleum refineries.

Toxicity: is adsorbed by undamaged skin; has an overall toxic and irritating effect. LD_{50} for rats - 2.1 g.kg [55,70,86].

o-Diethyl phthalate (ethyl phthalate); $C_{12}H_{14}O_4$ (222.23)

Oily liquid; m.p. -3°C, b.p. 294-295°C, f.p.125°C, CLE 0.75-4.25%, ρ 1.118; soluble in ethanol; slightly soluble in water (0.15%).

Present in wastes from the production of some basic organic chemicals, articles made of rubber, perfumes, denatured ethanol, cellulose esters, poly(vinyl acetate), resins, pesticides, plastics, explosives, some dyes, and camphor; also in wastes of metallurgical plants.

Toxicity: causes vomiting, dysfunctioning of the liver and the kidneys, and irritation of mucous membranes; has a narcotic effect. LD_{50} for rats and rabbits - 1 g/kg and for mice 6.2 g/kg [5,55,70,86].

Diethyl succinate (ethyl succinate); $C_8H_{14}O_4$ (174.2)

Liquid with a slight odor; m.p. -21°C, b.p. 217.7°C, f.p. 100°C, ρ 1.040; soluble in ethanol and diethyl ether; insoluble in water.

Present in wastes from the production of some plasticizers, foodstuffs, and perfumes.

Toxicity: slightly toxic; has an irritating effect. LD_{50} for rats - 8.5 g/kg [55,70,86].

Diethyl sulfate (ethyl sulfate); $C_4H_{10}O_4S$ (154.19)

Oily liquid with an odor of pepper; m.p. -24.4°C, b.p. 208°C(d.), f.p. 109°C, ρ 1.180; soluble in ethanol and diethyl ether; insoluble in water.

Present in wastes from the production of ethylene and ethylating agents.

Toxicity: is adsorbed by undamaged skin; has an overall toxic and irritating effect; is carcinogenic and mutagenic [37]. LD_{50} for rats - 880 mg/kg [55,70].

Diethyl sulfide; $C_4H_{10}S$ (90.19)

Liquid with an odor of ether; m.p. -102°C, b.p. 92.93°C, ρ 0.837; soluble in ethanol and diethyl ether; insoluble in water.

Present in wastes from the production of solvents for anhydides of mineral salts; also in wastes of galvanic processes involving the use of precious metals.

Toxicity: very toxic; has a narcotic effect. LD_{50} for mice - 1.16 g/kg; MPC established in FRG - 20 mg/m^3, with the concentration of diethyl sulfide in the ejections being $\leq$0.1 kg/hr [55,70,86].

Difluorodichloromethane (freon 12); CCl_2F_2 (120.92)

Gas with an odor of ether; m.p. -158°C, b.p. -29°C, ρ 1.484; soluble in organic solvents; insoluble in water.

Present in wastes from the production of refrigerants and aerosols.

Toxicity: has an irritating effect, and at high concentrations - a narcotic effect [55,70,86].

m-Dihydroxybenzene (resorcinol; 1,3-dihydroxybenzene); $C_6H_6O_2$ (110.11)

White crystals; m.p. 110°C, b.p. 281°C, f.p. 127°C, s.-i.p. 608°C, ρ 1.28; soluble in organic solvents and water.

Present in wastes from the production of some plastics, tanning agents, dyes, textiles, and perfumes.

Toxicity: has an overall toxic and irritating effect; is allergenic; is adsorbed by undamaged skin [55,70].

Removal: with the aid of filters made of nylon, polypropylene, polyesters, poly(vinyledene chloride), or teflon [34].

Determination: spectrophotometric analysis [382].

o-Dihydroxybenzene (pyrocatechin; 1,2-dihydroxybenzene); $C_6H_6O_2$ (110.11)

White crystals; m.p. 105°C, b.p. 240°C, f.p. 127°C, ρ 1.37; soluble in organic solvents and water.

Present in wastes from the production of some dyes, photographic chemicals, and analytical reagents.

Toxicity: has an overall toxic and irritating effect; is allergenic; is adsorbed by undamaged skin. LD_{50} for rats - 3.89 g/kg [68,75].

2,2'-Dihydroxydiethyl ether (diethylene glycol; glycol ether); $C_4H_{10}O_3$ (106.12)

Liquid; m.p. -8°C, b.p. 245°C, f.p. 135°C, ρ 1.116; soluble in water, ethanol, and diethyl ether.

Present in wastes from the production of some plastics, antifreeze, wool, cotton, viscose, silk, dyes, cork, glue, gelatin, and casein.

Toxicity: causes vomiting, respiratory depression, cramps, and dysfunctioning of the kidneys; has an irritating effect. LD_{50} for rats and guinea pigs - 20.7 and 13.2 g/kg respectively [55,70,72].

Diisopropyl amine ; $C_6H_{15}N$ (101.9)

Liquid with an odor of amine; m.p. -96.3°C, b.p. 84.1°C, f.p. -7°C, CLE 1.1-7.1%, ρ 0.722; soluble in water and organic solvents.

Present in wastes from the production of some organic compounds and pesticides.

Toxicity: causes vomiting, cramps, and irritation of the skin, mucous membranes, and the respiratory tracts. LD_{50} for rats - 770 mg/kg [5,55,70,86].

Diisopropyl ether (2-isopropoxypropane); $C_6H_{14}O$ (102.17)

Liquid with an odor of ether; m.p. -86.2°C, b.p. 68.5°C, f.p. -28°C, s.-i.p. 443°C, CLE 1.4-21%, ρ 0.726; soluble in organic solvents; slightly soluble in water (0.2%).

Present in wastes from the production of some solvents, varnishes and paints, and acetic acid; also in wastes of petroleum refineries and radiochemical plants.

Toxicity: causes paralysis of the respiratory center; and irritation of mucous membranes, the skin, and the respiratory tracts. MPC established in FRG - 300 mg/m^3 at an ejection rate of not more than 6 kg/hr [51,55,70].

Diketene (ß-isocrotyl lactone); $C_4H_4O_2$ (84.07)

Liquid with a sharp odor; m.p. -8°C, b.p. 127.4°C, f.p. 34°C, ρ 1.093; soluble in organic solvents; insoluble in water.

Present in wastes from the production of some basic organic compounds, plasticizers, dyes, acetoacetic acid and its esters; some solvents, corrosion inhibitors, and pharmaceuticals.

Toxicity: has an overall toxic effect; causes irritation of the respiratory organs. SC_{odor}, $SC_{irritation}$, and TC_{odor} - 0.011, 0.011, and 0.019 mg/m^3 respectively [49,290]; MPSDC - 0.007 mg/m^3, toxicity classification - II [20,21]; MPSDC established in Bulgaria, GDR, and Yugoslavia - 0.007 mg/m^3; MPADC established in GDR - 0.002 mg/m^3 [44,45,55].

Determination: see [27].

N,N-Dimethylacetamide (acetdimethylamide); C_4H_9NO (87.12)

Liquid with an odor of ammonia; m.p. -20°C, b.p. 165.5°C, s.-i.p. 77°C, CLE 2.0-11.5%, ρ 0.936; soluble in water and organic solvents.

Present in wastes from the production of synthetic rubber, some plastics, petrochemicals, and solvents.

Toxicity: is adsorbed by undamaged skin; has an overall toxic effect (causes cramps, partial paralysis, hallucination) and an irritating effect; is teratogenic. LD_{50} for rats and mice - 3.56 and 4.2 g/kg respectively [5,55,86]; MPSDC and MPADC - 0.2 and 0.006 mg/m^3 respectively; toxicity classification - II [96].

Determination: gas chromatographic analysis (sensitivity 10-12 μg/ml) [294], colorimetric analysis (concentrations from 0.5 to 100.0 mg/m^3 [295].

N-(Dimethylamidosulfonyl)-N-(fluorodichloromethylthio) aniline (euparen); $C_9H_{11}Cl_2$-$FN_2O_2S_2$ (333.3)

White crystals with a slight odor; m.p. 95-97°C, soluble in xylene (70 g/ℓ); insoluble in water.

Present in wastes from the production of some pesticides.

Toxicity: slightly toxic: causes impairment of the central nervous system. LD_{50} for rats and mice - 1850-2500 mg/kg; and for cats, rabbits, and guinea pigs - 1000 mg/kg [5].

N,N-Dimethylamine; C_2H_7N (45.09)

Gas with a sharp odor; m.p. -92.2°C, b.p. 6.9°C, f.p. 400°C, CLE 2.8-14.4%, ρ 0.654; soluble in water and organic solvents.

Present in wastes from the production of some basic organic compounds, synthetic rubber, plastics, articles made of rubber, soap, varnishes and paints, paper, pharmaceuticals, and surface-active agents.

Toxicity: causes irritation of mucous membranes, the skin, and the respiratory tracts [47,79]; is carcinogenic [79]. TC_{odor}, $TC_{irritation}$, and $TC_{c.n.s.}$ - 0.0025 [5]-1.1 [32] mg/m^3, 0.05 mg/ℓ, and 0.00093-.00033 mg/ℓ respectively [5]; MPSDC and MPADC - 0.005 mg/m^3; toxicity classification - II [20,21].

Determination: colorimetric analysis (sensitivity 0.5-5.0 mg/m^3) [291].

N,N-Dimethylaminobenzene (dimethylphenylamine); $C_8H_{11}N$ (121.18)

Liquid with an odor of pitch; m.p. 2.5°C, b.p. 194.15°C, f.p. 37°C, s.-i.p. 400°C, ρ 0.955; soluble in organic solvents; slightly soluble in water.

Present in wastes from the production of varnishes and paints, photographic chemicals, dyes, and some explosives.

Toxicity: is adsorbed by undamaged skin. $TC_{irritation}$ (for rats) - 0.011 mg/ℓ [5]; MPSDC and MPADC - 0.0055 mg/m^3; toxicity classification - II [20,21].

Removal: by adsorption on activated carbon [46].

Determination: see [20,27,293].

<u>bis</u>-(Dimethylamino)fluorophosphine oxide (dimefox, pestox); $C_4H_{12}FN_2OP$ (154.13)

Liquid with a fishy odor; b.p. 67°C, ρ 1.115; soluble in water and chloroform.

Present in wastes from the production of some pesticides.

Toxicity: has an overall toxic effect; causes vomiting, ptyalism, cramps; does damage to the spinal cord and peripheral nerves; suppresses the functioning of enzymes. LD_{50} for rats - 5 mg/kg [47,70].

O,O-Dimethyl-S-(5,6-benzo-3,4-dihydro-4-hydroxy-1,2,3-triazinyl-3-methyl) dithiophophate (methyltriazothione; P-1582); $C_{10}H_{12}N_3O_3PS$ (317.3)

White crystals; m.p. 73-74°C, ρ 1.44; soluble in most organic solvents; slightly soluble in water: 33 mg/ℓ of H_2O at 18°C.

Present in wastes from the production of some pesticides.

Toxicity: LD_{50} for rats and mice - 80 and 20 mg/kg respectively [47]; MPADC established in USA - 0.2 mg/m^3 [40].

N-Dimethylbenzylamide; $C_9H_{13}N$ (135.21)

Liquid with a slight odor; b.p. 180-182°C, ρ 0.894; soluble in ethanol and diethyl ether and in hot water.

Present in wastes from the production of some basic organic compounds, polymers, and corrosion inhibitors.

Toxicity: causes degenerative changes in the liver and the kidneys; irritation of the skin and the respiratory tracts. LC_{50}, $TC_{chronic}$, and $TC_{irritation}$ - 1.2, 0.1-0.2, and 0.03-0.04 mg/ℓ respectively [5]

Determination: see [5].

N,N,-Dimethyl-N'-(3,4-dichlorophenyl) urea (diuron); $C_9H_{10}Cl_2N_2O$ (233.1)

White crystals; m.p. 158-159°C; soluble in water (42 mg/ℓ) and in acetone (5.3%).

Present in wastes from the production of some pesticides.

Toxicity: has an overall toxic effect. LD_{50} for rats - 3400 mg/kg [54]; MPSDC and MPADC - 0.05 mg/m^3 [24].

O,O-Dimethyl-O(2,2-dichlorovinyl phosphate) (vapona; nuvan); $C_4H_7Cl_2O_4P$ (221.0)

Liquid with a sharp odor; m.p. 35°C(≅6.7 Pa), ρ 1.415; soluble in organic solvents; slightly soluble in water (1%).

Present in wastes from the production of some pesticides.

Toxicity: is adsorbed by undamaged skin; has an overall toxic and irritating effect. $TC_{irritation}$ - 1-3 mg/m^3 [52,67,70]; MPSDC and MPADC - 0.007 and 0.002 mg/m^3 respectively [21,24].

4,4-Dimethyl-1,3-dioxane (dimethyldioxane); $C_6H_{12}O$ (116.16)

Liquid with a sharp odor; m.p. -88.5°C, b.p. 133.1°C, f.p. 30°C, s.-i.p. 351°C, ρ 0.964; soluble in organic solvents and in water (17.95%).

Present in wastes from the production of some basic organic compounds, synthetic rubber, and isoprene.

Toxicity: has an overall toxic effect: does damage to the cardiac muscle, the kidneys, and the liver [5,56]. $SC_{chronic}$ and TC_{odor} - 0.004 and 0.34 mg/m^3 respectively [296]; MPSDC - 0.004 mg/m^3; toxicity classification - II [20,21].

N,N-Dimethylformamide (dimethylformamide); C_3H_7NO (73.09)

Liquid with an odor of ammonia; m.p. -61°C, b.p. 153°C, f.p. 58°C, s.-i.p. 420°C, CLE 2.2-15.2%, ρ 0.944; soluble in water and organic solvents.

Present in wastes from the production of some synthetic fibers and basic organic compounds.

Toxicity: is adsorbed by undamaged skin; suppresses the functioning of enzymes; causes irritation of the eyes and the respiratory tracts; does damage to embryos; is carcinogenic and mutagenic [79]. TC_{odor} and $TC_{irritatiton}$ - 0.14 and 0.5 [297]-0.8 [5] mg/m^3 respectively. LD_{50} for rats and mice - 4.6 and 4.2 g/kg respectively [5]; MPSDC and MPADC - 0.03 mg/m^3; toxicity classification - II [20, 21]; MPSDC and MPADC established in Bulgaria and Yugoslavia - 0.03 mg/m^3, and in GDR - -.03 and 0.01 mg/m^3 respectively [44,51]; MPC established in FRG - 150 mg/m^3 [51].

Determination: gas chromatographic analysis (sensitivity 0.05 μg) [298].

2,6-Dimethylheptanone-4 (diisobutyl ketone); $C_9H_{18}O$ (142.24)

Oily liquid with a peculiar odor; m.p. 5.9°C, b.p. 168.1°C, f.p. 49°C, ρ 0.809; soluble in organic solvents; slightly soluble in water (500 mg/ℓ).

Present in wastes from the production of some solvents and pesticides.

Toxicity: is adsorbed by undamaged skin; has a narcotic effect [55]. LD_{50} for rats - 5.8 g/kg [86].

N,N-Dimethylhydrazine (uns-dimethylhydrazine); $C_2H_8N_2$ (60.10)

Liquid with a fishy odor; m.p. -58°C, b.p. 64.9°C, ρ 0.791; soluble in water and organic solvents.

Present in wastes from the production of special fuels.

Toxicity: has an overall toxic effect; causes irritation of mucous membranes, the eyes, and the skin; is carcinogenic and teratogenic. LD_{50} for rats and mice - 122 and 265 mg/kg respectively [5,55,70,79].

O,O-Dimethyl-S-(N-methylcarbamoylmethyl dithiophosphate (rogor); $C_5H_{12}NO_3PS_2$ (229.2)

White crystals; m.p. 49-51°C, f.p. 160°C, s.-i.p. 372°C; soluble in organic solvents and in water (3.9%).

Present in wastes from the production of some pesticides.

Toxicity: very toxic. SC_{odor}, $SC_{c.n.s.}$, and TC_{odor} - 0.01, 0.0032, and 0.015 mg/m^3 respectively [299]; LD_{50} for laboratory animals - 100-230 mg/kg [67]. MPSDC and MPADC - 0.003 mg/m^3; toxicity classification - II [23].

O,O-Dimethyl-O-(N-methyl-cis-crotonanid-3-yl) phosphate (CD-9129); $C_7H_{14}NO_5$ (223.17)

Brown liquid; m.p. 25-30°C, b.p. 125°C; soluble in water.

Present in wastes from the production of some pesticides.

Toxicity: very toxic. LD_{50} for rats - 1 mg/kg [47]; recommended MPSDC - 0.001 mg/m^3 [46].

o-Dimethyl phthalate; $C_{10}H_{10}O_4$ (194.19)

Liquid with a slight odor; b.p. 282°C, f.p. 132°C, ρ 1.190; soluble in organic solvents; insoluble in water.

Present in wastes from the production of some pesticides, cellulose esters, poly(vinyl acetate), articles made of rubber, and some flotation agents.

Toxicity: is adsorbed by undamaged skin; causes irritation of mucous membranes; is allergenic [55]. $TC_{chronic}$ - 0.002 mg/ℓ [5].

Removal: by using filters made of polyethylene, poly(vinyl chloride), or polypropylene [34].

Dimethyl sulfoxide; C_2H_6OS (78.13)

Liquid; m.p. 18.5°C, b.p. 189°C, f.p. 95°C, ρ 1.01; soluble in organic solvents and water.

Present in wastes from the production of some basic organic chemicals, petrochemicals, pharmaceuticals, cellulose sulfate, pesticides, synthetic fibers, and antifreeze.

Toxicity: has an overall toxic effect: causes vomiting, excitement followed by depression of the central nervous system, chills, insomnia, cramps , dystrophic changes in the brain and internal organs; is teratogenic; does damage to embryos. $TC_{chronic}$ - 1 mg/ℓ; LD_{50} for rats and mice - 19.7-28.3 and 16.5-24.0 g/kg respectively [5,60,70,313]; MPC established in FRG - 300 mg/m^3 [51].

Dimethyl terephthalate; $C_{10}H_{10}O_4$ (194.18)

White crystals; m.p. 140-141°C, b.p. 288°C.

Present in wastes from the production of polyethyleneterephthalate fibers.

Toxicity: has an overall toxic effect: causes dystrophy of internal organs; causes irritation of mucous membranes and the upper respiratory tracts. $TC_{irritation}$ and $TC_{electric\ sensitivity}$ - 0.2 and 0.07 mg/m^3 respectively; LD_{50} for rats - >3.2 g/kg [5,86].

<u>bis</u>(O,O-Dimethylthiophosphoryl-o-phenyl-4)sulfide (abate); $C_{16}H_{20}O_6P_2S_3$ (466.5)

White crystals (technical grade - brown liquid); m.p. 30-30.5°C, ρ 1.320; soluble in acetonitrile, carbontetrachloride, and toluene; insoluble in water.

Present in wastes from the production of some pesticides.

Toxicity: is adsorbed by undamaged skin. $TC_{irritation}$ - 25 mg/m^3; LD_{50} for rats - 20 mg/kg [47,67]; MPSDC established in USA - 10 mg/m^3 [40].

2,4-Dinitrochlorobenzene (dinitrochlorobenzene); $C_6H_3ClN_2O_4$ (202.56)

Yellow crystals; m.p. 53°C, b.p. 315°c, ρ 1.498; soluble in organic solvents; insoluble in water.

Present in wastes from the production of some dyes and explosives.

Toxicity: has an overall toxic effect; causes irritation of mucous membranes and the skin; is allergenic [5,55,70]; ASL - 0.002 mg/m^3 [18].

4,6-Dinitro-o-cresol; $C_7H_6N_2O_5$ (199.1)

Yellow crystals; m.p. 85.5!C; soluble in organic solvents; slightly soluble in water (0.013%).

Present in wastes from the production of some pesticides.

Toxicity: very toxic; is adsorbed by undamaged skin; does damage to the cardiac muscle, the liver and the kidneys; causes irritation of the upper respiratory tracts; does damage to embryos and gonads. $TC_{irritation}$ - 0.3-2.9 mg/m^3; LD_{50} for rats, mice, and cats - 85, 47, and 50 mg/kg respectively [5,67,70]; MPSDC and MPADC - 0.003 and 0.0008 mg/m^3 respectively [24]; MPSDC established in USA - 0.2 mg/m^3 [40].

2,4-Dinitrophenol (dinitrophenol); $C_6H_4N_2O_5$ (184.11)

Yellow crystals with an odor of phenol; m.p. 114-115°C, ρ 1.681; soluble in organic solvents; slightly soluble in water (0.56%).

Present in wastes from the production of some varnishes and paints; also in wastes of wood-processing plants.

Toxicity: is adsorbed by undamaged skin; has an overall toxic and irritating effect; is allergenic; a case has been reported where death followed inhalation of air containing 40 mg/m^3 of 2,4-dinitrophenol [5,55]; recommended MPSDC - 0.01 [46]; ASL - 0.004 mg/m^3 [18].

2,4-Dinitrotoluene (dinitrotoluene); $C_7H_6N_2O_4$ (182.14)

Pale yellow crystals; m.p. 70.5°C, b.p. 300°C(d.), ρ 1.32; soluble in pyridine; slightly soluble in water.

Present in wastes from the production of some varnishes and paints, basic organic chemicals, and plasticizers.

Toxicity: very toxic; is adsorbed by undamaged skin; has an irritating effect [5,55]; recommended MPSDC - 0.01 mg/m^3 [46]; ASL - 0.004 mg/m^3 [18].

1,4-Dioxane (diethylene oxide); $C_4H_8O_2$ (88.10)

Liquid with a slight odor; m.p. 13°C, b.p. 101.7-101.8°C, f.p. 5°C, CLE 1.97-22.5%, ρ 1.033; soluble in water and organic solvents.

Present in wastes from the production of some solvents.

Toxicity: is adsorbed by undamaged skin; has an overall toxic effect: causes degenerative changes in the liver and the kidneys, and general depression; causes irritation of mucous membranes and the respiratory tracts. TC_{odor} and TC_{acute} - 0.1 and 0.7 mg/m^3 respectively [5,55,70].

Removal: by adsorption on activated carbon [46].

Diphenyl (biphenyl; xenene); $C_{12}H_{10}$ (154.21)

Colorless crystals; m.p. 71°C, b.p. 255.9°C, f.p. 105°C, CLE 0.6-5.8%, ρ

0.990; soluble in organic solvents; slightly soluble in water (0.8).

Present in wastes from the production of some pesticides and organic chemicals.

Toxicity: has an overall toxic effect: causes general depression, cramps, paralysis; causes irritation of the respiratory tracts. LD_{50} for rats - 2.2 g/kg [47,55, 70].

Removal: by using filters made of nylon or teflon [34].

Diphenyl ether (phenyl ether; diphenyl oxide); $C_{12}H_{10}O$ (170.20)

Liquid with an odor of geranium; m.p. 26.84°C, b.p. 259°C, f.p. 115°C, s.-i. p. 618°C, CLE 0.8-7.5%, ρ 0.073; soluble in benzene; insoluble in water.

Present in wastes from the production of some chemicals and perfumes.

Toxicity: has an overall toxic effect. LD_{50} for rats - 3.99 g/kg [70,72].

N,N'-Diphenylguanadine (melaniline; diphenylguanadine); $C_{13}H_{13}N_3$ (211.27)

White crystals; m.p. 151.5°C, d.p. 170°C, f.p. 150°C, s.-i.p. 646°C, lower CLE 12.6 g/m^3, ρ 1.13; soluble in ethanol; slightly soluble in water.

Present in wastes from the production of articles made of rubber, sheilac, and some analytical reagents.

Toxicity: has an overall toxic effect: causes incoordination of voluntary movements, cramps; causes irritation of mucous membranes and the respiratory tracts. LD_{50} for rats and mice - 375 and 290 mg/kg respectively [5,55,70].

Diphenylmethane-4,4'-diisocyanate; $C_{15}H_{10}N_2O_2$ (250.26)

Yellow crystals; m.p. 40.5°C, b.p. 156-158°C, ρ 1.85; soluble in organic solvents; reacts with water.

Present in wastes from the production of some synthetic fibers,polymers, and synthetic rubber.

Toxicity: has an overall toxic effect; causes retrosternal pains; causes coughing is allergenic: causes bronchial asthma. $TC_{irritation}$ - 0.005 mg/ℓ; ASL - 0.001 mg per m^3 [91].

Determination: see [302].

Diphenyl sulfide (phenyl sulfide); $C_{12}H_{10}S$ (186.27)

Liquid with an odor of burned rubber; m.p. -40°C, b.p. 296°C, ρ 1.118; soluble in diethyl ether; insoluble in water.

Present in wastes from the production of some lubricants, petrochemicals, latexes, medicinal preparations, dyes, detergents, and solvents.

Toxicity: has an overall toxic effect: causes excitement followed by depression, and functional disturbances of internal organs; causes irritation of mucous membranes and the respiratory tracts [5,70,86].

Determination: gas chromatography [94].

N,N-Dipropylamine (dipropyl amine); $C_6H_{15}N$ (101.2).

Liquid with an odor of ammonia; m.p. -39.6°C, b.p. 110.7°C, f.p. 17.4°C, s.-i.p. 280°C, lower CLE 2.1%, ρ 0.738; soluble in water and organic solvents.

Present in wastes from the production of some basic organic chemicals.

Toxicity: has an overall toxic effect; causes excitement followed by depression, internal bleeding, dystrophy of the liver and the cardiac muscle; causes severe irritation of the respiratory tracts, mucous membranes, and the skin. LD_{50} for rats - 0.2-0.4 g/kg [5,55,70,86].

Dipropyl phthalate; $C_{14}H_{18}O_4$ (250.28)

Liquid; b.p. 129-132°C(≅133 Pa), ρ 1.071; soluble in organic solvents.

Present in wastes from the production of some basic organic chemicals and plastics.

Toxicity: has an overall toxic, irritating, and narcotic effect; is allergenic [55, 86].

Diquat (1,1-ethylene-2,2'-bipyridildibromide); $C_{12}H_{12}Br_2N_2$ (360.1)

White crystals; m.p. 335-340°C(d.); soluble in water (700g/ℓ); insoluble in hydrophobic solvents.

Present in wastes from the production of some pesticides.

Toxicity: has an overall toxic effect: causes dysfunctioning of the liver and kidneys. LD_{50} for rats, rabbits, and guinea pigs - 100, 228, and 123 mg/kg respectively [5]; MPSDC established in USA - 0.5 mg/m^3 [40].

N,N'-Dithiodithiomorpholine (dimorpholinodisulfate); $C_8H_{16}N_2O_2S_2$ (236.0)

Pale yellow crystals; m.p. 125-126°C, f.p. 140°C, s.-i.p. 290°C, lower CLE 20.5 g/m^3, ρ 1.36; soluble in organic solvents; insoluble in water.

Present in wastes from the production of articles made of rubber and some pesticides.

Toxicity: has an overall toxic effect: causes functional disturbances of the central nervous system; has an irritating effect. LD_{50} for mice 2.7 g/kg [5,70].

Dodecylguanidino acetate (dodine); $C_{15}H_{23}N_3O_2$ (287.4)

White crystals; m.p. 136°C soluble in ethanol; slightly soluble in water (0.063%).

Present in wastes from the production of some pesticides.

Toxicity: causes irritation of the respiratory tracts. LD_{50} for rats, mice, cats, rabbits, and guinea pigs - 1120, 266, 266, 535, and 176 mg/kg respectively [5]; MPSDC and MPADC - 0.001 and 0.002 mg/m^3 respectively [24].

Endosulfan (thiodan; 1,2.3,4,7,7-hexachlorobicyclo-[2.2.1]-hepto-2-en-5,6-bis(oxymethyl)sulfite); $C_9H_6Cl_6O_3S$ (406.9)

White crystals; m.p. 108-109°C (type I); m.p. 296-298°C (type II); soluble in organic solvents; insoluble in water.

Present in wastes from the production of some pesticides.

Toxicity: very toxic; causes ptyalism, cramps. $TC_{chronic}$ - 0.5 mg/m^3; LD_{50} for rats and mice - 40-100 and 32075 mg/kg respectively [55,67,70]; MPSDC and MPADC - 0.005 and 0.001 mg/m^3 respectively [24].

Endrin (1,2,3,4,10,10-hexachloro-exo-6,7-epoxy-1,4,4a,5,6,7,8,8a-octahydro-1,4-endo,endo-5,8-dimethanonaphthalene); $C_{18}H_8Cl_6O$ (380.9)

White crystals; m.p. 226-230°C,(d.); soluble in organic solvents; insoluble in water.

Toxicity: very toxic; causes dysfunctioning of the central nervous system, the kidneys, and the respiratory organs, and cramps [70]. The use of this pesticide is banned in the USSR, Bulgaria, Hungary, and Poland [85].

Epichlorohydrin (3-chloro-1,2-epoxypropane); C_3H_5ClO (92.53)

Liquid with an odor of chloroform; m.p. -57°C, b.p. 116.1°C, f.p. 26°C, s.-i.p. 410°C, CLE 2.3-49.0%, ρ 1.18; soluble in organic solvents and in water (6.55%).

Present in wastes from the production of some organic chemicals, synthetic rubber, and solvents.

Toxicity: has an overall toxic effect: causes dysfunctioning of the kidneys; is adsorbed by undamaged skin [55,70]. SC_{odor} and TC_{odor} - 0.2 and 0.3 mg/m^3 respectively; $SC_{c.n.s.}$ and $TC_{c.n.s.}$ - 0.2 and 0.3 mg/m^3 respectively; $SC_{chronic}$ and $TC_{chronic}$ - 0.2 and 2 mg/m^3 respectively [435]; MPASDC and MPADC - 0.2 mg/m^3; toxicity classification - II [20,21]; MPSDC and MPADC established in Bulgaria and Yugoslavia - 0.2 mg/m^3, and in GDR - 0.2 and 0.06 mg/m^3 respectively; MPC established in FRG - 20 mg/m^3 [44,51].

Removal: by using filters made of polyethylene, polypropylene, nylon, polyesters, or teflon [34].

Determination: gas chromatographic analysis (sensitivity 0.1 mg/m^3) [20,27,436].

Eptam (EPTC; N,N-dipropyl-S-ethylthiocarbamate); $C_9H_{19}NOS$ (189.3)

Oily liquid with an unpleasant odor; b.p. 232°C, ρ 0.954; soluble in organic solvents and in water (375 mg/ℓ).

Present in wastes from the production of some pesticides.

Toxicity: has an overall toxic effect: causes headache, asthenia, nausea; dysfunctioning of the central nervous system, the liver, and other internal organs; is harmful to embryos. $TC_{irritation}$ - 0.02 mg/ℓ; LD_{50} for mice and cats - 750 and 150 mg/kg respectively [67]; MPSDC and MPADC - 0.02 mg/m^3 [24].

Erbon (ß- [(2,4,5-trichlorophenoxy)ethyl]-α ,α-dichloropropionate); $C_{11}H_9Cl_5O_3$ (366.48).

White crystals; m.p. 36-50°C, b.p. 161-164°C(≅65 Pa); soluble in organic solvents; insoluble in water.

Toxicity: has an overall toxic effect. LD_{50} and $LD_{chronic}$ for mice - 0.90-0.96 g/kg and 1.25 mg/kg respectively [5].

Erythritol (1,2,3,4 -tetrahydroxybutane; butantetrol-1,2,3,4); $C_4H_{10}O_4$ (122.12)

White crystals; m.p. 121.5°C, b.p. 329-331°C, ρ 1.45; soluble in water.

Present in wastes from the production of some surface-active substances, lubricants, explosives, and pharmaceuticals.

Toxicity: has an overall toxic effect. LD_{50} for dogs - 5 g/kg [70].

Ethane; C_2H_6 (30.07)

Gas; m.p. -182.23°C, b.p. 88.63°C, f.p. -152°C, s.-i.p. 472°C, CLE 2.9-15.0%, ρ 1.04 (air); slightly soluble in water.

Present in wastes of plants processing natural gas and of oil refineries; also in wastes from the production of some plastics and petrochemicals.

Toxicity: causes irritation of mucous membranes and the respiratory tracts; at high concentrations has a narcotic effect [55,70].

Removal: by adsorption on activated carbon [46]; by thermal catalytic oxidation (efficiency 100%) [89].

Determination: infrared spectroscopy [437]; gas chromatography (automatic methods) [438].

Ethanolamine (aminoethyl alcohol; 2-hydroxyethylamine; 2-aminoethanol); C_2H_7NO (61.09)

Oily liquid with an odor of ammonia; m.p. 10.5°C, b.p. 171°C, f.p. 85°C, ρ 1.022; soluble in water and ethanol.

Present in wastes from the production of some organic chemicals, leather, mineral fertilizers, surface-active agents, and some pharmaceuticals.

Toxicity: causes irritation of the skin. LD_{50} for rats - 2.1-2.74 g/kg [70,86].

1-Ethoxydiethyleneglycol acetate; $C_8H_{16}O_4$ (176.22)

Liquid with a slight odor; m.p. -11°C, b.p. 218.5°C, ρ 1.009; soluble in water.

Present in wastes from the production of some basic organic chemicals.

Toxicity: has an overall toxic and irritating effect. LD_{50} for rats - 11 g/kg [55,70].

Ethyl acetate; $C_4H_8O_2$ (88.09)

Liquid with a fruity odor; m.p. -83.6°C, b.p. 77.15°C, f.p. -3°C, s.-i.p. 426.7°C, CLE 2.18-11.4%, ρ 0.901; soluble in organic solvents and water.

Present in wastes from the production of acetic acid, some solvents, synthetic rubber, fats, waxes, explosives, essential oils, and acetyl acetone.

Toxicity: has an overall toxic effect: does damage to the liver, the kidneys, the pancreatic glands, and the spleen; causes conjunctivitis, dermatitis, eczema, and does damage to the cornea; has a narcotic effect [5,55,70]. SC_{odor} and TC_{odor} - 0.5 and 0.6 mg/m^3 respectively; $SC_{sensitivity\ to\ light}$ and $TC_{sensitivity\ to\ light}$ - 0.18 and 0.3 mg/m^3 respectively [442]; MPSDC and MPADC - 0.1 mg/m^3; toxicity classification - IV [20,21]; MPSDC and MPADC established in Bulgaria and Yugoslavia - 0.1 mg/m^3, and in GDR - 0.1 and 0.3 mg/m^3 respectively; MPC established in FRG - 300 mg/m^3 [44,51].

Removal: by adsorption on activated carbon [34,46,49]; by using filters made of polyethylene, polyprppylene, nylon, polyesters, or teflon [34].

Ethyl acetate; $C_5H_8O_2$ (100.11)

Colorless liquid with a sharp odor; m.p. -71.2°C, b.p. 99.5°C, f.p. 8°C, CLE 1.6-9.6%, ρ 0.91; soluble in organic solvents; slightly soluble in water (1.51%).

Present in wastes from the production of Plexiglas, gas masks, textiles, some plastics, artificial leather, dyes, and solvents; also in wastes of wood-processing plants, and from the treatment of cables and wires, the coating of chemical equipment, and the waterproofing of textiles.

Toxicity: has an overall toxic effect: causes ptyalism, lacrimation, respiratory depression, degenerative changes in the cardiac muscle, the liver, the kidneys, and the spleen: causes irritation of the skin; at high concentrations has a narcotic effect; is adsorbed by undamaged skin.

Removal: by adsorption on activated carbon [46]; by combustion [439].

Ethyl adipate (diethyl adipate); $C_{10}H_{18}O_4$ (202.24)

Liquid; m.p. -21.4°C, b.p. 245°C, ρ 1.009.

Present in wastes from the production of synthetic rubber and some intermediate compounds and plastics.

Toxicity: is mutagenic [37].

Ethylamine (aminoethane); C_2H_7N (45.09)

Liquid; m.p. -81°C, b.p. 16.6°C, f.p. -39°C, s.-i.p. 384°C, CLE 3.5-14.0%, ρ 0.68; soluble in water.

Present in wastes from the production of some organic chemicals, articles made of rubber, pharmaceuticals, cosmetic preparations, detergents; emulsifying agents, and pesticides; also in wastes of metal-working plants.

Toxicity: has an overall toxic effect: causes pathological changes in the cardiac muscle, the brain, the kidneys, and the lungs; causes irritation of mucous membranes, the respiratory tracts, and the skin; is mutagenic [5,79]. TC_{odor}, $TC_{sensitivity\ to\ light}$, and $TC_{chronic}$ - 0.05, 0.05, and 0.056 mg/m^3 respectively [440,441].

Removal: by adsorption on activated carbon [46].

Ethyl-n-amyl ketone (octanone-3); $C_8H_{16}O$ (128.22)

Liquid; b.p. 161°C, f.p. 58.9°C, ρ 0.83; soluble in organic solvents; insoluble in water.

Present in wastes from the production of some plastics, solvents, and perfumes.

Toxicity: has an overall toxic and irritating effect. LD_{50} for mice and guinea pigs - 3.8 and 2.5 g/kg respectively [86].

Ethylbenzene (phenylethane); C_8H_{10} (106.17)

Liquid; m.p. -95°C, b.p. 136.19°C, f.p. 20°C, s.-i.p. 420°C, CLE 0.9-3.9%, ρ 0.86; soluble in organic solvents; slightly soluble in water (0.014%).

Present in wastes from the production of some plastics, petrochemicals, synthetic rubber, and man-made fibers.

Toxicity: has an overall toxic effect: causes pathological changes in the lungs and the brain; causes irritation of mucous membranes and the respiratory tracts; at high concentrations has a narcotic effect; is adsorbed by undamaged skin [55,70]. MPSDC and MPADC - 0.02 mg/m^3; toxicity classification - III [20,21].

Removal: by adsorption on activated carbon [46]; by using filters made of nylon, polyesters, or teflon [34].

Determination: gas chromatographic analysis (sensitivity 0.1 μg) [27,60].

N,N-Ethylbenzylaniline (benzylethylaniline); $C_{15}H_{17}N$ (211.29)

Light yellow oily liquid; b.p. 286°C, ρ 1.03; soluble in organic solvents; insoluble in water.

Present in wastes from the production of some organic chemicals and dyes.

Toxicity: is adsorbed by undamaged skin; has an overall toxic effect: causes headache, supppression of the functioning of the central nervous system, methemoglobinemia, anemia, cyanosis, and damage to the cardiac muscle; has an irritating effect; is allergenic [55,70]; ASL - 0.01 mg/m^3 [91].

Ethylenebromohydrin (glycol bromohydrin; 2-bromoethanol); C_2H_5BrO (124.98)

Liquid; b.p. 149-150°C, ρ 1.790; soluble in water.

Present in wastes from the production of some solvents.

Toxicity: causes irritation of mucous membranes and the eyes; is mutagenic [70,447].

Ethylenechlorohydrin (2-chloroethanol-1); C_2H_5ClO (80.52)

Liquid with an odor of ether; m.p. -62.6°C, b.p. 128.7°C, f.p. 55°C, s.-i.p. 425°C, CLE 4.9-15.9%, ρ 1.205; soluble in water and organic solvents.

Present in wastes from the production of some organic chemicals.

Toxicity: has an overall toxic effect: causes vomiting, respiratory depression, pulmonary edema, degenerative changes in the kidneys and the liver; causes irritation of mucous membranes; at high concentrations has a narcotic effect; is adsorbed by undamaged skin. LD_{50} for rats - 95 mg/kg [55,70].

Ethylenecyanohydrin (ß-hydroxypropionitrile); C_3H_5NO (71.08)

Liquid; m.p. -46°C, b.p. 227°C(d.), f.p. 129°C, ρ 1.04; soluble in organic solvents and water.

Present in wastes from the production of acrylic acid and some solvents and inorganic compounds.

Toxicity: very toxic; causes death of animals 5 minutes after trace amounts have got into their eyes; causes inccordination of voluntary movements, cramps, dystrophic changes in the liver, the kidneys and the cardiac muscle, and pulmonary edema; has an irritating effect. TC_{acute} - 0.165 mg/ℓ [5,55,70].

Ethylenediamine (1,2-diaminoethane); $C_2H_6N_2$ (60.10)

Liquid with an odor of ammonia; m.p. 9.95°C, b.p. 117.2°C, f.p. 33°C, s.-i.p. 403°C, CLE lower limit - 2.42%, ρ 0.908; soluble in water and ethanol.

Present in wastes from the production of some surface-active agents, solvents, articles made of rubber, pesticides, and some organic chemicals.

Toxicity: has an overall toxic and irritating effect; is allergenic. LD_{50} for rats, mice, and guinea pigs - 1.16, 0.448, and 0.448 g/kg respectively [5,55,70].

Ethylene (ethene); C_2H_4 (28.05)

Gas; m.p. -169.15°C, b.p. -103.7°C, f.p. -136.1°C, s.-i.p. 540°C, CLE 3.34%, ρ 0.56; soluble in water (25.6 ml/100 ml).

Present in wastes from the production of some organic chemicals, pharmaceuticals, and alcohols; also in wastes of metal-working plants.

Toxicity: has an irritating effect; at high concentrations has a narcotic effect; is mutagenic [38,55,70,79]. TC_{odor} and $TC_{sensitivity\ to\ light}$ - 20 and 11 mg/m^3 respectively [443]. MPSDC and MPADC - 3 mg/m^3; toxicity classification - III [20, 21]; MPSDC and MPADC established in Bulgaria and Yugoslavia - 3 mg/m^3, and in

GDR - 3 and 2 mg/m^3 respectively [44,51].

Removal: by adsorption on activated carbon [46]; by combustion [49]; by thermal catalytic oxidation [89].

Determination: gas chromatographic analysis [60]; see also [444-446].

Ethylene oxide (1,2-epoxyethane); C_2H_4O (44.05)

Gas; m.p. -112.5°C, b.p. 10.7°C, CLE 2.8-100%, ρ 0.871; soluble in water and organic solvents.

Present in wastes from the production of some surface-active agents, pesticides, textiles, foodstuffs, organic chemicals, ethylene glycol, and acrylonitrile.

Toxicity: has an overall toxic and irritating effect; is teratogenic and mutagenic [37,55,70]. TC_{odor}, $TC_{sensitivity\ to\ light}$, $TC_{c.n.s.}$, and $SC_{chronic}$ - 1.5, 1, 0.65 and 0.3 mg/m^3 respectively [449]; MPSDC and MPADC - 0.3 and 0.03 mg/m^3 respectively; MPSDC and MPADC established in Bulgaria, GDR, and Yugoslavia - 0.3 and 0.03 mg/m^3 respectively [20,21]; MPC established in FRG - 20 mg/m^3.

Removal: by adsorption on activated carbon [46]; by thermal catalytic oxidation [89].

Determination: see [27,450,451].

Ethylene sulfide (thiirane); C_2H_4S (60.12)

Liquid with an unpleasant odor; m.p. -109°C, b.p. 55-56°C, ρ 1.004; slightly soluble in ethanol and diethyl ether.

Present in wastes from the production of some plastics and articles made of rubber.

Toxicity: very toxic; causes cramps, hemorrhage, and pulmonary edema; irritation of mucous membranes, the respiratory tracts and the skin; has a narcotic effect; is adsorbed by undamaged skin. LD_{50} for mice - 35.6 mg/kg [5,55]; MPSDC - 0.5 mg/m^3; toxicity classification - I [20,21].

Ethyl alcohol (ethanol; alcohol); C_2H_6O (46.07)

Liquid with a peculiar odor; m.p. -114.15°C, b.p. 78.39°C, f.p. 13.0°C, s.-i. p. 422.8°C, CLE 3.28-18.95%, ρ 0.793; soluble in water and organic solvents.

Present in wastes from the production of starch by the fermentation process; some pharmaceuticals, organic chemicals, and perfumes; also in wastes of sugar beet processing plants.

Toxicity: is adsorbed by undamaged skin; has an overall toxic effect: causes excitement followed by depression, incoordination of voluntary movements, drowsiness, nausea, and vomiting; is mutagenic [55,70,454]. SC_{odor} and TC_{odor} - 6.97 and 7.1 mg/m^3 respectively; $SC_{c.n.s.}$ and $TC_{c.n.s.}$ - 4.9 and 6.1 mg/m^3 respectively; $TC_{sensitvity\ to\ light}$ - 6.1 mg/m^3 [453]; MPSDC and MPADC - 5 mg/m^3; toxicity classification - IV [20,21].

Removal: by using filters made of nylon, polyesters, polyethylene, poly(vinylidene chloride), polypropylene, or teflon [34]; by adsorption on activated carbon [46]; by using scrubbers [445]; and by thermal catalytic oxidation [456].

Ethyl chloride (chloroethane); C_2H_5Cl (64.52)

Liquid with an odor of ether; m.p. -138.3°C, b.p. 12.27°C, f.p. -50°C, s.-i.p. 494°C, CLE 3.8-15.4%, ρ 0.88; soluble in organic solvents; slightly soluble in water (0.574%).

Present in wastes from the production of lead tetraethyl, some solvents, and organosilicon compounds.

Toxicity: has an overall toxic, irritating, and at high concentrations also a narcotic effect [55,70]. ASL - 0.10 mg/m^3 [90]; MPC established in FRG - 300 mg per m^3 [51].

Removal: by adsorption on activated carbon [46]; by using filters made of nylon, poyesters, polyethylene, poly(vinylidene chloride), teflon, or polypropylene [34].

Ethyl chloroacetate; $C_4H_7ClO_2$ (122.55)

Liquid with a strong fruity odor; b.p. 143.6°C, ρ 1.15; soluble in organic solvents; insoluble in water.

Present in wastes from the production of some organic chemicals.

Toxicity: has an overall toxic and a strong irritating effect [55,70].

Ethyl formate; $C_3H_6O_2$ (74.08)

Liquid with a slight odor; m.p. -80.5°C, b.p. 54.3°C, f.p. -20°C, s.-i.p. 440°C, CLE 2.8-16%, ρ 0.916; soluble in ethanol, diethyl ether, and water (11.8%9.

Present in wastes from the production of some pesticides, organic chemicals, nitrocellulose solvents, foodstuffs, alcoholic bevarages, and formamide.

Toxicity: has an overall toxic, irritating, and at high concentrations also a narcotic effect [5,55,70].

Removal: by adsorption on activated carbon [46].

bis(2-Ethylhexyl) phthalate (diethylhexyl phthalate); $C_{24}H_{38}O_4$ (390.57)

Liquid; m.p. -46°C, b.p. 231°C(≅1.65 kPa), f.p. 206°C, ρ 0.986; soluble in organic solvents; insoluble in water.

Present in wastes from the production of synthetic rubber and some polymeric materials.

Toxicity: causes irritation of mucous membranes and the respiratory tracts.

LD_{50} for rats and mice - 3.7 and 6.54 ml/kg respectively [5].

Ethyl isothiocyanate; C_3H_5NS (87.15)

Oily liquid with an odor of mustard; m.p. -5.9°C, b.p. 131-132°C, ρ 1.004; soluble in organic solvents; insoluble in water.

Present in wastes from the production of ethylamine and mercuric chloride.

Toxicity: has an overall toxic effect; causes irritation of the eyes, the respiratory tracts, and the skin. TC_{odor} - 6 mg/m^3 [5,70].

Ethyl mercaptan (ethanethiol); C_2H_6S (62.13)

Liquid with an unpleasant odor; m.p. -147.89°C, b.p. 190°C, CLE 2.8-18.2%, ρ 0.919; soluble in organic solvents and in water (6.76 g/ℓ).

Present in wastes from the production of some plastics, antioxidants, pharmaceuticals, and pesticides; also in wastes of natural gas processing plants.

Toxicity: has an overall toxic, irritating, and at high concentrations also a narcotic effect [55,70]. TC_{odor} - 0.0007 [64]-0.0026 [54] mg/m^3.

Removal: by adsorption on activated carbon [46,49]; by thermal catalytic oxidation (efficiency 90-95%) [89].

Determination: see [452].

Ethyl urethane (ethyl-N-ethyl carbamate); C_3H_7NO (89.09)

White crystals; m.p. 49°C, b.p. 180°C, ρ 0.986; soluble in water and organic solvents.

Present in wastes from the production of some organic chemicals, pesticides, amines, and pharmaceuticals.

Toxicity: has an overall toxic effect: causes nausea, vomiting, degenerative changes in the brain and the liver, pathological changes in the bones, and supression of the nerves in the brain and spinal cord; has a narcotic effect [47,55,70].

Removal: by using filters made of polyethylene, polypropylene, nylon, polyesters, poly(vinylidene chloride), or teflon [34].

Ferbam (fermate; ferric dimethyldithio carbamate); $C_9H_{18}FeN_3S_6$ (416.50)

Dark powder; m.p. 180°C(d.); soluble in organic solvents and in water (120 mg/ℓ).

Present in wastes from the production of some pesticides.

Toxicity: has an overall toxic effect: causes headache, vomiting, cramps, and dysfunctioning of the kidneys; causes dermititis. LD_{50} for rats and mice - 1.13 and 6.9 g/kg respectively [5,70].

Fluorodichloromethane freon 21); $CHCl_2F$ (102.93)

Liquid; m.p. - 135°C, b.p. 8.9°C, ρ 1.425; soluble in ethanol and diethyl ether; insoluble in water.

Present in wastes from the production of refrigerants and some solvents.

Toxicity: has an irritating effect [55,86].

Folithion (sumithion; O,O,-dimethyl-O-(3-methyl-4-nitrophenyl) thiophospahte); $C_9H_{12}NO_5PS$ (277.2)

Liquid with a peculiar odor; b.p. 109°C(≅13 Pa), ρ 1.308; soluble in organic solvents; slightly soluble in water (30 mg/ℓ).

Present in wastes from the production of some pesticides.

Toxicity: very toxic; is adsorbed by undamaged skin; causes nausea, vomiting, ptyalism, respiratory depression, cramps [47]. $TC_{irritation}$ - 1 mg/m^3; LD_{50} for rats and mice - 410-516 and 329-715 mg/kg respectively [67]; MPSDC and MPADC - 0.005 and 0.001 mg/m^3 respectively [24].

Determination: polarographic analysis (sensitivity 5 μg) [63].

Folpet (phthalan; N-trichloromethylthiophthalimide); $C_9H_4Cl_3NO_2S$ (296.6)

White crystals; m.p. 177°C; insoluble in water.

Present in wastes from the production of some pesticides.

Toxicity: has an overall toxic effect: causes dyspnea; irritation of mucous membranes, is harmful to embryos [73]. LD_{50} for mice and guinea pigs - 1.5 and 0.98 g/kg respectively [5,67].

Formaldehyde (methanal); CH_2O (30.03)

Gas; m.p. -118°C, b.p. -19°C, CLE 7.0-72.0%, ρ 0.82; soluble in water.

Present in wastes from the production of some organic chemicals, construction materials, linoleum, asphaltic roofing paper, roof-sheeting materials, artificial parchment paper, foamed plastics, and synthetic fatty acids.

Toxicity: has an overall toxic effect: causes dysfunctioning of the central nervous system, the eyes, the liver and kidneys; has a strong irritating effect; is allergenic, carcinogenic, and mutagenic [5,37,45,52,55,67,70]. SC_{odor}, TC_{odor}, $TC_{sensitivity\ to\ light}$, and $TC_{c.n.s.}$ - 0.005, 0.07, 0.084, and 0.098 mg/m^3 respectively; recommended MPSDC for plants - 0.02 mg/m^3 [23]; MPSDC and MPADC - 0.035 and 0.012 mg/m^3 respectively [21]. The maximum permissible concentration (mg/m^3) established in some countries is as follows:

Country	MPSDC	MPADC
Bulgaria	0.035	0.012
GDR	0.035	0.012
Hungary	0.035	0.012

Yugoslavia	0.035	0.012
Romania	0.03	0.01
Poland	0.05	0.02
Czechoslovakia	0.05	0.015

Removal: by catalytic oxidation [48]; filtering [34]; adsorption on activated carbon [423].

Determination: colorimetric analysis (sensitivity 0.2 μg) [29]; photometric analysis (sensitivity 0.02 $\mu g/m^3$) [63]; automatic methods [374,419,420-422].

meta-Formaldehyde (α-trioxymethylene); $C_3H_6O_3$ (90.08)

White crystals with an odor of ether; m.p. 62°C, b.p. 114.5°C(d.), s.-i.p. 415°C, ρ 1.17; soluble in water, ethanol, and diethyl ether.

Present in wastes from the production of some organic chemicals.

Toxicity: has an overall toxic and irritating effect; is allergenic [55].

Formic acid (methanoic acid); CH_3O_2 (46.03)

Liquid with a sharp odor; m.p. 8.3°C, b.p. 100.8°C, f.p. 60°C, s.-i.p.504°C, CLE 14.3-33.3%, ρ 1.22; soluble in water, ethanol, and diethyl ether.

Present in wastes from the production of some basic organic chemicals, textiles, leather, and foodstuffs.

Toxicity: has an overall toxic effect: causes dystrophic changes in the liver and kidneys, pulmonary hemorrhage, corneal opacity; causes chemical burns of mucous membranes [5,55,70].

Removal: by adsorption on activated carbon [46]; by using filters made of polyesters, polyethylene, poly(vinylidine chloride), polypropylene, or teflon [34].

Determination: paper chromatography (sensitivity 0.5 $\mu g/m^3$) [29].

Formothion (antio; O,O-dimethyl-S-(N-methyl-N-formylcarbamoylmethyl) dithiophosphate); $C_6H_{12}NO_4PS_2$ (257.3)

Yellowish liquid or crystalline substance; m.p. 25-26°C, ρ 1.361; soluble in organic solvents; slightly soluble in water (0.1%).

Present in wastes from the production of some pesticides.

Toxicity: is adsorbed by undamaged skin. LD_{50} for rats and mice - 350-400 and 92.5 mg/kg respectively; MPSDC and MPADC - 0.01 and 0.006 mg/m^3 respectively [24].

Determination: chromatographic analysis (sensitivity 0.005 mg/m^3) [418].

Furacillin (nitrofural); $C_6H_6N_4O_4$ (198.2)

Yellow crystals; m.p. 230-236°C; insoluble in water.

Present in wastes from the production of some pharmaceuticals and foodstuffs.

Toxicity: has an overall toxic effect: causes morphological changes in the bronchis, liver, and kidneys; is allergenic [5].

2-Furfural (furfural); $C_5H_4O_2$ (96.09)

Liquid; m.p. -36.5°C, b.p. 161.7°C, f.p. 61°C, s.-i.p. 260°C, CLE 1.8-3.4%, ρ 1.159; soluble in organic solvents and in water (8.3%).

Present in wastes from the production of polyfurfurolphenol, some varnishes and paints, articles made of rubber, petrochemicals, plastics, abrasives; also in wastes from some hydrolysis processes.

Toxicity: has an overall toxic effect: causes difficulty in breathing, cramps, paraplegia; causes irritation of the respiratory tracts and mucous membranes. SC_{odor} and TC_{odor} - 0.05 and 1 mg/m^3 respectively; $SC_{c.n.s.}$ and $TC_{c.n.s.}$ - 0.05 and 0.084 mg/m^3 respectively; $SC_{sensitivity\ to\ light}$ and $TC_{sensitivity\ to\ light}$ - 0.05 and 0.031 mg/m^3 respectively; $SC_{chronic}$ and $TC_{chronic}$ - 0.05 and 0.33 mg per m^3 respectively [428]; LD_{50} for rats, mice, and guinea pigs - 126, 425, and 541 mg/kg respectively [5,55,70]; MPSDC and MPADC - 0.05 mg/m^3; toxicity classification - III [20,21]; MPSDC and MPADC established in Bulgaria and Yugoslavia - 0.05 mg/m^3, and in GDR and Romania - 0.15 and 0.05 mg/m^3 respectively; MPC established in FRG - 20 mg/m^3 [44,51].

Removal: with the aid of filters made of polyethylene, polypropylene, nylon; polyesters, or teflon [34].

Determination: gas chromatographic analysis (sensitivity 1 μg) [20,27,29,60].

Furfuran (furan); C_4H_4O (68.07)

Liquid with an odor of chloroform; m.p. -85.68°C, b.p. 31.83°C, f.p. -40°C, CLE 2.3-14.3%, ρ -.936; soluble in organic solvents; slightly soluble in water (≅1%).

Present in wastes from the production of some dyes, fats and oils, tetrahydrofuran, and maleic anhydride.

Toxicity: is adsorbed by undamaged skin; has an overall toxic effect: causes dyspnea, cramps, and paralysis of the respiratory center; causes irritation of the eyes and respiratory tracts; has a cumulative effect. $TC_{chronic}$ - 0.005-0.01 mg/ℓ; LC_{50} for mice - 208 mg/ℓ [5,70].

Furfuryl alcohol (α-furyl carbinol); $C_5H_6O_2$ (98.10)

Liquid; m.p. -31°C, b.p. 171°C, ρ 1.129; soluble in organic solvents (except for alkanes) and water.

Present in wastes from the production of some solvents; also in wastes of cement and metal-working plants.

Toxicity: has an overall toxic and irritating effect. LD_{50} for rats - 275 mg/kg

[55,70].

Determination: chromatographic analysis [427].

Gasoline, predominantly C_4-C_{12} (≅200)

Liquid; b.p. 35-195°C, CLE 0.79-5.48%, ρ 0.700-0.780.

Present in wastes of oil-extracting installations and some machine-building plants; also in wastes from the production of varnishes and paints, petrochemicals, leather, articles made of rubber, and textiles.

Toxicity: is carcinogenic [233]. SC_{odor} and TC_{odor} - 0.15 and 0.3-3.12 mg per m^3 respectively [49]; $SC_{chronic,\ irritation}$ and $TC_{chronic,sensitivity\ to\ light}$ - 21.8 and 56 mg/m^3 respectively [32]; $SC_{chronic}$ and $TC_{chronic}$ - 20 and 100 mg/m^3 respectively [234]. The maximum permissible concentration (recalculated as mg C/m^3) and toxicity classification [22,24] are as follows:

Gasoline source	MPSDC	MPADC	Toxicity classification
Petroleum with low S content	5.0	1.5	III
Shale	0.05	0.05	IV

The maximum permissible concentration (mg/m^3) established in some countries is as follows (gasoline with a low concentration of sulfur):

Country	MPSDC	MPADC
Bulgaria	5.0	1.5
Hungary	5.0	1.5
GDR	5.0	1.5
Poland	2.5	0.75
Romania	6.0	2.0
Yugoslavia	5.0	1.5
(gasoline with a high concentration of sulfur)		
Bulgaria	0.05	0.05
GDR	0.05	0.03
Yugoslavia	5.0	5.0

Removal: with the aid of filters made of polyesters, poly(vinylidine chloride), teflon, or polypropylene [34]; by adsorption on activated carbon, activated silica gel, or activated aluminum [235]; the efficiency of removing gasoline with the aid of scrubbers is 60-90%; of cyclones - 84.2%; of multicyclones - 93.8%; of electrostatic filters - 99.0%; and of various fiber filters - 99.7-99.9% [235].

Determination: chromatographic analysis (sensitivity 0.6 μg) [21,27].

Glycide (glycidol; epihydrin alcohol; 1,2-epoxy-3-propanol); $C_3H_6O_2$ (74.08)

Liquid; b.p. 162-163°C, ρ 1.165; soluble in water and organic solvents; slightly soluble in xylene.

Present in wastes from the production of dyes, varnishes and paints, and some basic organic compounds.

Toxicity: is adsorbed by undamaged skin; has an overall toxic effect: at first stimulates and then suppresses the nerves in the brain and spinal cord; has an irritating effect. LD_{50} for rats and mice - 0.45 and 0.85 g/kg respectively [70,72].

Glycol dibutyrate (ethylene dibutyrate); $C_{10}H_{18}O_4$ (202.24)

Liquid; b.p. 240°C, ρ 1.024; soluble in organic solvents; insoluble in water.

Present in wastes from the production of some solvents, lacquers, leather, cosmetics, and household chemicals.

Toxicity: TC_{odor} and $TC_{chronic}$ - 4 and 1 mg/m^3 respectively; LD_{50} for rats and mice - 4500 and 6650 mg/kg respectively; toxicity classification - IV [448].

Glycolic acid (hydroxyacetic acid); $C_2H_4O_3$ (76.05)

White crystalline powder; m.p. 79-80°C, ρ 1.27; soluble in water and organic solvents.

Present in wastes from the production of textiles, leather, and dyes; also in wastes of metal-working and machine-building plants.

Toxicity: causes irritation of mucous membranes and the skin. LD_{50} for rats - 1.6-3.2 g/kg [67,70,86].

Granosan (ceresan; ethylmercuric chloride); C_2H_5ClHg (264.9)

White crystals with a peculiar odor; m.p. 192.5°C; soluble in hot acetone; slightly soluble in water.

Present in wastes from the production of some pesticides.

Toxicity: very toxic; is adsorbed by undamaged skin; causes pathological changes in the brain, and chemical burns of the skin; is allergenic and mutagenic; is harmful to embryos and gonads; has a cumulative effect [55,70,73,282]. MPSDC and MPADC - 0.0003 and 0.0001 mg/m^3 respectively [24].

Heptachlor (3a,4,7,7a-tetrahydro-4,7-methano-1,4,5,6,7,8,8-heptachlorodiene); C_{10}-H_5Cl_7 (373.3)

White crystals with an odor of comphor; m.p. 95-96°C, b.p. 117-126 (6.7 Pa); soluble in organic solvents; insoluble in water.

Present in wastes from the production of some pesticides.

Toxicity: is adsorbed by undamaged skin; has an overall toxic effect: causes impairment of locomotive reflexes and cramps [47,70]; is carcinogenic and mutagenic [282,283]. MPSDC and MPADC - 0.001 and 0.0002 mg/m^3 respectively [21,24].

Heptafluorobutyric acid (perfluorobutyric acid); $C_4HF_7O_2$ (214.05)

Liquid; m.p. 17.5°C, b.p. 120°C.

Present in wastes from the production of some types of synthetic rubber and some pesticides.

Toxicity: has an irritating effect. TC_{odor} and $TC_{irritation}$ - 0.0005 and 0.05 mg/ℓ respectively [5].

n-Heptane; C_7H_{14} (100.20)

Liquid; m.p. -90.59°C, b.p. 90.43°C, f.p. -4°C, s.-i.p. 223°C, CLE 1.1-6.0%, ρ 0.684; soluble in organic solvents; insoluble in water.

Present in wastes of oil refineries and petrochemical plants.

Toxicity: has an irritating effect, and at high concentrations - a narcotic effect [55]. MPC established in FRG - 300 mg/m^3 [51].

Removal: by thermal catalytic oxidation (efficiency 100%) [89].

Heptanone-1 (heptanoic aldehyde; oenanthole); $C_7H_{16}O$ (116.20)

Liquid; m.p. -34°C, b.p. 176.3°C, f.p. 74°C, s.-i.p. 255°C, CLE 0.9-7.0%, ρ 0.823; soluble in organic solvents; slightly soluble in water (1 g/ℓ).

Present in wastes from the production of some basic organic chemicals.

Toxicity: has an overall toxic and irritating effect [70].

Heptanone-2 (n-amylmethyl ketone); $C_7H_{14}O$ (114.18)

Liquid with a sharp fruity odor; b.p. 151.5°C, ρ 0.822; soluble in organic solvents; slightly soluble in water (4.3 g/ℓ).

Present in wastes from the production of some solvents and aromatic substances.

Toxicity: is adsorbed by undamaged skin; causes irritation of mucous membranes; at high concentrations has a narcotic effect [55,70].

Heptene-2 (cis- and trans-isomers); C_7H_{14} (98.18)

Liquid; m.p. -109.2°C (trans-); b.p. 98.2°C, f.p. -2.2°C, ρ 0.709; soluble in organic solvents; insoluble in water.

Present in wastes from the production of some basic organic chemicals.

Toxicity: has an overall toxic and irritating effect. TC_{odor} - 1.2-1.5 mg/m^3; recommended MPSDC - 0.4 mg/m^3 [55].

Herban (norea; N-(3a,4,5,6,7a-hexahydro-4,7-metanoindan-5-yl)N,N'-dimethyl urea); $C_{13}H_{22}N_2O$ (221.3)

White crystals; m.p. 168-169°C, ρ 1.16; soluble in organic solvents; slightly soluble in water (20 mg/ℓ).

Present in wastes from the production of some pesticides.

Toxicity: has an overall toxic effect. MPC and LD_{50} for rats - 0.1 and 6.58 mg/kg respectively; LD_{50} for mice - 2.98 mg/kg [5,70].

Hexachlorobenzene (perchlorobenzene); C_6Cl_6 (284.80)

White leaves; m.p. 231°C, b.p. 322°C, ρ 2.044; soluble in benzene; slightly soluble in ethanol; insoluble in water.

Present in wastes from the production of some organic chemicals and pesticides.

Toxicity: has an overall toxic effect; causes irritation of the respiratory organs and the skin [55,70]. ASL - 0.13 mg/m^3 [18].

Removal: with the aid of filters made of polyethylene, polypropylene, or poly(vinylidene chloride) [34].

1,2,3,4,5,6-Hexachlorocyclohexane (mixture of 8 isomers); $C_6H_6Cl_6$ (290.8)

White crystalline powder with a musty odor; m.p. 88-309°C (depending on composition); soluble in organic solvents and in water (5-25 mg/ℓ).

Present in wastes from the production of some pesticides.

Toxicity: is adsorbed by undamaged skin; has an overall toxic and cumulative effect; is allergenic and mutagenic [55,57,282]; is harmful to gonads and embryos [73]. $TC_{chronic}$ - 1 mg/kg [55,67]; LD_{50} for warm-blooded animals - 300-500 mg/kg [73]; MPSDC and MPADC - 0.03 mg/m^3; toxicity classification - I [4,20,21]; MPSDC and MPADC in GDR - 0.03 and 0.01 mg/m^3 respectively [55]; recommended MPSDC - 0.001 mg/m^3 [46].

Hexachloro-1,3-cyclopentadiene (perchlorocyclopentadiene); C_5Cl_6 (275.0)

Yellow liquid with a sharp odor; m.p. 12°C, b.p. 238°C, ρ 1.708; soluble in ethanol; insoluble in water.

Present in wastes from the production of some pesticides.

Toxicity: has an overall toxic effect [55,86].

Hexachloroethane (carbon hexachloride; perchloroethane); C_2Cl_6 (236.76)

White crystals with an odor of camphor; m.p. 187°C, b.p. 185.5°C(777 mm Hg), ρ 2.09; soluble in organic solvents; insoluble in water.

Present in wastes from the production of articles made of rubber, some pharmaceuticals, and solvents for pharmaceutical preparations.

Toxicity: has an overall toxic effect: causes pathological changes in the liver; causes irritation of mucous membranes and the skin; at high concentrations has a narcotic effect [55,70].

Hexahydroazepine (hexamethyleneimine); $C_6H_{13}N$ (99.18)

Yellowish liquid with an odor of ammonia; m.p. -37°C, b.p. 138°C, ρ 0.879;

soluble in water and organic solvents.

Present in wastes from the production of some basic organic chemicals and pharmaceuticals.

Toxicity: is adsorbed by undamaged skin; has an overall toxic effect: does damage to the central nervous system and blood; causes irritation of mucous membranes, the skin, and the respiratory tracts. LD_{50} for rats - 22.4 mg/kg [279]; MPSDC and MPADC - 0.1 and 0.02 mg/m^3 respectively; toxicity classification - II [23].

Determination: see [20].

Hexamethylenediamine adipate; $C_{14}H_{30}N_2O_4$ (290.02)

Colorless oily liquid; b.p. 196-197°C; soluble in water (0.5:1).

Present in wastes from the production of polyamides.

Toxicity: has an overall toxic effect. LD_{50} for rats, mice, and rabbits - 5.9, 1.61-3.6, and 5 g/kg respectively [5].

1,6-Hexamethylenediisocyanate; $C_8H_{12}N_2O_2$ (169.06)

Liquid with a sharp odor; m.p. -67°C, b.p. 127°C(1.33 kPa), ρ 1.046; soluble in organic solvents; reacts with water and alcohols.

Present in wastes from the production of textiles, leather, and polyuretanes.

Toxicity: has an overall toxic effect: causes dysfunctioning of the central nervous system and the liver; causes irritation of mucous membranes, the skin and the respiratory tracts. $TC_{irritation}$ - 1 mg/m^3; LD_{50} for rats, mice, and rabbits - 0.913, 0.885, and 0.57 ml/kg respectively [5,46].

Hexamethylenetetramine (urotropine; formin; metheneamine); $C_6H_{12}N_4$ (140.19)

White crystals or powder; b.p. 263°C, ρ 1.270; soluble in water; insoluble in diethyl ether.

Present in wastes from the production of synthetic rubber, articles made of rubber, and some plastics and pharmaceuticals.

Toxicity: causes irritation of the skin and mucous membranes [5,70]; suspected of being carcinogenic [52].

n-Hexane; C_6H_{14} (86.17)

Liquid with a slight odor; m.p. -95°C, b.p. 69.95°C, f.p. -20°C, s.-i.p. 234°C, CLE 1.1-6.7%, ρ 0.659; insoluble in water; soluble in diethyl ether and ethanol.

Present in wastes from the production of petrochemicals, some thermometers, and varnishes and paints; also in wastes of metallurgical works.

Toxicity: causes irritation of the respiratory organs; at high concentrations has a narcotic effect [55,70]. MPSDC - 60 mg/m^3; toxicity classification - IV [20,21].

MPC established in FRG - 300 mg/m^3 [51].

Removal: with the aod of filters made of poly(vinylidene chloride), nylon, polyesters, or teflon [34]; by thermal catalytic (MoO_2) oxidation (efficiency 80-90%) [89]; by adsorption on activated carbon [46].

Hexanol-1 (n-hexyl alcohol); $C_6H_{14}O$ (102.17)

Liquid; m.p. -51.6°C, b.p. 157.47°C, f.p. 62°C, s.-i.p. 310°C, ρ 0.818; soluble in organic solvents; slightly soluble in water.

Present in wastes from the production of some pharmaceuticals.

Toxicity: has an overall toxic and narcotic effect. LD_{50} for rats - 4.9 g/kg [55,71,86].

1-Hexene; C_6H_{12} (84.16)

Liquid; m.p. -139.82°C, b.p. 63.49°C, f.p. -26.12°C, ρ 0.673; soluble in organic solvents; insoluble in water.

Present in wastes from the production of some basic organic chemicals.

Toxicity: has an overall toxic and irritating effect.

Hydrocyanic acid (prussic acid); HCN (27.03)

Liquid with an odor of bitter almond; m.p. -13.3°C, b.p. 25.7°C, f.p. -18°C, s.-i.p. 538°C, CLE 4.9-39%, ρ 0.68; soluble in water and organic solvents.

Present in wastes from the production of some pesticides and plastics.

Toxicity: very toxic; causes headache, nausea, vomiting, respiratory depression, asphyxia, cramps, death; is adsorbed by undamaged skin [55,70]. MPADC - 0.01 mg/m^3 [20,21]; established MPSDC and MPADC in GDR - 0.015 and 0.005 mg/m^3 respectively [44,51].

Removal: by adsorption on activated carbon [46].

Determination: photometric analysis (sensitivity 0.1 μg) [63]; colorimetric analysis (sensitivity 0.05 μg) [20,29,383].

p-Hydroxydiphenylamine (4-anilinophenol); $C_{12}H_{11}NO$ (185.17)

White crystals; m.p. 70°C, b.p. 330°C; soluble in organic solvents; slightly soluble in water.

Present in wastes from the production of synthetic rubber and cellulose esters; also in wastes of petroleum refineries.

Toxicity: is adsorbed by undamaged skin; has an overall toxic and irritating effect. TC_{acute} and $TC_{chronic}$ - 10 mg/ℓ and 4 mg/m^3 respectively [5,55, 284].

8-Hydroxyquinoline (oxine); C_9H_7NO (145.15)

White crystals; m.p. 76°C, b.p. 266.6°C; soluble in most organic solvents; slightly soluble in water and diethyl ether.

Present in wastes from the production of some pesticides, pharmaceuticals, and analytical reagents; also in wastes of metallurgical works.

Toxicity: has an overall toxic effect: causes dysfunctioning of all the sentient organs; is carcinogenic. LD_{50} for guinea pigs - 1.2 g/kg [5,47,55,70].

Indole (benzopyrrole); C_8H_7N (117.15)

White crystals with a sharp odor; m.p. 52-53°C, b.p. 253-254°C, ρ 1.09; soluble in organic solvents and hot water.

Present in wastes from the production of some perfumes and pharmaceuticals.

Toxicity: causes headache and nausea. TC_{odor} and $TC_{irritation}$ - 0.45 and >1 mg/m^3 respectively [5,70].

Isobutyl acetate; $C_6H_{12}O_2$ (116.16)

Liquid with an odor of ether; m.p. 98.9°C, b.p. 116.5°C, f.p. 18-23°C, s.-i.p. 420°C, ρ 0.858; soluble in organic solvents; slightly soluble in water (0.63%).

Present in wastes from the production of some aromatic substances, solvents, plastics, and varnishes and paints.

Toxicity: has an overall toxic effect; causes irritation of mucous membranes [55, 70].

Isobutyl alcohol (2-methylpropanol-1); $C_4H_{10}O$ (74.12)

Liquid with a sharp odor; m.p. -108°C, b.p. 108.5°C, f.p. 28°C, s.-i.p. 390°C, CLE 1.68-7.3%, ρ 0.802; soluble in ethanol, diethyl ether, and water (11.1%).

Present in wastes from the production of petrochemicals, polyethylene, butyl acetate, some alcohols, plastics, and varnishes and paints.

Toxicity: has an overall toxic and irritating effect. SC_{odor}, TC_{odor}, $SC_{chronic}$, $TC_{chronic}$ - 0.24, 0.40, 0.1, and >0.1 mg/m^3 respectively [315]; LD_{50} for rats - 2.46 g/kg [70]; MPSDC and MPADC - 0.1 mg/m^3; toxicity classification - IV [23]; MPC established in FRG - 300 mg/m^3 [51].

Isobutylene (2-methylpropene-1); C_4H_8 (56.10)

Gas; m.p. -140.35°C, b.p. 7.01°C, f.p. -76.1°C, s.-i.p. 465°C, CLE 1.7-9.0%, ρ 0.618; soluble in organic solvents; insoluble in water.

Present in wastes from the production of some antioxidants, foodstuffs, plastics, and isobutylene derivatives.

Toxicity: has a narcotic effect; causes asphyxia [55,70]; ASL - 0.40 mg/m^3 [90].

Removal: by thermal catalytic oxidation (efficiency 100%) [89].

Isobutyl mercaptan (2-methylpropanethiol-1); $C_4H_{10}S$ (90.19)

Liquid with an unpleasant odor; m.p. -144.86°C, b.p. 88.52°C, f.p. -19.5°C, ρ 0.834; soluble in ethanol and diethyl ether; very slightly soluble in water.

Present in wastes from the production of some basic organic chemicals, cellulose, and paper.

Toxicity: has a narcotic effect; causes vomiting [70].

Isolan (1-isopropyl-3-methylpyrozol-5-yl-dimethyl carbamate); $C_{10}H_{17}N_3O_2$ (211.27)

Liquid; b.p. 105°C(40 Pa), ρ 1.07; soluble in water and organic solvents.

Present in wastes from the production of some pesticides.

Toxicity: very toxic; suspected of being carcinogenic. LD_{50} for rats - 54 mg per kg [47,70].

Isooctyl alcohol (a mixture of dimethylhexanols and methylheptanols); $C_8H_{18}O$ (130.23)

Liquid with a sharp odor; b.p. 185-190°C, f.p. 74°C, ρ 0.831; soluble in organic solvents; insoluble in water.

Present in wastes from the production of plasticizers, oil-additives, surface-active agents, and some antibiotics.

Toxicity: $SC_{irritation}$ and $TC_{irritation}$ - 0.15 and 1.5 mg/m^3 respectively [316]; MPSDC and MPADC - 0.15 mg/m^3; toxicity classification - IV [21].

Isopropyl acetate; $C_5H_{10}O_2$ (102.13)

Liquid; m.p. -73.4°C, b.p. 89°C, f.p. 4°C, ρ 0.869; soluble in organic solvents; slightly soluble in water (2.9%).

Present in wastes from the production of plastics, oils and fats, some perfumes and foodstuffs, cellulose, and varnishes and paints.

Toxicity: has an overall toxic effect: causes pathological changes in the liver; causes irritation of mucous membranes; at high concentrations has a narcotic effect [5,55,70].

Removal: by adsorption on activated carbon [46].

Isopropyl alcohol (propanol-2); C_3H_8O (60.09)

Liquid with an odor of alcohol; m.p. -89.5°c, b.p. 82.4°C, f.p. 13°C, lower CLE 2.23%, ρ 0.785; soluble in water and organic solvents.

Present in wastes from the production of some basic organic chemicals, pharmaceuticals, acetone, isopropyl ether, isopropyl acetate, varnishes and paints, petrochemicals, and some plastics.

Toxicity: has an overall toxic effect; causes irritation of mucous membranes and the respiratory organs; at high concentrations has a narcotic effect [317]. LD_{50} for rats - 5.8 g/kg [55,70]; MPSDC and MPADC - 0.6 mg/m^3; toxicity classification - III [20,21]; MPC established in FRG - 300 mg/m^3 [51].

Removal: by adsorption on activated carbon [46].

Determination: colorimetric analysis (sensitivity 2 μg) [29].

Isopropyl amine (2-aminopropane); C_8H_9N (59.11)

Liquid with an odor of ammonia; m.p. -95.2°C, b.p. 31.9°C, f.p. -37°C, s.-i.p. 402°C, lower CLE 2.15%, ρ 0.688; soluble in water and organic solvents.

Present in wastes from the production of surface-active agents, articles made of rubber, some chemicals used in the textile industry, and some pharmaceuticals and pesticides.

Toxicity: causes irritation of mucous membranes, the eyes, and the skin. LD_{50} for rats - 820 mg/kg [55,70]; ASL - 0.01 mg/m^3 [90].

Isopropylbenzene hydroperoxide (cumene hydroperoxide); $C_9H_{12}O_2$ (152.0)

Liquid; b.p. 153°C, f.p. 60°C, ρ 1.061; soluble in organic solvents; slightly soluble in water.

Present in wastes from the production of some basic organic chemicals.

Toxicity: is adsorbed by undamaged skin; has an overall toxic and irritating effect [55]. MPSDC and MPADC - 0.007 mg/m^3; toxicity classification - II [20,21]; MPSDC established in Bulgaria - 0.007 mg/m^3; MPSDC and MPADC established in GDR - 0.007 and 0.02 mg/m^3 respectively [44,51].

Determination: colorimetric analysis (sensitivity 1.5 μg) [29].

Isopropylcyanide (isopropylcarbylamine); C_4H_9N (69.11)

Liquid with an odor of bitter almond; b.p. 107-108°C, ρ 0.793; soluble in organic solvents; slightly soluble in water.

Present in wastes from the production of some organic chemicals.

Toxicity: very toxic; causes dizziness, nausea, vomiting, loss of consciousness, cramps; rats die when exposed for 1 hour at 0.2 mg/ℓ; LD_{50} for rats, mice, and rabbits - 50-100, 5-10, and 14 mg/kg respectively [5].

2-Isopropylthiopropane (diisopropyl sulfide; isopropyl sulfide); $C_6H_{14}S$ (118.23)

Liquid with an odor of ether; m.p. -78.06°C, b.p. 120.04°C, ρ 0.814; soluble in organic solvents; insoluble in water.

Present in wastes from the production of petrochemicals, lubricants, latexes, some medicinal preparations, detergents, and solvents.

Toxicity: has an overall toxic effect: causes dystrophic changes in the cardiac muscle and the liver and kidneys, and loosening of vascular walls. LC_{50} for mice - 33.7 mg/ℓ [5,86].

Determination: gas chromatographic analysis (sensitivity 0.001 mg/m^3) [94].

Lactonitrile (ethylidenecyanohydrin; α-hydroxypropionitrile); C_3H_5NO (71.08)

Liquid; m.p. -40°C, b.p. 182-184°C(d.), ρ 0.987; soluble in water and organic solvents.

Present in wastes from the production of some basic organic chemicals.

Toxicity: very toxic; has an overall toxic and irritating effect. LD_{50} for warm-blooded animals - 1 mg/kg (skin tests) [5,55].

Lead tetraethyl (tetraethyl lead); $C_8H_{20}Pb$ (323.5)

Liquid; b.p. 200°C(d.), ρ 1.652; soluble in organic solvents; insoluble in water.

Present in wastes from the production of gasoline additives and some dyes and organic chemicals; also in wastes of petrochemical plants.

Toxicity: is adsorbed by undamaged skin; is suspected of being carcinogenic [52,55,70].

Removal: see [388].

Determination: colorimetric analysis [64]; gas chromatographic analysis (sensitivity 0.025 μg) [389].

Lead tetramethyl (tetramethyl lead); $C_4H_{12}Pb$ (267.33)

Liquid; m.p. -30°C, b.p. 110°C(≅1.3 kPa), ρ 1.995; soluble in organic solvents; insoluble in water.

Present in wastes from the production of gasoline additives; also in wastes of petroleum refineries.

Toxicity: has an overall toxic and irritating effect. LD_{50} for rats - 80-109 mg per kg [84].

Linuron (N-methyl-N-methoxy-N'-(3,4-dichlorophenyl) urea); $C_9H_{10}Cl_2N_2O_3$ (249.11)

White crystals; m.p. 93-94°C; slightly soluble in water (75 mg/ℓ).

Present in wastes from the production of some pesticides.

Toxicity: has an overall toxic effect: stimulates and then depresses the nerves in the brain and spinal cord; causes respiratory depression, incoordination of voluntary movements, and cramps. $TC_{chronic}$ - 4.38 mg/m^3; LD_{50} for rats and mice - 1500-2170 and 2420 mg/kg respectively [5].

Malathion (carbophos; O,O-dimethyl-S-[1,2-bis(ethoxycarbinol)ethyl] dithiophosphate); $C_{10}H_{19}O_6PS$ (330.4)

Liquid with a peculiar odor; b.p. 156-157°C(≅93 Pa), ρ 1.23; soluble in organic solvents and in water (150 mg/ℓ).

Present in wastes from the production of some pesticides.

Toxicity: has an overall toxic effect: causes ptyalism, vomiting, diarrhea, and dyspnea; has an irritating effect. LD_{50} for rats and mice - 400-1400 mg/kg [67]; MPSDC - 0.015 mg/m^3; toxicity classification - II [20,21].

Determination: see [27].

Malazide (3-hydroxy-6-pyridazinone); $C_4H_4N_2O_2$ (112.09)

White crystals; m.p. 296°C; slightly soluble in water (0.6%).

Present in wastes from the production of herbicides and pyridine.

Toxicity: very toxic; causes dysfunctioning of the central nervous system and the liver; is carcinogenic and mutagenic [37,70,93].

Maleic anhydride (2,5-dihydrofurandione); $C_4H_2O_3$ (98.06)

Colorless crystals; m.p. 52.85°C, b.p. 202°C, f.p. 70°C, s.-i.p. 376°C, CLE 2.2-13%, ρ 1.48; soluble in organic solvents and water.

Present in wastes from the production of some basic organic chemicals, pharmaceuticals, dyes, and articles made of rubber.

Toxicity: is adsorbed by undamaged skin; has an overall toxic effect: causes pathological changes in internal organs and in the composition of the blood; causes irritation of mucous membranes and the respiratory tracts; suspected of being carcinogenic [52]. SC_{odor}, TC_{odor}, $SC_{c.n.s.}$, $TC_{irritation}$, $TC_{chronic}$ - 0.9, 1.3, 0.85, 1, and 0.08 mg/m^3 respectively [329]; MPSDC and MPADC - 0.2 and 0.05 mg/m^3 respectively; MPSDC and MPADC established in Bulgaria, Yugoslavia, and GDR - 0.2 and 0.05 mg/m^3 respectively [51].

Removal: with the aid of scrubbers (efficiency 98%); by thermal catalytic oxidation (efficiency 98%) [330].

Determination: see [20].

Maneb (poly-N,N'-ethylene-bis(dithiocarbamate) manganese); $[C_4H_6MnN_2S_4]_n$ $(265.3)_n$

Yellow powder; d.p. 120°C, ρ 1.92; slightly soluble in water.

Present in wastes from the production of some pesticides.

Toxicity: has an overall toxic effect: causes incoordination of voluntary movements, paralysis; is allergenic and mutagenic; is harmful to gonads and embryos. $TC_{chronic}$ - 2.0-4.7 mg/m^3; LD_{50} for rats and mice - 3.6-7.5 and 2.9 g/kg respectively [5,73].

Melamine (cyanouramide; 2,4,6-triamino-1,3,5-triazine); $C_3H_6N_6$ (126.13)

White crystals; m.p. 354°C(d.), ρ 1.57; slightly soluble in water and organic solvents.

Present in wastes from the production of some polymers, varnishes and paints, electric insulating materials, tanning agents, and ion-exchange resins.

Toxicity: has an overall toxic effect. $TC_{chronic}$ - 0.08-0.1 mg/m^3; LD_{50} for mice - 1 g/kg [5,70,86].

Menazon (S-(4,6-diamino-1,3,5-triazinyl-2-methyl)-O,O-dimethyldithiophosphate); $C_6H_{12}N_5O_2PS_2$ (281.3)

White crystals with an unpleasant odor; m.p. 160-162°C; slightly soluble in water and organic solvents (except for methyl cellosolve).

Present in wastes from the production of some pesticides.

Toxicity: has an overall toxic effect: causes vomiting, respiratory depression, cramps, and disturbance of metabolism; has an irritating effect. LD_{50} for rats - 890 mg/kg [47,70].

2-Mercaptobenzothiazole (vulcafor; captax; 2(3H)-benzothiazolthione); $C_7H_5NS_2$ (167.26)

Yellow crystals; m.p. 179-181°C, s.-i.p. 515-520°C, lower CLE 26 g/m^3; soluble in organic solvents; insoluble in water.

Present in wastes from the production of synthetic rubber and some dyes.

Toxicity: in the case of prolonged exposure causes chronic upper respiratory disorders and conjunctivitis. $TC_{chronic}$ - 0.35-0.4 g/m^3; LD_{50} for rats, mice, and rabbits - 2.3-2.7 g/kg [5].

Mercuriphenyl acetate (phenylmercuric acetate); $C_8H_8HgO_2$ (336.75)

White crystals; m.p. 149-153°C; soluble in organic solvents; slightly soluble in water (4.3 g/ℓ).

Present in wastes from the production of some pesticides.

Toxicity: very toxic; causes nausea, vomiting, impairment of the central and the peripheral nervous system, tremors, carpopedal spasms; also causes irritation of the respiratory tracts, mucous membranes, and the skin [70].

Mercury diethyl (diethyl mercury); $C_4H_{10}Hg$ (258.73)

Liquid with a slight odor; b.p. 159°C, ρ 2.460; soluble in organic solvents; insoluble in water.

Present in wastes of some chemical plants.

Toxicity: very toxic; is adsorbed by undamaged skin; causes dysfunctioning of the central nervous system; has an irritating effect; is mutagenic and allergenic [70, 312].

Mesidine (2,4,6-trimethylaniline); $C_9H_{13}N$ (135.21)

Viscous liquid; b.p. 232-233°C, ρ 0.963; soluble in organic solvents; slightly soluble in water.

Present in wastes from the production of some dyes, pharmaceuticals, and explosives.

Toxicity: is adsorbed by undamaged skin; has an overall toxic effect: causes abnormal gait and tremors, and does damage to the composition of the blood, the retina, the liver and kidneys, and the cardiac muscle; causes irritation of mucous membranes and the eyes. SC_{odor}, $SC_{sensitivity\ to\ light}$, $SC_{irritation}$, and $TC_{irritation}$ (for rabbits) - 0.0095, 0.0055, 0.0030 [331], and 0.01 mg/m^3 respectively; LD_{50} for rats, mice, guinea pigs, and rabbits - 500, 372, 337.5, and 337.5 mg/kg respectivly [5]; MPSDC and MPADC - 0.003 mg/m^3; toxicity classification - II [20, 21]; MPSDC and MPADC established in GDR - 0.01 and 0.003 mg/m^3 respectively [51].

Determination: colorimetric analysis [332].

Mesitylene (1,3,5- trimethylbenzene); C_9H_{12} (120.2)

Colorless liquid with a peculiar odor; m.p. -44.7°C, b.p. 164.7°C, CLE 0.37-11.6%, ρ 0.865; soluble in organic solvents; very slightly soluble in water (0.002%).

Present in wastes from the production of some dyes, solvents, and varnishes and paints; also in wastes of petroleum refineries.

Toxicity: very toxic; causes pathological changes in the composition of the blood; has a narcotic effect [47]; ASL - 0.020 mg/m^3 [18].

Metaldehyde (polymer of acetaldehyde); $[C_2H_4O]_n$

White crystals; m.p. 246°C; soluble in organic solvents; insoluble in water.

Present in wastes from the production of some basic organic chemicals.

Toxicity: is adsorbed by undamaged skin; has an overall toxic and irritating effect. LD_{50} for laboratory animals - 175-290 mg/kg [67]; MPSDC and MPADC - 0.003 mg/m^3; toxicity classification - II [20,21].

Methacrylic acid (2-methylpropenoic acid); $C_4H_6O_2$ (86.09)

White crystals or liquid; m.p. 16°C, b.p. 163°C, ρ 1.015; soluble in organic solvents and water.

Present in wastes from the production of some polymers, glues, synthetic rubber, and Plexiglas.

Toxicity: has an overall toxic and irritating effect [55,70]. $TC_{irritation}$ - 0.44 mg/m^3 [333].

Methacrylonitrile; C_4H_5N (67.09)

Liquid; m.p. -35.8°C, b.p. 30.3°C, f.p. 12.8°C, ρ 0.800; soluble in organic solvents and in water (2.83%).

Present in wastes from the production of synthetic rubber, some polymers, and Plexiglas.

Toxicity: is adsorbed by undamaged skin; has an overall toxic and irritating effect; is mutagenic; does harm to gonads and emryos [73]. TC_{odor} and $TC_{irritation}$ - 0.038-0.06 and 0.0055-0.038 mg/ℓ respectively; LD_{50} for rats and mice - 240 and 15 mg/kg respectively [5].

Methallyl chloride (2-chloro-2-methylpropene-1); C_4H_7Cl (90.55)

Liquid; m.p. <-80°C, b.p. 72.17°C, f.p. 6°C, s.-i.p. 478°C; CLE 2.1-11.6%, ρ 0.926; soluble in organic solvents; slightly soluble in water (0.1%).

Present in wastes from the production of some basic organic chemicals and pesticides.

Toxicity: has an overall toxic effect: causes pathological changes in organs; has an irritating effect and, at high concentrations, a narcotic effect. LD_{50} for laboratory animals - 538-1350 mg/kg [5,47,55,70]; MPSDC and MPADC - 0.02 and 0.01 mg per m^3 respectively [24].

Methane (marsh gas; mine methane); CH_4 (16.04)

Gas; m.p. -182.49°C, b.p. -161.56°C, f.p. -187.8°C, s.-i.p. 537.8°C, CLE 5-15%; soluble in ethanol, diethyl ether, and water.

Present in wastes of coke-tar plants; also in wastes from the production of some basic organic chemicals, acetylene, formaldehyde, ammonia, hydrocyanic acid, and hydrogen [70].

Toxicity: slightly toxic; at high concentrations causes asphyxia [55,70].

Removal: by thermal catalytic oxidation (efficiency 100%) [89].

Determination: see [64,334-341].

Methoxychlor (1,1-bis(4-methoxyphenyl)-2,2,2-trichloroethane); $C_{16}H_{15}Cl_3O_2$ (445.5)

White crystals; m.p. 89°C, ρ 1.02; soluble in organic solvents; practically insoluble in water.

Present in wastes from the production of some pesticides.

Toxicity: has an overall toxic effect: causes pathological changes in internal organs; causes irritation of mucous membranes; is allergenic. LD_{50} for rats - 6 g/kg [47,55,70]; MPSDC established in USA - 10 mg/m^3 [40]; recommended MPSDC in USA - 0.15 mg/m^3 [46].

o-Methoxyphenol (guaiacol); $C_7H_8O_2$ (124.13)

Viscous liquid; m.p. 32°C, b.p. 205°C, f.p. 91°C, s.-i.p. 385°C, ρ 1.128; soluble in organic solvents and in water (1.7%).

Present in wastes from the production of aromatic substances, papaverine, catechol (o-dihydroxybenzene), and some other pharmaceuticals.

Toxicity: has an overall toxic effect: causes vomiting, palpitation, cramps [72,86]. ASL - 0.015 mg/m^3 [18].

Methyl acetate; $C_3H_6O_2$ (74.08)

Liquid with a sharp odor; m.p. -98.5°C, b.p. 57.2°C, f.p. -9.4°C, s.-i.p. 470°C, CLE 3.15-15.6%, ρ 0.939; soluble in organic solvents and in water (31.9%).

Present in wastes from the production of cellulose, celluloid, oils and fats; and artificial leather.

Toxicity: is adsorbed by undamaged skin; has an overall toxic effect: causes irritation of the eyes and respiratory tracts; has a narcotic effect. SC_{odor} and TC_{odor} - 0.4 and 0.5 mg/m^3 respectively; $SC_{sensitivity\ to\ light}$ and $TC_{sensitivity\ to\ light}$ - 0.12 and 0.18 mg/m^3 respectively; $SC_{c.n.s.}$ and $TC_{c.n.s.}$ - 0.07 and 0.08 mg/m^3 respectively; LD_{50} for rats, mice, rabbits, and guinea pigs - 2.9, 2.4, 2.4, and 3.6 g/kg respectively [5,55,70,345]; MPSDC and MPADC - 0.07 mg/m^3; toxicity classification - IV [20,21]; MPSDC and MPADC established in Bulgaria and Yugoslavia - 0.07 mg/m^3, and in GDR - 0.2 and 0.07 mg/m^3 respectively; MPC established in FRG - 150 mg/m^3 [24,51].

Removal: with the aid of filters made of polyethylene, poly(vinylidene chloride), polypropylene, nylon, polyesters, or teflon [34]; by adsorption on activated carbon [46].

Methyl alcohol (methanol; wood alcohol); CH_4O (32.04)

Liquid; m.p. -97.9°C, b.p. 64.5°C, CLE 6.7-36.5%, ρ 0.791; soluble in water and organic solvents.

Present in wastes from the production of some solvents, wood chemicals, methyl esters and ethers, organic and inorganic acids, varnishes and paints, formaldehyde, cellulose sulfate, cinematic film, and some pharmaceuticals, and polymeric materials; also in wastes of metal-working plants and furniture-making plants, and from oil- and fat-processing.

Toxicity: is adsorbed by undamaged skin; has an overall toxic effect: does damage to the central nervous system, the retina, and internal organs; has an irritating and narcotic effect [5,47,55,70]. SC_{odor} and TC_{odor} - 3.7 and 4.1 mg/m^3 respectively; $SC_{sensitivity\ to\ light}$ and $TC_{sensitivity\ to\ light}$ - 2.4 and 3.5 mg/m^3 respectively [355]; $SC_{irritation}$ and $TC_{irritation}$ - 1.01 and 4.5 mg/m^3 respectively [356]; MPSDC and MPADC - 1.0 and 0.5 mg/m^3 respectively [20,21]; MPSDC and MPADC established in Bulgaria, GDR, Hungary, Czechoslovakia, and Yugoslavia - 1.0 and

0.5 mg/m^3 respectively, and in Romania - 3.0 and 1.0 mg/m^3 respectively; MPC established in FRG - 300 mg/m^3 [51].

Removal: with the aid of scrubbers and then by adsorption on activated carbon (efficiency 100%); by thermal catalytic oxidation [357]; with the aid of filters made of nylon, polyesters, polyethylene, poly(vinylidene chloride), teflon, or polypropylene [34]; by adsorption on activated carbon [87].

Determination: colorimetric analysis [29], photometric analysis (sensitivity 1 μg); gas chromatographic analysis (sensitivity 0.03 mg/m^3) [63].

Methyl acrylate; $C_4H_6O_2$ (86.09)

Liquid; m.p. -75°C, b.p. 80.2°C, f.p. -5°C, CLE 2-13%, ρ 0.950; soluble in organic solvents and in water (5.2%).

Present in wastes from the production of some polymers, synthetic rubber, lacquers, Plexiglas, textiles, paper, carton, bandages, and electrotechnical goods.

Toxicity: has an overall toxic effect (at high concentrations causes cramps and death); has an irritating effect: causes lacrimation, ptyalism, dermatitis, pyrodermia. $TC_{chronic}$ - 0.0001-0.02 mg/ℓ; LD_{50} for rabbits - 1.3 mg/kg [5,70,342]; MPSDC - 0.01 mg/m^3; toxicity classification - IV [20,21,23]; MPSDC established in Bulgaria and Yugoslavia - 0.01 mg/m^3; MPSDC and MPADC established in GDR - 0.03 and 0.01 mg/m^3 respectively; MPC established in FRG - 20 mg/m^3 [44,51].

Removal: by adsorption on activated carbon [46]; by combustion [87].

Determination: gas chromatographic analysis (sensitivity 0.003 mg/m^3) [20,343, 344].

Methylal (dimethoxymethane); $C_3H_8O_2$ (76.09)

Liquid; m.p. -104.8°C, b.p. 42.3°C, f.p. -21°C, s.-i.p. 237°C, ρ 0.851; soluble in ethanol, diethyl ether, and in water (32%).

Present in wastes from the production of resins and perfumes.

Toxicity: has an overall toxic effect: causes pathological changes in the liver and kidneys; has an irritating and, at high concentrations, a narcotic effect [55,70]. ASL - 0.15 mg/m^3 [90].

Methylamine (aminomethane); CH_5N (31.06)

Gas; m.p. -92.5°C, b.p. -6.5°C, f.p. -25°C, s.-i.p. 410°C, CLE 4.9-20.7%, ρ 0.662; soluble in water and organic solvents.

Present in wastes from the production of surface-active agents, some pharmaceuticals, solvents, dyes, alkaloids, tanning agents, and pesticides.

Toxicity: has an overall toxic and irritating effect; is premutagenic. TC_{odor}, $TC_{irritation}$, and $TCDC_{acute}$ - 0.0005-0.001, 0.01, and 0.05 mg/ℓ respectively [5,70].

N-Methylaniline (N-methylaminobenzene; N-methylphenylamine); $C_7H_{10}N$ (107.15)

Liquid; m.p. -57°C, b.p. 195.7°C, ρ 0.986; soluble in organic solvents; slightly soluble in water.

Present in wastes from the production of some explosives and dyes.

Toxicity: is adsorbed by undamaged skin; has an overall toxic effect: causes pathological changes in the composition of the blood and bone marrow; causes irritation of the respiratory tracts; is allergenic. $TCDC_{acute}$ - 0.008 mg/ℓ [5,55]; MPSDC and MPADC - 0.04 mg/m^3 [20,21,44].

Determination: see [20].

Methyl bromide (bromomethane); CH_3Br (94.95)

Colorless gas; m.p. -93.7°C, b.p. 3.6°C, CLE 13.5-14.5%, ρ 1.732; soluble in ethanol and diethyl ether; slightly soluble in water.

Present in wastes from the production of some pesticides and fire-extinguishing compositions.

Toxicity: very toxic; is adsorbed by undamaged skin; does damage to the central nervous system, the kidneys, and the lungs; causes chemical burns [67]. LC_{50} for rats - 21 mg/m^3 [47]; MPSDC and MPADC - 0.02 and 0.01 mg/m^3 respectively [24].

2-Methylbutadiene-1,3 (isoprene); C_5H_8 (68.11)

Liquid; m.p. -146°C, b.p. 34.07°C, f.p. -48°C, CLE 1.7-11.5%, ρ 0.881; soluble in organic solvents; insoluble in water.

Present in wastes from the production of synthetic rubber and some perfumes and pharmaceuticals.

Toxicity: causes irritation of mucous membranes and the respiratory tracts; at high concentrations has a narcotic effect. LD_{50} for mice - 144 mg/kg [47,55,70]; ASL - 0.04 mg/m^3 [90].

Removal: by adsorption on activated carbon [46].

2-Methylbutanol-4 (isoamyl alcohol); $C_5H_{12}O$ (88.15)

Liquid with a peculiar odor; m.p. -117.2°C, b.p. 132°C, f.p. 43°C, s.-i.p. 340°C, CLE 1.2-9%, ρ 0.813; soluble in ethanol and diethyl ether; slightly soluble in water (2.6%).

Present in wastes from the production of solvents for fats, resins, alkaloids, pyroxylin, artificial silk, and lacquers.

Toxicity: has an overall toxic effect; causes irritation of mucous membranes [70].

<u>tert</u>-Methylbutyl ketone (pinacolin; 3,3-dimethylbutanone-2); $C_6H_{12}O$ (100.16)

Liquid with an odor of camphor; m.p. -52.5°C, b.p. 106.2°C, CLE 1.27-8.0% [70], ρ 0.820; soluble in organic solvents and in water (2.5%).

Toxicity: has an overall toxic and irritating effect.

Methyl chloride (chloromethane); CH_3Cl (50.49)

Gas; m.p. -96.7°C, b.p. -23.76°C, f.p. 0.0°C, s.-i.p. 632°C, CLE 7.6-19%, ρ 1.0; soluble in organic solvents; slightly soluble in water (0.9%).

Present in wastes from the production of some basic organic chemicals, synthetic rubber, and solvents.

Toxicity: very toxic; causes pathological changes in the liver and kidneys, and impairment of the cardiovascular system; is mutagenic; at high concentrations has a narcotic effect [38,47,55,70]. ASL - 0.06 mg/m^3 [90].

Removal: by adsorption on activated carbon [46].

4-Methyl-3-chloro-2-methylvaleranilide (solan); $C_{13}H_{18}ClNO$ (239.8)

White crystals; m.p. 85-86°C; soluble in organic solvents; slightly soluble in water (8 mg/ℓ).

Present in wastes from the production of some pesticides.

Toxicity: slightly toxic. LD_{50} for rats and mice - 5 and 1.8 mg/kg respectively [67]; MPSDC and MPADC - 0.07 mg/m^3 [24].

2-Methyl-4-chlorophenoxyacetic acid (methoxone); $C_9H_9ClO_3$ (200.6)

White crystals with a slight odor; m.p. 120-120.2°C; soluble in ethanol; slightly soluble in water (1.5 g/ℓ).

Present in wastes from the production of some pesticides.

Toxicity: has an overall toxic and irritating effect; is mutagenic. LD_{50} for rats - 700 mg/kg [5,70,73].

4-Methyldihydropyran (4-methyl-2,3-dihydro-γ-pyran); C_6H_9O (98.15)

Liquid with a sharp odor; b.p. 119.2°C, f.p. 19°C, s.-i.p. 235°C, ρ 0.916; soluble in organic solvents; slightly soluble in water.

Present in wastes from the production of some basic organic chemicals and synthetic rubber.

Toxicity: is adsorbed by undamaged skin; has an overall toxic, irritating, and narcotic effect. TC_{odor} and TC_{acute} - 0.015 and 0.12 mg/ℓ respectively; LD_{50} for mice - 1.95 g/kg [5].

Methyl disulfide (dimethyl disulfide); $C_2H_6S_2$ (94.2)

Liquid with a sharp odor; m.p. -84.7°C, b.p. 110°C, ρ 1.062.

Present in wastes from the production of paper.

Toxicity: causes irritation of the respiratory tracts. MPSDC - 0.7 mg/m^3; toxicity classification - IV [20,21,346].

Methyl formate (methyl methanoate); $C_2H_4O_2$ (60.05)

Liquid; m.p. -99.8°C, b.p. 31.8°C, f.p. -21°C, s.-i.p. 456°C, CLE 5.5-21.8%, ρ 0.97; soluble in ethanol, diethyl ether, and in water (30.4%).

Present in wastes from the production of some basic organic chemicals and pesticides.

Toxicity: has an overall toxic and irritating effect; at a concentration of 1% by volume in the air causes death within 2.5 hours, and of 5% - after 0.5 hour [55,70].

Methyl lactate; $C_4H_8O_3$ (104.10)

Liquid; m.p. -68°C, b.p. 144.8°C, f.p. 49°C, s.-i.p. 385°C, ρ 1.118; soluble in organic solvents and water.

Present in wastes from the production of cellulose esters and some pharmaceuticals.

Toxicity: is adsorbed by undamaged skin; has an overall toxic, irritating, and, at high concentrations, a narcotic effect [70].

Methyl mercaptan (methanethiol); CH_4S (48.10)

Gas with an unpleasant odor; m.p. -122.97°C, b.p. 5.95°C, ρ 0.865; soluble in ethanol, diethyl ether, and in water (23.2 g/ℓ).

Present in wastes from the production of some ethers, pesticides, and basic organic chemicals.

Toxicity: has an overall toxic, irritating, and, at high concentrations, a narcotic effect. TC_{odor} - 0.00002-0.0001 mg/m^3 [5]; MPSDC - $9 \cdot 10^{-6}$ mg/m^3 [20,21].

Removal: by adsorption on activated carbon [348]; by oxidation [68,349].

Determination: chromatographic analysis [350]; chemisorption [351].

Methylmercaptophos (demeton; 7:3 mixture of O,O-dimethyl-O-(2-ethylthioethyl) thiophosphate and O,O-dimethyl-S-(2-ethylthioethyl) thiophosphate); $C_6H_{15}O_3PS_2$ (230.3)

Liquid; m.p. 93-102°C(53-67 Pa); soluble in organic solvents; slightly soluble in water (330 mg/ℓ).

Present in wastes from the production of some pesticides.

Toxicity: very toxic; is adsorbed by undamaged skin; causes vomiting, cramps; has a cumulative effect [70].

Methyl methacrylate; $C_4H_6O_2$ (100.11)

Liquid with an odor of ether; m.p. -48.2°C, b.p. 100.6°C, f.p. 8°C, CLE 1.47-

12.5%, ρ 0.943; soluble in water (1.56%).

Present in wastes from the production of Plexiglas and some other polymers.

Toxicity: has an overall toxic effect; causes irritation of mucous membranes and respiratory tracts; does damage to embryos; is teratogenic [5,55]. SC_{odor} and TC_{odor} - 0.1 and 0.2 g/m^3 respectively; $SC_{sensitivity\ to\ light}$ and $TC_{sensitivity\ to\ light}$ - 0.1 and 0.25 mg/m^3 respectively; $SC_{c.n.s.}$ and $TC_{c.n.s.}$ - 0.1 and 0.15 mg/m^3 respectively [352,354]; MPSDC and MPADC - 0.1 mg/m^3; MPSDC and MPADC established in Bulgaria and Yugoslavia - 0.1 mg/m^3, and in GDR - 0.3 and 0.1 mg/m^3 respectively; MPC established in FRG - 150 mg/m^3 [44,51].

Removal: by adsorption on activated carbon (efficiency 90%); by combustion [87].

Determination: polarographic analysis [27,353].

3-Methyl-1-(γ-methylbutylthiobutane (diisoaminosulfide; isoaminosulfide); $C_{10}H_{22}S$ (174.34)

Liquid with a sharp odor; b.p. 216°C, ρ 0.843; soluble in organic solvents; insoluble in water.

Toxicity: has an overall toxic effect: causes functional disturbances of the nervous system, dystrophic changes in internal organs, and irritation of mucous membranes and the respiratory tracts. TC_{odor} - 0.022 μg/ℓ; LD_{50} for mice - 17 mg/kg [5,70].

2-Methyl-1-(ß-methylpropylthio)propane (isobutyl sulfide; diisobutyl sulfide); $C_8H_{18}S$ (146.30).

Liquid; m.p. -82.2°C, b.p. 176.1°C, ρ 0.838; soluble in organic solvents; insoluble in water.

Present in wastes from the production of petrochemicals, lubricants, latexes, some medicinal preparations, dyes, detergents, and solvents.

Toxicity: very toxic; causes vascular disorders, and dystrophic changes in the cardiac muscle and the liver and kidneys [5,70].

Determination: gas chromatographic analysis (sensitivity 0.001 $\mu g/m^3$) [94].

α-Methylnaphthalene (1-methylnaphthalene); $C_{11}H_{10}$ (142.19)

Liquid; m.p. -30.8°C, b.p. 245°C, ρ 1.016; soluble in organic solvents; insoluble in water.

Present in wastes from the production of petrochemicals, solid fuels, phthalic anhydride, petroleum asphalt, some basic organic chemicals, and pesticides.

Toxicity: has an overall toxic effect [47].

4-Methyl-2-pentanone; $C_6H_{12}O$ (100.16)

Liquid with a slight odor of camphor; m.p. -83.5°C, b.p. 117-118°C, ρ 0.801; soluble in organic solvents and in water (1.8%).

Present in wastes from the production of some polymers, varnishes and paints, solvents, and antibiotics; also in wastes of petroleum refineries.

Toxicity: is adsorbed by undamaged skin; causes irritation of the eyes and the respiratory tracts; at high concentrations has a narcotic effect [55,70]. ASL - 0.10 mg/m^3 [91]; MPC established in FRG - 300 mg/m^3 [51].

Determination: photometric analysis (sensitivity 10 μg) [347].

N-Methylpiperazine; $C_5H_{12}N$ (100.17)

Liquid; m.p. -6.4°C, b.p. 138°C, ρ 0.9; soluble in water.

Present in wastes from the production of some basic organic chemicals.

Toxicity: has an overall toxic effect: causes brain edema, paralysis; also causes irritation of the respiratory tracts and mucous membranes. $TC_{chronic}$ - 0.025-0.035 mg/ℓ; LC_{50} for rats and mice - 1.40 and 1.45 mg/kg respectively [5,55].

2-Methylpropanal (isobutyraldehyde); C_4H_8O (72.10)

Liquid with a sharp odor; m.p. -65.9°C, b.p. 63.5°C, f.p. -22°C, s.-i.p. 176°C, CLE 2.2-2.9%, ρ 0.793; soluble in ethanol, diethyl ether, and in water (10%).

Present in wastes from the production of some basic organic chemicals, textiles, plastics, perfumes, and articles made of rubber.

Toxicity: is adsorbed by undamaged skin; has an overall toxic effect; causes irritation of mucous membranes. LD_{50} for rats - 3.7 g/kg [55,70].

2-Methylpropanethiol-2 (tert-butyl mercaptan); $C_4H_{10}S$ (90.18)

Liquid with a sharp odor; m.p. 1.1°C, b.p. 64.2°C, f.p. -26°C, ρ 0.800; soluble in organic solvents; slightly soluble in water.

Present in wastes from natural gas processing plants and from the production of cellulose and paper.

Toxicity: has an overall toxic and irritating effect [55,70].

Methylpyrrolidone (1-methyl-2-pyrrolidone); C_5H_9NO (99.13)

Liquid; m.p. -24°C, b.p. 202°C, ρ 1.027; soluble in organic solvents; insoluble in water.

Present in wastes from the production of some basic organic chemicals and synthetic fibers.

Toxicity: has an overall toxic effect: causes respiratory and vascular disorders, and impairment of the central nervous system. $TC_{irritation}$ - 0.10-0.15 mg/ℓ; LD_{50} for rats, mice, rabbits, and guinea pigs - 7.9, 5.4, 3.5, and 4.4 g/kg respectively [5,55].

Methyl sulfate (dimethyl sulfate); $C_2H_6O_4S$ (126.13)

Liquid; m.p. -26.8°C, b.p. 188.5°C(d.), f.p. 83°C, ρ 1.351; soluble in ethanol; slightly soluble in water.

Toxicity: very toxic; is adsorbed by undamaged skin; has an overall toxic effect: causes prostration, cramps, paralysis, dysfunctioning of the kidneys, liver, and heart, death; also causes irritation of mucous membranes and the respiratory tracts; is carcinogenic and mutagenic [55,70,79]; in USA recommended MPSDC - 0.01 mg/m^3 [46]; MPC established in FRG - 20 mg/m^3 [51].

Methyl sulfide (dimethyl sulfide); C_2H_6S (62.13)

Liquid with a sharp odor; m.p. -83°C, b.p. 37.5°C, f.p. -17.7°C, s.-i.p. 205°C, CLE 2.2-19.7%, ρ 0.845; soluble in ethanol; insoluble in water.

Present in wastes of petrochemical plants and cellulose and paper mills.

Toxicity: is adsorbed by undamaged skin; has an overall toxic effect. TC_{odor} and $TC_{chronic}$ - 0.00037 and 0.026 mg/ℓ respectively; MPSDC - 0.08 mg/m^3; toxicity classification - IV [20,21]; MPSDC established in Bulgaria, GDR, and Yugoslavia - 0.008 mg/m^3; MPADC established in GDR - 0.03 mg/m^3 [54]; MPC established in FRG - 20 mg/m^3 [51].

Determination: gas chromatographic analysis (sensitivity 0.001 $\mu g/m^3$) [94].

Molinate (ordram; yalan; N-hexamethylen-S-ethylthio carbamate); $C_9H_{17}NOS$ (187.3)

Oily liquid; b.p. 137°C(≅1.3 kPa); soluble in organic solvents and in water (880 mg/ℓ).

Present in wastes from the production of some pesticides.

Toxicity: has an overall toxic effect: causes flaccidity, abnormal gait, dyspnea, and emaciation; does damage to gonados and embryos. $TC_{chronic}$ - 4.6 mg/m^3; LD_{50} for rats and mice - 0.66 and 0.53 g/kg respectively [5,67]; MPSDC and MPADC - 0.01 and 0.006 mg/m^3 respectively [24].

Monolinuron (aresin; N-methyl-N-methoxy-N'-(4-chlorophenyl) urea); $C_9H_{11}ClN_2O_2$ (214.7)

White crystals; m.p. 79-80°C; soluble in organic solvents and in water (580 mg per ℓ).

Present in wastes from the production of some pesticides.

Toxicity: does damage to emryos; is teratogenic [5]. LD_{50} for rats and mice - 2.0-2.5 g/kg [54]; MPSDC and MPADC - 0.005 and 0.001 mg/m^3 respectively [24].

Monuron (N,N-dimethyl-N'-chlorophenyl) urea); $C_9H_{14}ClN_2O$ (198.7)

White crystalline powder; m.p. 176-177°C; soluble in chlorohydrocarbons and in water (230 mg/ℓ).

Present in wastes from the production of some pesticides.

Toxicity: has an overall toxic effect: causes pathological changes in the composition of the blood; is carcinogenic. LD_{50} for rats and guinea pigs - 3.5-3.7 and 1.5-3.7 g/kg respectively [5,47,67,70,73]; MPSDC and MPADC - 0.02 mg/m^3 [20,21].

Morpholine (tetrahydro-1,4-oxazine); C_4H_9NO (87.12)

Liquid; m.p. -4.9°C, b.p. 128°C, ρ 1.00; soluble in water and organic solvents.

Present in wastes from the production of some emulsifiers, solvents, dyes, casein, plastics, and waxes.

Toxicity: is adsorbed by undamaged skin; has an overall toxic effect: causes excitement, dyspnea, and degenerative and necrotic changes in the liver and kidneys; also causes irritation of mucous membranes, the respiratory tracts, and the skin. $TC_{irritation}$ and $TC_{chronic}$ - 0.008 and 0.07 mg/ℓ respectively; LD_{50} for rats - 1.6-1.7 g/kg [5,55,70].

Determination: colorimetric analysis (sensitivity 2 μg) [358].

Muritox (O,O-diethyl-S-(N-carboethoxy-N-methylcarbamylmethyl) dithiophosphate); $C_{10}H_{20}NO_5PS_2$ (329.37)

Yellow oily liquid; m.p. 9°C, b.p. 113°C(≅13 Pa); slightly soluble in organic solvents and in water (0.1%).

Present in wastes from the production of some pesticides.

Toxicity: LD_{50} for rats and mice - 36 and 106 mg/kg respectively [47].

Nabam(dithan C-31; sodium-N,N'-ethylene-bis(dithiocarbamate); $C_4H_6Na_2S_4$ (256.3)

White crystals; m.p. 160°C(d.); soluble in water (≅20%); insoluble in organic solvents.

Present in wastes from the production of some pesticides.

Toxicity: has an overall toxic, irritating, and, at high concentrations, a narcotic effect. LD_{50} for rats and mice - 1.37 and 1.36 g/kg respectively [5,70].

Naphthalene; $C_{10}H_8$ (128.16)

White leaves; m.p. 80.3°C, b.p. 217.9°C, CLE 1.7-8.2%, ρ 1.16; soluble in organic solvents; insoluble in water.

Present in wastes from the production of household chemicals, some dyes, plastics, pesticides, and pharmaceuticals, phthalic acid, celluloid, smokeless gunpowder, and tetralin.

Toxicity: is adsorbed by undamaged skin; has an overall toxic effect: causes vomiting, headache, anemia, cramps, and necrosis of the liver; also causes irritation of the respiratory tracts. MPSDC and MPADC - 0.003 mg/m^3; toxicity classification - IV [20,21]; MPC established in FRG - 150 mg/m^3 [51].

Determination: see [359].

Naphthalene-ß-sulfonic acid (naphthalene-2-sulfonic acid); $C_{10}H_8O_3S$ (208.23)

White leaves; m.p. 90°C; soluble in organic solvents and water.

Present in wastes from the production of some basic organic chemicals.

Toxicity: is adsorbed by undamaged skin; has an overall toxic and irritating effect. LD_{50} for rats and mice - 4.4 and 1.5 g/kg respectively [5].

Naphthaquinone (1,4-naphthaquinone); $C_{10}H_6O_2$ (158.15)

Greenish yellow crystals; m.p. 123-126°C, subl.p. 100°C, ρ 1.42; soluble in organic solvents; slightly soluble in water.

Present in wastes from the production of some dyes, medicinal preparations, basic organic chemicals, and pesticides.

Toxicity: has an overall toxic effect: causes pathological changes in internal organs and the blood composition; also causes irritation of mucous membranes, the skin, and the respiratory tracts. LD_{50} for rats and guinea pigs - 190 and 400 mg/kg respectively [5,55,70]. MPSDC and MPADC - 0.005 mg/m^3; toxicity classification - I; MPSDC and MPADC established in Bulgaria and Yugoslavia - 0.005 mg/m^3, and in GDR - 0.005 and 0.002 mg/m^3 respectively [54].

Removal: by catalytic oxidation [48,360]; with the aid of scrubbers (efficiency 98%) [360].

ß-Naphthol (2-naphthol; 2-hydroxynaphthalene); $C_{10}H_8O$ (144.2)

White crystals; m.p. 122°C, b.p. 295°C, f.p. 153°C, ρ 1.22; soluble in organic solvents; insoluble in water.

Present in wastes from the production of some pharmaceuticals, articles made of rubber, perfumes, and basic organic chemicals.

Toxicity: is adsorbed by undamaged skin; has an overall toxic and irritating effect; at high concentrations causes anemia, vomiting, inflammation of the kidneys, and cramps. LD_{50} for rabbits - 3.8 g/kg [55,70]; MPSDC and MPADC - 0.006 and 0.003 mg/m^3 respectively; toxicity classification - II [23].

Naphthyl-α-amine (1-aminonaphthalene); $C_{10}H_9N$ (143.19)

White leaves; m.p. 50°C, b.p. 301°C, ρ 1.13; soluble in organic solvents; slightly soluble in water.

Present in wastes from the production of azo dyes, 1-naphthol, photographic chemicals, articles made of rubber, and textiles.

Toxicity: is mutagenic and carcinogenic [5,37,38,70,86,93]. ASL - 0.003 mg/m^3 [93].

Removal: with the aid of filters made of nylon, polyesters, or teflon [34].

N-(1-Naphthyl)thiourea; $C_{11}H_{10}N_2S$ (202.3)

White crystals (technical grade - grayish blue); m.p. 198°C; slightly soluble in water (0.06%).

Present in wastes from the production of some pesticides.

Toxicity: has an overall toxic and irritating effect; causes pleurisy and pulmonary edema. LD_{50} for rats - 6-8 mg/kg [5,55,70]; recommended MPSDC - 0.05 [5], 0.03 mg/m^3 [46]; MPSDC established in USA - 0.3 mg/m^3 [40].

Nemagon (fumazone; 1-chloro-2,3-dibromopropane); $C_3H_3Br_2Cl$ (236.36)

Brown liquid with a sharp odor; b.p. 196°C, ρ 2.09; soluble in organic solvents.

Present in wastes from the production of some fumigants and nematocides.

Toxicity: has an irritating and, at high concentrations, a narcotic effect [70]. MPSDC - 0.00003 mg/m^3 [21].

Neozone A (N-phenyl-α-naphthylamine); $C_{16}H_{13}N$ (219.29)

White crystals; m.p. 62°C, b.p. 335°C(≅34.4 kPa), lower CLE 54 g/m^3; soluble in water and organic solvents.

Present in wastes from the production of articles made of rubber, synthetic rubber, and some dyes and plastics.

Toxicity: very toxic; causes methemoglobinemia; has an irritating effect; is allergenic. LD_{50} for mice - 1.8 g/kg [5].

Neozone D (N-phenyl-ß-naphthylamine); $C_{16}H_{13}N$ (219.29)

White crystals; m.p. 108°C, b.p. 399.5°C, ρ 1.2; soluble in chloroform; insoluble in water.

Present in wastes from the production of articles made of rubber and some dyes and polymers.

Toxicity: is adsorbed by undamaged skin; has an overall toxic effect: causes degenerative changes in the liver and kidneys, sclerosis of the vascular walls, ulcers, and pigmentation of the hair, nails, and the skin; also has an irritating effect; is oncogenic and carcinogenic. $TC_{irritation}$ - 0.016 mg/ℓ; LD_{50} for rabbits - 1 g/kg [5,55,86].

Determination: photometric analysis (sensitivity 0.004 mg/ℓ) [361].

L-Nicotine (1-methyl-2-(3-pyridilpyrrolidone); $C_{10}H_{14}N_2$ (162.24)

Oily liquid; b.p. 247.3°C, ρ 1.01; soluble in organic solvents and water.

Present in wastes from the production of tobacco.

Toxicity: very toxic, is adsorbed by undamaged skin; causes dysfunctioning of the nervous system and the cardiovascular system. LD_{50} for rats - 55 mg/kg [5,70]; in USA recommended MPSDC - 0.0005 mg/m^3 [46].

Determination: gas chromatographic and colorimetric analysis [362].

Nicotinic acid (pyridine-3-carboxylic acid; provitamin PP); $C_6H_5NO_2$ (123.11)

White crystals; m.p. 235.5-236.5°C; soluble in water (1.3%)

Present in wastes from the production of some pharmaceuticals.

Toxicity: has an irritating effect: causes sneezing, dermatitis; is allergenic. LD_{50} for rats and mice - ≅7 g/kg [5,70].

Nicotinic sulfate (3-(1-methyl-2-pyrrolidyl)pyridine sulfate); $C_{20}H_{30}N_4O_4S$ (422.56)

Solid; m.p. 80°C, b.p. 247.3°C, ρ 1.01.

Present in wastes from the production of some pesticides.

Toxicity: very toxic; causes headache, dizziness, nausea, vomiting, functional disturbance of the heart; has an irritating effect. LD_{50} for rats and mice - 56.7 and 8.5 mg/kg respectively [67]; MPSDC and MPADC - 0.005 and 0.001 mg/m^3 respectively [24].

2-Nitroaniline (o-nitroaniline); $C_6H_6N_2O$ (138.12)

Yellow crystals; m.p. 71.5°C, b.p. 270°C(d.), s.-i.p. 505°C, lower CLE 39.6 g/m^3, ρ 1.44; slightly soluble in hot water.

Present in wastes from the production of some plastics and textile dyes.

Toxicity: is adsorbed by undamaged skin; has an overall toxic effect: causes incoordination of voluntary movements, tremors, cramps; methemoglobinemia, cyanosis; dysfunctioning of the liver and the cardiovascular and the central nervous system [5,55,70]. LD_{50} for rats - 3.52 g/kg; ASL - 0.006 mg/m^3 [18].

3-Nitroaniline (azoamine Orange K); $C_6H_6N_2O$ (138.12)

Yellow crystals; m.p. 114°C, b.p. 305.7°C(d.), ρ 1.43; soluble in organic solvents; slightly soluble in water.

Present in wastes from the production of some plastics and textile dyes.

Toxicity: is adsorbed by undamaged skin; has an overall toxic effect: causes incoordination of voluntary movements, tremors, cramps; methemoglobinemia, cyanosis; dysfunctioning of the liver, the cardiovascular and the central nervous system [5,55, 70]. ASL - 0.010 mg/m^3 [18].

4-Nitroaniline (azoamine Red Y); $C_6H_6N_2O$ (138.12)

Yellow crystals; m.p. 148°C, b.p. 366°C(d.), s.-i.p 414°C, lower CLE 28.8 g/m^3; soluble in organic solvents.

Present in wastes from the production of some plastics and textile dyes.

Toxicity: is adsorbed by undamaged skin; has an overall toxic effect: causes incoordination of voluntary movements, tremors, cramps, methemoglobinemia, cyanosis; dysfunctioning of the liver, the cardiovascular and the central nervous system [5,55,70]. LD_{50} for rats - 1.41 g/kg; ASL - 0.0006 mg/m^3 [18].

Nitrobenzene (oil of mirbane); $C_6H_5NO_2$ (123.11)

Greenish yellow liquid; m.p. 5.8°C, b.p. 210.8°C, f.p. 83°C, s.-i.p. 482°C, lower CLE 1.84%, ρ 1.208; soluble in organic solvents; slightly soluble in water (0.19%).

Present in wastes from the production of some basic organic chemicals, dyes, and perfumes, soaps, and aniline.

Toxicity: very toxic; is adsorbed by undamaged skin; causes headache, dizziness, nausea, vomiting, methemoglobinemia, cyanosis; also has an irritating effect. SC_{odor} and TC_{odor} - 0.008 and 0.0182 mg/m^3 respectively; $SC_{sensitvity\ to\ light}$ and $TC_{sensitivity\ to\ light}$ - 0.08 and 0.0157 mg/m^3 respectively; $SC_{c.n.s.}$ and $TC_{c.n.s.}$ - 0.008 and 0.0129 mg/m^3 respectively; $SC_{chronic}$ and $TC_{chronic}$ - 0.008 and 0.08 mg/m^3 respectively [363]; MPSDC and MPADC - 0.008 mg/m^3; toxicity classification - II [20,21]; MPSDC and MPADC established in Hungary, Yugoslavia, and Bulgaria - 0.008 mg/m^3 [44], and in GDR - 0.010 and 0.005 mg/m^3 respectively [54]; MPC established in FRG - 20 mg/m^3 [51].

Removal: with the aid of filters made of polypropylene, nylon, polyacrylates, or teflon [34].

Determination: nitration (sensitivity 1 μg) [27].

Nitrochloroform (chloropicrin; trichloronitromethane); CCl_3NO_2 (164.37)

Liquid with a sharp odor; m.p. -64°C, b.p. 112.3°C, ρ 1.65; soluble in organic solvents; insoluble in water. Present in wastes from the production of some pesticides and dyes.

Toxicity: has an overall toxic effect: causes pathological changes in the heart, liver and kidneys, small cerebral hemorrhage, pulmonary edema; also has an irritating effect: causes lacrimation and ptyalism. TC_{odor} - 0.6 mg/m^3 [5,55,67,70]; MPSDC and MPADC - 0.01 and 0.007 mg/m^3 respectively [24]; in USA recommended MPSDC - 0.005 mg/m^3 [46].

Removal: by adsorption on activated carbon [46].

Nitrocyclohexane; $C_6H_{11}NO_2$ (129.16)

Liquid; m.p. -34°C, b.p. 205.5°C(d.), ρ 1.10; soluble in organic solvents and in water (2%).

Present in wastes from the production of some basic organic chemicals.

Toxicity: causes depression, respiratory disorders, incoordination of voluntary movements, dysfunctioning of the central nervous system and the liver; cramps. $TC_{chronic}$ - 0.5 mg/m^3; LD_{50} for mice - 54.2 mg/kg [5,55].

Nitroethane; $C_2H_5NO_2$ (75.07)

Liquid with an unpleasant odor; m.p. -89.5°C, b.p. 114.1°C, lower CLE 4%, ρ 1.050; soluble in organic solvents and in water (4.5%).

Present in wastes from the production of some solvents and basic organic chemicals.

Toxicity: has an overall toxic effect: causes dysfunctioning of the liver and kidneys; also has an irritating effect. LD_{50} for rats and mice - 1.1 and 0.86 g/kg [5,55,70].

Determination: colorimetric analysis (sensitivity 0.1 mg/m^3) [364].

Nitroglycerin (glyceryl ether trinitrate); $C_3H_5N_3O_9$ (227.09)

Colorless liquid; m.p. 13.3°C, b.p. 160°C(15 mm Hg), ρ 1.69; soluble in acetone; slightly soluble in water.

Present in wastes from the production of some explosives and pharmaceutical preparations.

Toxicity: has an overall toxic effect: causes a slowing-down of the pulse, cyanosis, nausea, vomiting; also has an irritating effect; is adsorbed by undamaged skin [5,70].

Nitromethane; CH_3NO_2 (61.04)

Liquid; m.p. -28.5°C, b.p. 101.2°C, lower CLE 7.3%, ρ 1.138; soluble in organic solvents (except for alkanes); soluble in water (9.1%).

Present in wastes from the production of some plastics and esters, cellulose, chloropicrin, lubricants, and diesel fuel.

Toxicity: has an overall toxic effect: causes nausea, vomiting, cramps, methemoglobinemia; also irritation of the respiratory tracts; at high concentrations has a narcotic effect. LD_{50} for rats and mice - 940 and 950 mg/m^3 respectively [5,55,70].

Removal; by adsorption on activated carbon [46].

Determination: coulometric analysis (sensitivity 0.1 mg/m^3) [364].

Nitrophen (4-nitro-2',4'-dichlorodiphenyl ether); $C_{12}H_7Cl_2NO_3$ (284.1)

Light yellow crystals; m.p. 70-71°C; soluble in water.

Present in wastes from the production of some pesticides.

Toxicity: causes irritation of mucous membranes, the eyes, and the skin. MPSDC and MPADC - 0.02 and 0.01 mg/m^3 respectively [24].

Nitropropane (2-nitropropane; ß-nitropropane); $C_3H_7NO_2$ (89.09)

Liquid with an unpleasant odor; m.p. -91.3°C, b.p. 120.2°C, lower CLE 2.6%, ρ 0.988; soluble in organic solvents; slightly soluble in water.

Present in wastes from the production of some solvents, cellulose esters, nitro alcohols, and textiles.

Toxicity: has an overall toxic effect: causes vomiting, and does damage to the liver and kidneys; also causes irritation of the respiratory tracts. $TC_{irritation}$ - 0.7-0.16 mg/m^3 [5,55,70].

Removal: by adsorption on activated carbon [46].

Determination: coulometric analysis (sensitivity 0.1 mg/m^3) [364].

N-Nitrosodimethylamine (dimethylnitrosoamine); $C_2H_6N_2O$ (74.06)

Yellow liquid; b.p. 152-153°C, ρ 0.005; soluble in water and organic solvents.

Present in wastes from the production of some solvents.

Toxicity: has an overall toxic effect: causes pathological changes in the liver, death; is carcinogenic and mutagenic [5,37,38,70].

n-Nonyl alcohol (nonaol-1); $C_9H_{10}O$ (144.25)

Liquid with an odor of flowers; m.p. -5°C, b.p. 213.47°C, f.p. 96°C, ρ 0.827; soluble in organic solvents; insoluble in water.

Present in wastes from the production of surface-active agents, plasticizers, and lubricant additives.

Toxicity: has an overall toxic effect: causes dysfunctioning of the central nervous system and the liver; also has a narcotic effect [70,86].

Octamethyl (octamethyltetramide of pyrophosphoric acid); $C_8H_{24}N_4O_3P_2$ (286.3)

Liquid; b.p. 126°C(≅130 Pa); soluble in water.

Present in wastes from the production of some insecticides and pharmaceuticals.

Toxicity: very toxic; causes respiratory disorders, nausea, vomiting, ptyalism, cramps, inhibition of the functioning of enzymes; has a cumulative effect. LD_{50} for rats - 25 mg/kg [55,70].

Oleic acid; $C_{18}H_{34}O_2$ (282.45)

Oily liquid; m.p. 14°C, b.p. 286°C, ρ 0.895; soluble in organic solvents; insoluble in water.

Present in wastes from the production of soaps, textiles, articles made of rubber, cellulose sulfate, flotation agents, plasticizers, and lubricants.

Toxicity: has an irritating effect [5].

Removal: with the aid of filters made of nylon, polyacrylates or other polyesters, poly(vinylidene chloride), or teflon [34,55,70,86].

Ovex (ovotran; O-(4-chlorophenyl)-4'-chlorobenzosulfonate); $C_{12}H_8Cl_2O_3$ (303.2)

White crystals; m.p. 86.5°C; soluble in organic solvents; insoluble in water.

Present in wastes from the production of acaricides (a substance or preparation that kills mites).

Toxicity: has an over all toxic effect: causes ptyalism, cramps, loss of weight; at high concentrations - impairment of blood formation and dysfunctioning of the liver. $TC_{chronic}$ - 6-15 mg/m^3 [5,67]; MPSDC and MPADC - 0.02 mg/m^3 [24].

Oxalic acid (ethandioic acid); $C_2H_2O_4$ (90.06)

White crystals; m.p. 189.5°C, forms dihydrate ($C_2H_4O_4 \cdot 2H_2O$; m.p. 101.5°C); soluble in water, ethanol, and diethyl ether.

Present in wastes from the production of some organic chemicals; also in wastes of wood-working, metal-working, and leather-processing plants.

Toxicity: has an overall toxic effect: causes asthenia, nose bleeding, coughing, headache, dysfunctioning of the heart, cramps; also causes irritation of mucous membrane of the nose and skin ulceration.

Removal: with the aid of filters made of polypropylene, polyethylene, poly(vinylidene chloride), or polyesters [5,34,55,70].

Oxamide (oxalic acid amide); $C_2H_4N_2O_2$ (88.06)

White crystals; m.p. 419°C(d.); soluble in sulfuric acid; insoluble in water or diethyl ether.

Present in wastes from the production of cellulose nitrate.

Toxicity: has an overall toxic effect. LD_{50} for mice - 1220 mg/kg [70].

Paraldehyde (2,4,6-trimethyl-1,3,5-trioxane; par-acetaldehyde); $C_6H_{12}O_3$ (132.16)

Liquid with a peculiar odor; m.p. 12.6°C, b.p. 124.4°C, f.p. 35.5°C, s.-i.p. 237.7°C, ρ 0.994; soluble in organic solvents and in water (10.7%).

Present in wastes from the production of some organic compounds and pharmaceuticals.

Toxicity: has an overall toxic and irritating effect. LD_{50} for dogs - 3.5 g/kg [55,70].

Removal: with the aid of filters made of nylon, polyacrylates or other polyesters, polyethylene, poly(vinylidene chloride), teflon, or polypropylene [34].

Paraquat (1,1-dimethyl-4,4'-dipyrrine dimethylsulfate); $C_{12}H_{14}N_2$ (186.26)

White crystals; m.p. 300°C(d.); soluble in water.

Present in wastes from the production of some pesticides.

Toxicity: very toxic. LD_{50} for rats - 150 mg/kg [5].

Paraquatdichloride (1,1-dimethyl-4,4'-dipyridine dichloride); $C_2H_{14}N_2Cl_2$ (257.12)

White crystals; m.p. ≅300°C; soluble in water.

Present in wastes from the production of some pesticides.

Toxicity: very toxic; is adsorbed by undamaged skin. LD_{50} for rats and rabbits - 141 and 240 mg/kg respectively [47]; MPSDC established in USA - 0.5 mg/m^3 [47].

Parathion (metacide; folidol-80; O,O-dimethyl-O-(4-nitrophenyl) thiophosphate); $C_8H_{10}NO_5PS$ (263.2)

White crystals; m.p. 35-36°C, b.p. 158°C, ρ 1.35; soluble in ethanol and diethyl ether; slightly soluble in water (50 mg/ℓ).

Present in wastes from the production of some pesticides.

Toxicity: very toxic; has an overall toxic effect. $LD_{chronic}$ for cats, rats, mice, and rabbits - 0.5, 15-35, 15-35, and 100 mg/kg respectively [366]; MPSDC - 0.008 mg/m^3; toxicity classification - I [20,23].

Pelargonic acid (nonanoic acid); $C_9H_{18}O_2$ (158.23)

Oily liquid; m.p. 12.5°C, b.p. 253-255.6°C, ρ 0.907; soluble in organic solvents; slightly soluble in water (0.026%).

Present in wastes from the production of varnishes and paints.

Toxicity: has an irritating effect. TC_{odor} - 0.0055 mg/m^3 [55,70,86].

Pentachloroethane (pentalin); C_2HCl_5 (202.31)

Liquid; m.p. -29°C, b.p. 162°C, ρ 1.68; soluble in ethanol and diethyl ether; insoluble in water.

Present in wastes from the production of some organic chemicals and solvents.

Toxicity: has an overall toxic, irritating, and narcotic effect [55,70].

Pentachloronitrobenzene; $C_6Cl_5NO_2$ (295.36)

Yellow crystals; m.p. 146°C, b.p. 325°C(d.), ρ 1.72; soluble in organic solvents; insoluble in water.

Present in wastes from the production of some pesticides.

Toxicity: has an overall toxic effect: does damage to the vascular system, the lungs, liver, kidneys, heart, and the central nervous system; is carcinogenic and teratogenic. LD_{50} for warm-blooded animals - 1.5-1.7 g/kg [5,33,55,67,70]; MPSDC and MPADC - 0.01 and 0.006 mg/m^3 respectively [24].

2,3,4,5,6-Pentachlorophenol; C_6HCl_5O (266.35)

White crystals with an oder of phenol; m.p. 191°C, b.p. 310°C(d.), ρ 1.978; soluble in organic solvents; very slightly soluble in water.

Present in wastes from the production of some pesticides; also in wastes of wood-working plants.

Toxicity: very toxic; has an overall toxic and irritating effect. LC_{50} for rats and mice - 355 and 225 mg/kg respectively [67,70]; MPSDC and MPADC - 0.005 and 0.001 mg/m^3 respectively [24].

Determination: gas chromatographic analysis (sensitivity 0.51-0.61 μg) [369].

Pentane; C_5H_{12} (72.15)

Liquid; m.p. -129.72°C, b.p. 36.07°C, f.p. ≅-40°C, s.-i.p. 287°C, ρ 0.626; soluble in hydrocarbon solvents; insoluble in water.

Present in wastes from the production of some solvents; also in wastes of oil refineries.

Toxicity: has an irritating and, at high concentrations, a narcotic effect [5]. TC_{odor}, $TC_{c.n.s.}$, and $TC_{chronic}$ - 217, 130, and 116 mg/m^3 respectively [368]; MPSDC and MPADC - 100 and 25 mg/m^3 respectively; toxicity classification - IV [20, 21]; MPSDC and MPADC established in Bulgaria, GDR, and Yugoslavia - 100 and 25 mg/m^3 respectively [44]; MPC established in FRG - 300 mg/m^3 [51].

Removal: by thermal catalytic oxidation [89].

Phenanthrene; $C_{14}H_{10}$ (178.24)

White powder; m.p. 110°C, b.p. 340.1°C, ρ 0.98; soluble in organic solvents; insoluble in water.

Present in wastes from the production of stabilizers for explosive substances and smoke-forming compositions.

Toxicity: causes photosensitization of the skin; suspected of being carcinogenic [55,70,86].

Phenetidines (mixture of o-, m-, and para-phenetidine); $C_8H_{11}NO$ (137.18)

Liquid which darkens when exposed to the air; soluble in ethanol and diethyl ether; insoluble in water.

Present in wastes from the production of some dyes, pharmaceuticals, and foodstuffs.

Toxicity: has an overall toxic effect: causes general debility, headache, dizziness, paleness, cyanosis of integumentary tissues, dyspnea, palpitation, nausea, vomiting, anemia, methemoglobinemia, dysfunctioning of the liver and kidneys; also has an irritating effect; is carcinogenic (in the case of o-phenetidine). For p-phenetidine: $TC_{chronic}$ - 1 mg/ℓ; LD_{50} for rats and mice - 0.58 and 0.54 mg/kg respectively [5,55,70].

Phenkaton (S-(2,5-dichlorophenylthiomethyl)-O,O-diethyl dithiophsophate) $C_{11}H_{15}Cl_2$-

O_2PS_2(377.3)

Oily liquid; b.p. 120(≅0.13 Pa); soluble in organic solvents; slightly soluble in water.

Present in wastes from the production of some pesticides.

Toxicity: very toxic; causes nausea, vomiting, ptyalism, respiratory disorders, cramps. LD_{50} for rats and mice - 182 and 293 mg/kg respectively [47]; MPSDC and MPADC - 0.006 and 0.002 mg/m^3 respectively [24].

Phenol (carbolic acid); C_6H_5OH (94.11)

White crystals with a peculiar odor; m.p. 43°C, b.p. 182°C, f.p. 75°C, s.-i.p. 595°C, ρ 1.072; soluble in water and organic solvents.

Present in wastes from the production of some organic chemicals, petrochemicals, wood chemicals, coal-tar chemicals, pharmaceuticals, plastics, phenol, linoleum, asphaltic roofing paper, roof-sheeting materials, artificial parchment paper, foamed plastic and slag-cotton plates; also in wastes from crosstie-impregnation processes and of metallurgical plants.

Toxicity: is adsorbed by undamaged skin; has an overall toxic effect: causes nausea, vomiting, hurried breathing, cramps; and dysfunctioning of the central nervous system, the liver, kidneys, pancreas, and spleen; also causes severe irritation of mucous membranes and the respiratory tracts; is carcinogenic [47,51,55,70]. SC_{odor} and TC_{odor} - 0.0137 and 0.022 mg/m^3 respectively; $SC_{sensitivity\ to\ light}$ and $TC_{sensitivity\ to\ light}$ - 0.0137 and 0.0155 mg/m^3 respectively; $SC_{c.n.s.}$ and $TC_{c.n.s.}$ - 0.0137 and 0.0156 mg/m^3 respectively; $SC_{chronic}$ and $TC_{chronic}$ - 0.01 and 0.1 mg/m^3 respectively [412]. MPSDC and MPADC - 0.01 mg/m^3; toxicity classification - II [20,21]. Maximum permissible concentration (mg/m^3) established in different countries [44,51]:

Country	MPSDC	MPADC
Bulgaria	0.01	0.01
Hungary	0.01	0.01
GDR	0.03	0.01
Poland	0.02	0.01
Romania	0.75	0.25
Czechoslovakia	0.3	0.1
Yugoslavia	0.01	0.01
Italy	0.75	0.38
Finland	0.72	0.25
France	-	1.0
USA	0.08	0.5
FRG	0.75	0.5
Switzerland	0.75	0.1
Japan	0.26	-

Removal: by adsorption on activated carbon [46]; with the aid of filters made of polyethylene, poly(vinylidene chloride), teflon, or polypropylene [34]; by catalytic oxidation [413].

Determination: colorimetric analysis (sensitivity 1 μg) [29], thin-layer chromatography [414]; polarography (sensitivity 2 μg/ml) [63]; spectrophotometric analysis (sensitivity 0.004 $\mu g/m^3$) [63]; volumetric analysis [415]; and automatic methods [416].

Phenurone (N,N-dimethyl-N'-phenylurea); $C_9H_{12}N_2O$ (164.2)

White crystals; m.p. 136°C; soluble in organic solvents and in water (3850 mg per ℓ).

Present in wastes from the production of some pesticides.

Toxicity: slightly toxic. LD_{50} for rats and guinea pigs - 7.5 and 3.2 g/kg respectively [5,67,70].

N-Phenylaniline (diphenylamine; phenylamine); $C_{12}H_{11}N$ (169.22)

White crystals with an odor of flowers; m.p. 54°C, b.p. 302°C, ρ 1.16; soluble in organic solvents; insoluble in water.

Present in wastes from the production of some plastics, dyes, explosives, and chemical reagents.

Toxicity: is adsorbed by undamaged skin; has an overall toxic effect: is harmful to the cardiovascular and the central nervous system; causes irritation of mucous membranes and the skin. LD_{50} for rats and mice - 11.5 and 2.9 g/kg respectively [5,47,55,70].

Removal: with the aid of filters made of poly(vinylidene chloride), nylon, polyesters, or teflon [34].

p-Phenylaniline (4-aminodiphenyl); $C_{12}H_{11}N$ (169.23)

White crystals; m.p. 53°C, b.p. 302°C, ρ 1.160; soluble in diethyl ether; slightly soluble in water.

Present in wastes from the production of some basic organic chemicals.

Toxicity: is adsorbed by undamaged skin; causes irritation of the skin and mucous membranes; has an overall toxic effect; is carcinogenic and mutagenic [45,55,70,86,93].

1-Phenylbutane (n-butylbenzene); $C_{10}H_{14}$ (134.21)

Colorless liquid; m.p. -87.97°C, b.p. 183.27°C, ρ 0.860; soluble in organic solvents; insoluble in water.

Present in wastes from the production of petrochemicals, some basic organic chemicals, pesticides, plastics, surface-active agents, and asphalt.

Toxicity: is adsorbed by undamaged skin; causes irritation of the skin, dermatitis; has an overall toxic and narcotic effect [47,55,70].

Phenyl isocyanide (phenyl carbylamine); C_7H_5N (103.12)

Oily liquid with an odor of bitter almond; m.p. -12.9°C, b.p. 190.7°C, ρ 1.005; soluble in ethanol and diethyl ether; slightly soluble in water (0.2%).

Present in wastes from the production of synthetic rubber and some plastics.

Toxicity: very toxic; is adsorbed by undamaged skin; causes respiratory disorders, vomiting, loss of consciousness; has an irritating effect [55,71]. LD_{50} for rats - 800 mg/kg.

Phenylene diamines (mixture of all three isomers); $C_6H_8N_2$ (108.14)

Crystals which darken when exposed to the air; soluble in organic solvents and water.

Present in wastes from the production of some dyes, polymers, petrochemicals, and photohraphic chemicals, articles made of rubber, textiles, formaldehyde, and uretanes.

Toxicity: has an overall toxic and irritating effect; is allergenic [5,55,70,86].

Phenylhydrazine; $C_6H_8N_2$ (108.14)

Yellow liquid; m.p. 19.6°C, b.p. 243°C, ρ 1.09; soluble in organic solvents and in water (11.6%).

Present in wastes from the production of some dyes and pharmaceuticals.

Toxic: very toxic; causes dissolution of erythrocytes and leucocytes; methemoglobinemia, anemia; has an irritating effect; is allergenic. LD_{50} for rats, mice, and guinea pigs - 188, 175, and 80 mg/kg respectively [5,55,70,86].

Determination: see [411].

N-Phenylhydroxylamine; C_6H_7NO (109.13)

White crystals; m.p. 81-82°C; soluble in organic solvents and in water (2% - in cold H_2O; 10% - in hot H_2O).

Present in wastes from the production of some organic chemicals.

Toxicity: very toxic; is adsorbed bu undamaged skin; causes methemoglobinemia paralysis; has an irritating effect; is very allergenic [5,70].

N-Phenyl-α-naphthylamine; $C_{16}H_{13}N$ (219.29)

Crystals; m.p. 62°C; soluble in organic solvents and water.

Present in wastes from the production of articles made of rubber, polyethylene, motor fuels, olefins, and some dyes.

Toxicity: has an overall toxic effect; causes methemoglobinemia and severe irritation of the skin; is allergenic. LD_{50} for mice - 1.8 g/kg [5].

N-Phenyl-ß-naphthylamine; $C_{16}H_{13}N$ (219.29)

Crystals; m.p. 108°C; soluble in organic solvents; insoluble in water.

Present in wastes from the production of articles made of rubber, polyethylene, motor fuels, olefins, and some dyes.

Toxicity: has an overall toxic effect: causes degenerative changes in the liver and kidneys, and vascular sclerosis; also has an irritating effect [5,47,55].

o-Phenylphenol (2-hydroxydiphenyl; dowcide-I); $C_{12}H_{10}O$ (170.1)

White crystals; m.p. 57°C, b.p. 228°C; soluble in aromatic hydrocarbons; slightly soluble in water (0.7g/ℓ).

Present in wastes from the production of some pesticides and articles made of rubber.

Toxicity: very toxic: causes nausea, vomiting, cramps, collapse, dysfunctioning of the liver and kidneys; also has an irritating effect. LD_{50} for rats - 2.7 g per kg [55,70].

Phosalone (S-(6-chlorobenzoxasolinon-2-yl-3-methyl)-O,O-diethyldithiophosphate); $C_{12}H_{15}ClNO_4PS_2$ (367.8)

White crystals with an odor of garlic; m.p. 47.5-48°C; soluble in organic solvents; slightly soluble in water (10mg/ℓ).

Present in wastes from the production of some pesticides.

Toxicity: very toxic; causes severe irritation of mucous membranes and the respiratory tracts [67]. SC_{odor}, TC_{odor}, and $SC_{chronic}$ - 0.037, 0.049, and 0.012 mg/m^3 respectively [417]; LD_{50} for rats and mice - 84-108 mg/kg [70]; MPSDC and MPADC - 0.01 and 0.006 mg/m^3 respectively [24].

Phosdrin (mevinphos; O,O-dimethyl-O-(1-carbomethoxy-1-propen-2-yl) phosphate); $C_7H_3O_6P$ (224.16) .

Yellow liquid; b.p. 106-107°C; soluble in organic solvents and water.

Present in wastes from the production of some pesticides.

Toxicity: very toxic; causes nausea, vomiting, ptyalism, inhibition of the functioning of enzymes, cramps; has a cumulative effect. LD_{50} for rats and mice - 4 and 9 mg/kg respectively [55,70].

Phosgene (carbonyl chloride); $COCl_2$ (98.92)

Gas; m.p. -118°C, b.p. 8.2°C; soluble in organic solvents; slightly soluble in water.

Present in wastes from the production of some solvents, dyes, pharmaceuticals, and plastics.

Toxicity: extremely toxic; causes pulmonary edema, death; also causes irritation of the respiratory tracts and mucous membranes [55, 70,86]. ASL- 0.003 mg per m^3 [91].

p-Phthalic acid (terephthalic acid); $C_8H_6O_4$ (166.13)

White powder; m.p. 425°C (in sealed capillary); subl. p. >300°C; slightly soluble in hot ethanol; very slightly soluble in water; soluble in pyridine.

Present in wastes from the production of some polyesters, plastics, and analytical reagents.

Toxicity: slightly toxic. LD_{50} for rats - >6.4 g/kg [70,86].

Determination: see [387].

o-Phthalic anhydride; $C_8H_4O_3$ (148.11)

White crystals with an odor of bitter almond; m.p. 130.8°C, b.p. 284.5°C, f. p. 75-140°C, s.-i.p. 595°C; soluble in ethanol and in water (0.6%).

Present in wastes from the production of some plasticizers, dyes and plastics; benzoic acid, phthalates, and phthaleins.

Toxicity: has an overall toxic and irritating effect. SC_{odor} and TC_{odor} - 0.32 mg/m^3; $SC_{sensitivity\ to\ light}$ and $TC_{sensitivity\ to\ light}$ - 0.55 and 0.92 mg/m^3 respectively; $SC_{irritation}$ and $TC_{irritation}$ - 0.2 and 0.54 mg/m^3 respectively [55,425]. MPSDC and MPADC - 0.1 mg/m^3; toxicity classification - II [20,21].

Removal: with the aid of scrubbers (efficiency 98%); by thermal catalytic oxidation (efficiency 98-100%) [89,426].

Determination: polarographic analysis (sensitivity 1 μg/ml) [63].

Phthalophos (imidan; O,O-dimethyl-S-(N-phthalimidomethyl) dithiophosphate); $C_{11}H_{12}NO_4PS_2$ (317.3)

White crystals; m.p. 72-72.5°C; soluble in organic solvents; very slightly soluble in water (0.0025%).

Present in wastes from the production of some pesticides.

Toxicity: very toxic; causes inhibition of the functioning of enzymes; is harmful to embryos; is mutagenic. LD_{50} for laboratory animals - 37-210 mg/kg [67,374]; MPSDC and MPADC - 0.009 and 0.004 mg/m^3 respectively [24].

Determination: polarographic analysis (sensitivity 1 μg) [63,424].

Picene (1,2,7,8-dibenzophenanthrene); $C_{22}H_{14}$ (278.33)

Colorless leaves; m.p. 364°C, b.p. 520°C; slightly soluble in organic solvents; insoluble in water.

Present in wastes from the production of some plastics and petrochemicals; also in wastes of coal-tar chemical plants.

Toxicity: suspected of being carcinogenic [41,70].

Picloram (tordon; 4-amino-3,5,6-trichloropyridine-2 carbonic acid); $C_6H_3Cl_3N_2O_2$ (241.5)

White crystals with an odor of chlorine; d.p. >190°C; slightly soluble in water (0.43 g/ℓ).

Present in wastes from the production of some herbicides.

Toxicity: causes lacrimation and coughing. LD_{50} for rats, mice, and rabbits - 3.75, 1.5, and 2.0 g/kg respectively [5,67,70]. MPSDC and MPADC - 0.03 and 0.02 mg/m^3 respectively [24].

Polycarbacine

Light brown solid substance; m.p. 120°C, d.p. 140°C; soluble in weak alkaline solutions; insoluble in water.

Present in wastes from the production of some fungicides.

Toxicity: has an overall toxic effect: does damage to the liver and lungs; is harmful to embryos; is teratogenic [67]. MPSDC and MPADC - 0.002 and 0.0008 mg per m^3 respectively [24].

Prometryne (2-methylthio-4,6-bis(isopropylamino)-sym-triazine); $C_{10}H_{19}N_5$ (241.4)

White crystals; m.p. 118-120°C; soluble in organic solvents and in water (48 mg per ℓ).

Present in wastes from the production of some herbicides.

Toxicity: has an irritating effect. LD_{50} for rats and $TC_{chronic}$ - 3.12 g/kg and 50mg/kg respectively. MPSDC and MPADC - 0.04 mg/m^3 [24].

Propachlor (N-(isopropyl)chloroacetamide); $C_{11}H_{14}ClNO$ (211.7)

White crystals; m.p. 78-79°C, b.p. 110°C(≅4 Pa); slightly soluble in water.

Present in wastes from the production of some herbicides.

Toxicity: has an irritating effect. $TC_{irritation}$ - 0.002 mg/ℓ; LD_{50} for rats and mice - 1056 and 306 mg/kg respectively [5,67].

Determination: polarographic analysis [63,376].

Propanide (surcopur; N-(3,4-dichlorophenyl)propionamide); $C_9H_9Cl_2NO$ (218.1)

White crystals; m.p. 91-92°C; soluble in organic solvents; slightly soluble in water (0.2 g/ℓ).

Present in wastes from the production of some herbicides.

Toxicity: has an overall toxic effect: causes pathological changes in the composition of the blood; also causes irritation of the respiratory tracts and the skin; has a cumulative effect. TC_{odor} and $TC_{irritation}$ - 0.008 and 0.004 mg/m^3 respectively [375]; LD_{50} for rats and mice - 2.5 and 0.67 g/kg respectively [5,67]. MPSDC and MPADC - 0.005 and 0.001 mg/m^3 respectively [24].

Propazine (4,6-bis(isopropylamino)-2-chloro-sym-triazene); $C_3H_{16}ClN_5$ (229.7)

White crystals; m.p. 212-214°C; soluble in water (8.6 mg/ℓ).

Present in wastes from the production of some herbicides.

Toxicity: has an irritating effect; is mutagenic. LD_{50} for rats and mice - 3.84 and 3.18 mg/kg respectively [5,73]; MPSDC and MPADC - 0.04 mg/m^3 [24].

Propham (isopropyl-N-phenyl carbamate); $C_{10}H_{13}NO_2$ (179.22)

White crystals; m.p. 90°C, ρ 1.09; soluble in organic solvents and in water.

Present in wastes from the production of some pesticides.

Toxicity: $TC_{irritation}$ - 24.4 mg/m^3 [67]. MPSDC and MPADC - 0.02 mg/m^3[24].

ß-Propiolactone; $C_3H_4O_2$ (72.06)

Liquid; m.p. -33°C, b.p. 155°C, ρ 1.149; soluble in water and organic solvents.

Present in wastes from the production of organic compounds and pharmaceuticals.

Toxicity: very toxic; has an irritating effect; is carcinogenic and mutagenic; is adsorbed by undamaged skin. LD_{50} for rats - 50-100 mg/kg [55,70].

Propionic acid (propanoic acid); $C_3H_6O_2$ (74.08)

Liquid; m.p. -20.8°C, b.p. 141.1°C, f.p. 54.4°C, s.-i.p. 440°C, ρ 0.994; soluble in water and organic solvents.

Present in wastes from the production of some pesticides, organic chemicals, esters, aromatic substances, and cellulose.

Toxicity: has an irritating effect. SC_{odor} and TC_{odor} - 0.02 and 0.078 mg/m^3 respectively; $SC_{chronic}$ and $TC_{chronic}$ - 0.26 and 2.1 mg/m^3 respectively [381]; LD_{50} for rats and mice - 1.51 and 1.37 g/kg respectively [5,70]; MPSDC - 0.015 mg per m^3; toxicity classification - III [23].

Removal: by adsorption on activated carbon; with the aid of filters made of polyethylene, polypropylene, nylon, polyesters, poly(vynylidene chloride), or teflon.

Determination: paper chromatography (sensitivity 5 μg) [29].

Propionic aldehyde (propanal); C_3H_6O (58.1)

Liquid; m.p. -102°C, b.p. 48.8°C, f.p. <-20°C, s.-i.p. 227°C, CLE 2.5-13.6%, ρ 0.807; soluble in organic solvents and in water (16.7%).

Present in wastes from the production of organic compounds and fat substitutes.

Toxicity: has an overall toxic, irritating, and narcotic effect. SC_{odor}, TC_{odor}, $SC_{chronic}$, and $TC_{chronic}$ - 0.012, 0.048, 0.5, and 3.0 mg/m^3 respectively; LD_{50} for rats - 1.4 g/kg [55,70,381].

n-Propylacetate; $C_5H_{10}O_2$ (102.14)

Liquid with an odor of fruits; m.p. -92.5°C, b.p. 101.6°C, f.p. 14°C, s.-i.p. 435°C, CLE 1.8-9.6%, ρ 0.887; soluble in organic solvents and in water (1.89%).

Present in wastes from the production of some solvents, perfumes, plastics, and articles made of rubber.

Toxicity: has an overall toxic, irritating, and, at high concentrations, a narcotic effect; is adsorbed by undamaged skin [47,70].

Determination: gas chromatographic analysis (sensitivity 0.2 mg/m^3) [377].

n-Propyl alcohol (propanol-1); C_3H_8O (60.1)

Liquid; m.p. -127°C, b.p. 97.2°C, f.p. 23°C, s.-i.p. 371°C, CLE 2.1-13.5%, ρ 0.803; soluble in organic solvents and water.

Present in wastes from the production of some pesticides, solvents, plastics, and pharmaceuticals.

Toxicity: has an overall toxic and irritating effect. LD_{50} for rats - 1.87 g/kg [55,70]; MPSDC and MPADC - 0.3 $mg/^3$; toxicity classification - III [20,21]; MPADC established in Yugoslavia and GDR - 0.3 mg/m^3; MPSDC established in Bulgaria and GDR - 0.3 and 1.0 mg/m^3 respectively [50].

Removal: by adsorption on activated carbon [46].

n-Propylamine (1-aminopropane); C_3H_9N (59.11)

Liquid with an odor of ammonia; m.p. -83°C, b.p. 47.8°C, f.p. -37.1°C, s.-i.p. 329°C, lower CLE 2.0%, ρ 0.719; soluble in water and organic solvents.

Present in wastes from the production of some chemical reagents and pharmaceuticals.

Toxicity: causes irritation of mucous membranes and the skin; at high concentrations has a narcotic effect [5,47,70].

Propylene (propene); C_3H_6 (42.08)

Gas; m.p. -187.65°C, b.p. -47.75°C, f.p. -107.8°C, s.-i.p. 400°C, CLE 2.2-10.3%, ρ 0.58; soluble in ethanol and diethyl ether; slightly soluble in water.

Present in wastes from the production of some plastics; acetone, isopropylbenzene, isopropyl alcohol.

Toxicity: has an overall toxic effect; is precarcinogenic and mutagenic [79]. SC_{odor} and TC_{odor} - 3.3 and 17.3 $mg/^3$ respectively; $SC_{sensitivity\ to\ light}$ and $TC_{sensitivity\ to\ light}$ - 3.3 and 11.0 mg/m^3 respectively; $TC_{c.n.s.}$ - 3.3 mg/m^3 [378]; MPSDC and MPADC - 3 mg/m^3; toxicity classification - III [20,21]; MPSDC and MPADC established in Bulgaria - 3 mg/m^3, and in GDR - 3 and 2 mg/m^3 respectively [46].

Removal: by thermal catalytic oxidation [89].

Determination: paper chromatographic analysis [60,379,380].

n-Propyl mercaptan (propanthiol-1); C_3H_8S (76.16)

Liquid with an unpleasant odor; m.p. -113.3°C, b.p. 97.2°C, f.p. -20°C, ρ

0.841; soluble in ethanol, diethyl ether, and in water (1.96 g/ℓ).

Present in wastes from the production of some synthetic fibers; also in wastes of furnaces operating on natural gas and gas stoves (with n-propyl mercaptan added as an odorant).

Toxicity: has an overall toxic effect: causes respiratory disorders and inccordination of voluntary movements; is adsorbed by undamaged skin. TC_{odor} - $2 \cdot 10^{-4}$-$0.6 \cdot 10^{-5}$ mg/ℓ; LC_{50} and LD_{50} for mice - 40 mg/ℓ and 3.7 g/kg respectively [5].

Pyramine (phenazone; 5-amino-2-phenyl-4-chloro-3-pyridazone); $C_{10}H_8ClN_3O$ (221.7)

White crystals; m.p. 205-206°C; soluble in water (400 mg/ℓ).

Present in wastes from the production of some pesticides and herbicides.

Toxicity: has an overall toxic effect; is allergenic. $TC_{chronic}$ - 25 mg/m^3 [5].

Pyridine; C_5H_5N (79.10)

Liquid with a sharp odor; m.p. -41.6°C, b.p. 115.3°C, ρ 0.983; soluble in organic solvents and water.

Present in wastes from the production of some solvents, dyes, textiles, leather, pharmaceuticals, organic compounds, and pesticides; also in wastes of plants processing solid fuels.

Toxicity: has an overall toxic effect: causes pathological changes in the liver and kidneys; also has an irritating effect. TC_{odor}, $TC_{irritation}$, $TC_{sensitivity\ to\ light}$, $TC_{chronic}$ - 0.0002-0.003, 0.0016-0.005, 0.00009, and 0.001 mg/ℓ respectively [5]; MPSDC and MPADC - 0.08 mg/m^3; toxicity classification - II [20,21]; MPSDC and MPADC established in Bulgaria and Yugoslavia - 0.08 mg/m^3, in GDR - 0.08 and 0.03 mg/m^3 respectively, and in Romania - 0.15 and -.05 mg/m^3 respectively; MPC established in FRG - 20 ml/m^3 [44,51].

Removal: by adsorption [46,373]; with the aid of filters made of nylon, polypropylene, poly(vinylidene chloride), or polyethylene [34]; by combustion [373].

Determination: coulometric analysis (sensitivity 0.1 mg/m^3) [364].

Pyrrolidine (tetrahydropyrrole); C_4H_9N (71.12)

Liquid with an odor of ammonia; b.p. 88.5-89°C, ρ 0.857; soluble in organic solvents and water.

Present in wastes from the production of some pharmaceuticals.

Toxicity: very toxic; has an overall toxic and irritating effect. $TC_{chronic}$ - 0.0026 mg/m^3; LD_{50} for rats - 0.3 g/kg [5].

Quinoline (leucoline); C_9H_7N (129.16)

Oily liquid with a peculiar odor; m.p. 15.6°C, b.p. 237-238°C, ρ 1.09; soluble in water and organic solvents.

Present in wastes from the production of some organic chemicals, cyanine dyes, hydroquinoline, and some solvents, pesticides, and pharmaceuticals.

Toxicity: has an overall toxic and irritating effect [5,55,70].

Rotenone; $C_{23}H_{22}O_6$ (394.4)

White crystals; m.p. 163°C, ρ 1.27; soluble in acetone, chloroform, and in water (15 mg/ℓ).

Present in wastes from the production of some insecticides and plastics.

Toxicity: has an overall toxic effect: causes vomiting, cramps, adipose degeneration of the liver; also causes irritation of the eyes, the respiratory tracts, and the skin; is allergenic. LD_{50} for mice - 350 mg/kg [47,55,70].

Simazine (2-chloro-4,6-bis(ethylamino)-sym-triazene); $C_7H_{12}ClN_5$ (201.7)

White crystals; m.p. 227-228°C; soluble in organic solvents; slightly soluble in water (5 mg/ℓ).

Toxicity: has an overall toxic effect: causes paralysis and pathological changes in the liver. LD_{50} for rats and mice - 1390 and 4100 mg/kg respectively [5,67]; MPSDC and MPADC - 0.02 mg/m^3 [24].

Sodium and calcium cyanide mixture

Dark gray powder with an odor of bitter almond.

Present in wastes from the production of some pesticides.

Toxicity: very toxic; causes headache, dizziness, palpitation, and ptyalism; also causes irritation of mucous membranes and the respiratory tracts [67]; MPSDC and MPADC - 0.009 and 0.004 mg/m^3 respectively [24].

Sodium trichloroacetate; $C_2Cl_3NaO_2$ (185.4)

White crystals; d.p. ≅350°C; soluble in water.

Present in wastes from the production of some pesticides.

Toxicity: causes irritation of mucous membranes and the respiratory tracts; is adsorbed by undamaged skin. LD_{50} for rats, guinea pigs, rabbits, and mice - 6, 6, 6, and 5 g/kg respectively [5].

Styrene (phenylethylene; vinylbenzene); C_8H_8 (104.14)

Liquid; m.p. -30.6°C, b.p. 145.2°C, CLE 1.1-5.2%, ρ 0.906; soluble in organic solvents; slightly soluble in water (0.05%).

Present in wastes from the production of some plastics, synthetic rubber, and articles made of rubber.

Toxicity: has an irritating and, at high concentrations, a narcotic effect. SC_{odor} and TC_{odor} - 0.003 and 0.02 mg/m^3 respectively; $SC_{sensitivity\ to\ light}$ and $TC_{sensitivity\ to\ light}$ - 0.003 and 0.02 mg/m^3 respectively; $SC_{c.n.s.}$ and $TC_{c.n.s.}$ - 0.003 and 0.005 mg/m^3 respectively; $TC_{chronic}$ - 0.5 mg/m^3 [384]; MPSDC and MPADC - 0.003 mg/m^3; toxicity classification - III [20]: MPSDC and MPADC established in Bulgaria, Hungary, and Yugoslavoa - 0.003 mg/m^3, and in GDR - 0.003 and 0.01 mg/m^3 respectively [44]; MPC established in FRG - 150 mg per m^3 [51].

Determination: gas chromatographic analysis (sensitivity 0.003 mg/m^3) [385]; thin-layer chromatography (sensitivity 2 μg); photometric analysis (sensitivity 5 μg) [386]; paper chromatography (sensitivity 1 μg) [29].

Succinic acid (butandioic acid); $C_4H_6O_4$ (118.09)

White crystals; m.p. 185°C, b.p. 235°C(d.), ρ 1.56; soluble in ethanol, diethyl ether, and in water (6.8%).

Present in wastes from the production of some pharmaceuticals, varnishes and paints, perfumes, and photographic chemicals.

Toxicity: causes severe irritation of mucous membranes and the respiratory tracts [55,70].

Removal: with the aid of filters made of nylon, polyethylene, poly(vinylidene chloride), polypropylene, polyesters (including polyacrylates), or teflon [34].

Telodrin (1,3,4,5,6,7,8,8-octachloro-1,3,3a,4,7,7a-hexahydro-4,7-methanoisobenzofuran); $C_9H_4Cl_8O$ (411.79)

White crystals; m.p. 248-257°C; soluble in organic solvents and in water (1%).

Present in wastes from the production of some pesticides.

Toxicity: is harmful to embryos. LD_{50} for rats (intravenous) - 1.8 mg/kg [70,73].

1,1,2,2-Tetrabromoethane (sym-tetrabromoethane); $C_2H_2Br_4$ (345.68)

Yellowish liquid with an odor of camphor; m.p. 0.1°C, b.p. 239-242°C, ρ 2.964; soluble in organic solvents; slightly soluble in water (651 mg/ℓ).

Present in wastes of metallurgical works.

Toxicity: has an overall toxic effect: causes pathological changes in the liver and kidneys; also has an irritating and narcotic effect. LD_{50} for guinea pigs - 400 mg/kg [70].

1,3-Tetrachlorobutadiene (perchlorobutadiene); C_4Cl_4 (260.74)

Oily liquid with an odor of turpentine; m.p. -18.6°C, b.p. 215°C, s.-i.p.

585°C, ρ 1.679; soluble in organic solvents; slightly soluble in water (20 mg/ℓ).

Present in wastes from the production of some pesticides.

Toxicity: very toxic; is adsorbed by undamaged skin; has an overall toxic effect; is harmful to embryos. LD_{50} for rats, mice, and guinea pigs - 165, 51, and 90 mg/m^3 respectively; MPSDC and MPADC - 0.001 and 0.0002 mg/m^3 respectively [24].

Determination: see [58].

sym-Tetrachloroethane (acetylene tetrachloride; 1,1,2,2-tetrachloroethane); $C_2H_2Cl_4$ (167.86)

Liquid; m.p. -36°C, b.p.142°C, s.-i.p. 474°C, ρ 1.54; soluble in organic solvents; slightly soluble in water (0.13%).

Present in wastes from the production of trichloroethane and some solvents.

Toxicity: has an overall toxic effect: causes dysfunctioning of the liver, kidneys, and heart; also causes irritation of mucous membranes, the eyes, and the respiratory organs; at high concentrations has a narcotic effect; is adsorbed by undamaged skin [55,86]. MPSDC - 0.06 mg/m^3; toxicity classification - IV [96]; MPC established in FRG - 20 mg/m^3 [51].

Tetrachloroethylene (ethylene tetrachloride; perchloroethylene); C_2Cl_4 (165.85)

Liquid with an odor of chloroform; m.p. -22°C, b.p. 121°C, ρ 1.625; soluble in organic solvents; slightly soluble in water (0.04%).

Present in wastes from the production of some solvents and organic chemicals, and textiles.

Toxicity: has an overall toxic effect; causes irritation of mucous membranes, the eyes, and the respiratory tracts; is adsorbed by undamaged skin [55,70,86,98]. $TC_{chronic}$ - 19 mg/m^3 [371,372]; MPSDC - 0.06 mg/m^3; toxicity classification - II [23].

Tetracyanoethylene; C_6N_4 (128.10)

Yellow crystals; m.p. 198-200°C, b.p. 223°C; soluble in organic solvents; slightly soluble in water.

Present in wastes from the production of some reagents, dyes, and pesticides.

Toxicity: very toxic; has an overall toxic and irritating effect; is adsorbed by undamaged skin and mucous membranes. LD_{50} for mice - 28 mg/kg [5].

Tetraethyl pyrophosphate; $C_8H_{20}P_2O_7$ (290.20)

Liquid; b.p. 144-145°C(≅0.4 kPa), ρ 1.19; soluble in organic solvents; insoluble in water.

Present in wastes from the production of some pesticides and pharmaceuticals.

Toxicity: very toxic; has an overall toxic effect. LD_{50} for rats and rabbits - 1.12 and 5 mg/kg respectively [47,70].

Tetrahydro-3,5-dimethyl-1,3,5-thiodiazin-2-thion (thiazone; dazomet); $C_5H_{10}N_2S_2$ (162.2)

White (or gray) crystals; m.p. 104°C; soluble in chloroform and acetone; slightly soluble in water.

Present in wastes from the production of some pesticides.

Toxicity: has an overall toxic and irritating effect. LD_{50} for rats, mice, guinea pigs, and rabbits - 0.32-0.62, 0.18, 0.16, and 0.12 g/kg respectively [5,67]; MPSDC and MPADC - 0.07 mg/m^3 [24].

Tetrahydrofuran (tetramethylene oxide); C_4H_8O (72.10)

Liquid; m.p. -108.5°C, b.p. 65.5-65.8°C, f.p. -20°C, s.-i.p. 250°C, CLE 1.84-11.8%, ρ 0.889; soluble in organic solvents and water.

Present in wastes from the production of some varnishes and paints, plastics, synthetic fibers, and medicinal preparations.

Toxicity: has an overall toxic effect: causes dysfunctioning of the liver and kidneys; is adsorbed by undamaged skin; also causes irritation of mucous membranes; at high concentrations has a narcotic effect. $TC_{chronic}$ - 1 mg/ℓ; LD_{50} for rats, mice, and guinea pigs - 3, 2.3, and 2.3 g/kg respectively [5,55,70]; MPSDC and MPADC - 0.2 mg/m^3; toxicity classification - IV [20].

Determination: gas chromatographic analysis [29].

Tetrahydrofurfuryl alcohol; $C_5H_{10}O_2$ (102.13)

Liquid; b.p. 177°C, ρ 1.049; soluble in water and organic solvents.

Present in wastes from the production of some organic compounds, solvents, fats, and waxes.

Toxicity: has an overall toxic effect: causes dysfunctioning of the central nervous system and the liver; also causes irritation of mucous membranes and the skin. LD_{50} for rats and mice - 2.5-4.0 and 2.3 g/kg respectively [5,55,70].

Tetralin (1,2,3,4-tetrahydronaphthalene); $C_{10}H_{12}$ (132.20)

Colorless liquid; m.p. -35.79°C, b.p. 207.57°C, CLE 0.8-3.2%, ρ 0.870; soluble in organic solvents; insoluble in water.

Present in wastes from the production of some solvents, naphthalene, fats and oils, resins, waxes, and lacquers.

Toxicity: has an overall toxic effect: causes cataract, kidney disorders; also causes irritation of mucous membranes, the skin, and the eyes; at high concentrations has a narcotic effect. LD_{50} for rats - 2.86 g/kg [55,70]; ASL - 0.04 mg/m^3 [90].

2,5,7,10-Tetramethyl-6-thioundecene (diisoheptyl sulfide); $C_{12}H_{18}S$ (230.46)

Liquid with an odor of ether; b.p. 129-130°C, ρ 0.8362; soluble in organic solvents; insoluble in water.

Present in wastes from the production of some petrochemicals, lubricants, latexes, dyes, solvents, and detergents.

Toxicity: has an overall toxic effect: is harmful to the central nervous system, and causes functional disturbances of internal organs; also causes irritation of mucous membranes and the respiratory tracts [5].

Determination: gas chromatographic analysis (sensitivity 0.001 $\mu g/m^3$) [94].

Tetramethylthiuram disulfide; $C_6H_{12}N_2S_4$ (240.4)

White crystals; m.p. 155-156°C, ρ 1.29; soluble in chloroform; slightly soluble in water.

Present in wastes from the production of some pesticides and articles made of rubber.

Toxicity: has an overall toxic effect: causes headache, palpitation, difficult breathing, and impairment of the endocrine system, the liver, heart, and the central nervous system; also causes irritation of mucous membranes, the respiratory tracts, and the skin; is allergenic, teratogenic, carcinogenic, and, at high concentrations, mutagenic; is harmful to embryos. LD_{50} for rats, rabbits, and mice - 0.4-0.85, 0.2, and 1.2-3.5 g/kg respectively [5,47,70]; MPSDC and MPADC - 0.01 and 0.006 mg/m^3 respectively [24].

Tetranitromethane; CN_4O_8 (196.04)

Liquid with a sharp odor; m.p. 14.2°C, b.p. 125.7°C, ρ 1.639; soluble in organic solvents; insoluble in water.

Present in wastes from the production of some organic compounds and explosives.

Toxicity: has an overall toxic effect: causes cramps, pulmonary edema, disorders of the liver and kidneys; also causes irritation of the eyes and respiratory tracts; is adsorbed by undamaged skin. $TC_{c.n.s.}$ - 0.001-0.002 mg/ℓ [5,55,70].

Thiazone (3,5-dimethyltetrahydro-1,3,5-thiadiazin-2-thione); $C_5H_{10}N_2S_2$ (162.28)

White crystals; m.p. 105°C; soluble in organic solvents and in water (12%).

Present in wastes from the production of some pesticides.

Toxicity: has an overall toxic effect: causes ptyalism and cramps.

Thioacetamide (acetothioamide); C_2H_5NS (75.14)

Yellow crystals; m.p. 115°C; soluble in water and ethanol; slightly soluble in diethyl ether.

Present in wastes from the production of some reagents and synthetic rubber.

Toxicity: has an overall toxic effect: causes disorders of the liver and kidneys; also has an irritating effect; is carcinogenic. LD_{50} for rats - 0.2 g/kg [5,41,55].

Thiocron (O,O-dimethyl-S[N-(2-methoxyethyl)carbanyl]dithiophosphate); $C_7H_{16}NO_4$-PS_2 (273.31)

White crystals with a pleasant odor; m.p. 46°C; slightly soluble in water, ethanol, and diethyl ether.

Present in wastes from the production of some herbicides.

Toxicity: LD_{50} for rats - 420 mg/kg [47].

Thioglycolic acid (mercaptoacetic acid); $C_2H_4O_2S$ (92.11)

Liquid with a sharp odor; m.p. -16.5°C, b.p. 129°C(≅3.9 kPa), ρ 1.325; soluble in water and organic solvents.

Present in wastes from the production of some antibiotics, cosmetic preparations, perfumes, textiles, synthetic fibers, pesticides, analytical reagents, and thioglycolates.

Toxicity: has an irritating effect. $TC_{irritation}$ - 0.0003-0.0015 mg/ℓ; LD_{50} for rats and mice - 125 and 250 g/kg respectively [5,70].

ß-Thiolactic acid (2-mercaptopropionic acid); $C_3H_6O_2S$ (106.14)

Oily liquid; m.p. 16.8°C, b.p. 111°C(≅1.8 kPa), ρ 1.219; soluble in water and ethanol.

Present in wastes from the production of some antioxidants.

Toxicity: has an overall toxic and irritating effect [70]. ASL - 0.0001 mg/m^3 [91].

Thiophene (thiofuran); C_4H_4S (84.14)

Liquid; m.p. -38°C, b.p. 84.1°C, ρ 0.064; soluble in organic solvents; insoluble in water.

Present in wastes from the production of some plastics, antioxidants, dyes, pharmaceuticals, and pesticides.

Toxicity: has an overall toxic effect: causes excitement, cramps: also causes irritation of mucous membranes and the respiratory tracts; at high concentrations has a narcotic effect [5,55]. SC_{odor} and TC_{odor} - 1.2 and 2.1 mg/m^3 respectively; $SC_{sensitivity\ to\ light}$ and $TC_{sensitivity\ to\ light}$ - 0.8 and 1.0 mg/m^3 respectively; $SC_{c.n.s.}$ and $TC_{c.n.s.}$ - 0.6 and 0.8 mg/m^3 respectively; $SC_{chronic}$ - 0.6 mg/m^3 [390]; MPSDC - 0.6 mg/m^3; toxicity classification - IV [20,21]; MPSDC established in Bulgaria and Yugoslavia - 0.6 mg/m^3; MPSDC and MPADC established in GDR - 0.6 and 0.2 mg/m^3 respectively [54].

Determination: spectrophotometric analysis (sensitivity 2.5 μg) [27,391].

Thiosemicarbazide (aminothiourea); CH_5N_3S (91.13)

White crystals; m.p. 183°C; slightly soluble in water and organic solvents.

Present in wastes from the production of some organic compounds, pesticides, analytical reagents, and pharmaceuticals.

Toxicity: very toxic; has an overall toxic effect: causes cramps. LD_{50} for rats, cats, and dogs - 23, 20, and 10 mg/kg respectively [5,70].

Thiourea(thiocarbamide); CH_4N_2S (76.12)

White crystals; m.p. 180-182°C, ρ 1.40; soluble in water and organic solvents.

Present in wastes from the production of some plastics, synthetic fibers, organic compounds, and petrochemicals.

Toxicity: has an overall toxic effect: causes headache, drowsiness, xeroderma, a bitter taste in the mouth, paleness, swelling of the face; is carcinogenic; is adsorbed by undamaged skin. LD_{50} for mice - 8 g/kg [5,41,55].

Thiuram disulfide (antabuse; disulfiram; tetraethylthiuram disulfide); $C_{10}H_{20}N_2S_4$ (296.55)

White crystals; m.p. 73°C, lower CLE 7.8 g/m^3; soluble in organic solvents; insoluble in water.

Present in wastes from the production of some pharmaceuticals and articles made of rubber.

Toxicity: has an overall toxic effect [5].

Thiuram sulfide (tetramethylthiuram sulfide); $C_6H_{12}N_2S_3$ (208.38)

Yellow crystals; m.p. 110°C, f.p. 140°C, s.-i.p. 270°C, lower CLE 57.5 g/m^3; soluble in organic solvents; insoluble in water.

Present in wastes from the production of articles made of rubber.

Toxicity: has an overall toxic effect: causes internal bleeding, dystrophic changes in internal organs, pulmonary and kidney necrosis; also has an irritating effect. $TC_{irritation}$ - 400 mg/m^3; LD_{50} for rats and mice - 0.413 and 1.15 g/kg respectively [5].

Tillam (N-butyl-S-propyl-N-ethylthio carbamate); $C_{10}H_{21}NOS$ (203.4)

Liquid; b.p. 142°C(≅2.7 kPa); soluble in organic solvents and in water (92 mg per ℓ).

Toxicity: has an overall toxic effect: causes nausea and palpitation; also causes dermatitis. LD_{50} for rats and mice - 1.12 and 0.72 g/kg respectively [5,70]; MPSDC and MPADC - 0.02 and 0.01 mg/m^3 respectively [24].

Toluene (methylbenzene); C_7H_8 (92.13)

Liquid; m.p. -95°C, b.p. 110.6°C, CLE 1.3-6.7%, ρ 0.866; soluble in organic solvents; very slightly soluble in water (0.014%).

Present in wastes from the production of benzoic acid, benzaldehyde, and some explosives, dyes, and organic compounds.

Toxicity: has an overall toxic, irritating, and, at high concentrations, a narcotic effect; is carcinogenic and mutagenic; is adsorbed by undamaged skin [45,52, 55,70,394]. SC_{odor}, TC_{odor}, $SC_{c.n.s.}$, $TC_{c.n.s.}$, and $SC_{chronic}$ - 1.27, 2, 0.6, 1, and 0.6 mg/m^3 respectively [395]; MPSDC and MPADC - 0.6 mg/m^3; toxicity classification - III [20,21]; MPSDC and MPADC established in Bulgaria, GDR, Hungary, and Yugoslavia - 0.6 mg/m^3; MPC established in FRG - 150 mg/m^3 [44,51].

Removal: by adsorption on activated carbon [396]; by thermal catalytic oxidation (efficiency 92-100%) [397]; with the aid of filters made of polypropylene, nylon, polyesters including polyacrylates, or teflon [34,398,399].

Determination: gas chromatographic analysis (sensitivity 0.0005 mg/m^3) [29,63]; nitration (sensitivity 2 μg) [27]; polarographic analysis (sensitivity 2.5 μg/ml) [20, 63].

2,4-Toluylene diisocyanate; $C_9H_6N_2O_2$ (174.15)

Liquid; m.p. 21.8°C, b.p. 121°C(≅1.3 kPa), ρ 1.21; soluble in organic solvents; undergoes reaction with water.

Present in wastes from the production of polyuretans and polyester-based sealing agents.

Toxicity: has an overall toxic effect; causes irritation of mucous membranes, the skin, and the respiratory tracts; is allergenic [5,47,55,70]. SC_{odor}, TC_{odor}, $SC_{c.n.s.}$, $TC_{c.n.s.}$, and $TC_{chronic}$ - 0.15, 2, 0.05, 0.1, and 0.02 mg/m^3 respectively [392]; MPSDC and MPADC - 0.05 and 0.02 mg/m^3 respectively; toxicity classification - I [20,21]; MPSDC and MPADC established in Bulgaria, GDR, Romania, and Yugoslavia - 0.05 and 0.02 mg/m^3 respectively [44,51].

Determination: see [27,393].

Toxaphene; $C_{10}H_{10}Cl_8$ (413.8)

Brown wax-like substance with an odor of turpentine; m.p. 65-90°C; soluble in organic solvents; insoluble in water.

Present in wastes from the production of some pesticides.

Toxicity: has an overall toxic effect: causes vomiting, headache, dysfunctioning of the central nervous system; also causes irritation of mucous membranes; is mutagenic [374]. $SC_{irritation}$ and $TC_{irritation}$ - 0.0023-0.004 and 0.0002-0.0003 mg/ℓ respectively [90]; MPSDC established and recommended in USA - 0.5 and 0.001 mg/m^3 respectively [40].

Triallat (avadex; N,N-diisopropyl-S-(2,3,3-trichloroallyl) thiocarbamate); $C_{10}H_{16}O_3NOS$ (304.7)

White crystals with an unpleasant odor; m.p. 29.30°C, b.p. 148-149°C(≅1.2 kPa); soluble in organic solvents; slightly soluble in water (9 mg/ℓ).

Present in wastes from the production of some pesticides.

Toxicity: has an overall toxic effect: causes headache and vomiting; also causes irritation of mucous membranes and the respiratory tracts, lacrimation, and ptyalism; suspected of being carcinogenic; is adsorbed by undamaged skin. LD_{50} for rats, mice, and cats - 1.47, 0.93, and 1.0 g/kg respectively [5]; MPSDC and MPADC - 0.02 and 0.01 mg/m^3 respectively [24].

S,S,S-Tributyltrithio phosphate; $C_{12}H_{27}OPS_3$ (314.5)

Light yellow liquid with a peculiar odor; b.p. 150°C(≅40 Pa), ρ 1.148; soluble in water (≅50 mg/ℓ).

Present in wastes from the production of some pesticides.

Toxicity: is adsorbed by undamaged skin. $SC_{irritation}$, $TC_{irritation}$, $TC_{chronic}$, and TC_{odor} - 0.009, 8.4-13.0, 1.7, and 0.052 mg/m^3 respectively [54]; TC on the basis of the electric response of the cerebral cortex - 0.014 mg/m^3 [260]; MPSDC and MPADC - 0.01 mg/m^3; toxicity classification - II [4,20,21].

Determination: analysis with the use of ammonium molybdate reagent [27]; chromatography (sensitivity 0.005 mg/ℓ) [261].

Trichloroacetic acid; $C_2HCl_3O_2$ (163.4)

White crystals; m.p. 58°C, s.-i.p. 711°C, subl.p. 197.55°C, ρ 1.62; soluble in water and organic solvents.

Present in wastes from the production of some organic chemicals, pharmaceuticals, and specimens for microscopic investigations.

Toxicity: very toxic; has an overall toxic effect; also has an irritating effect: causes chemical burns, and irritation of the mucous membranes and the skin. LD_{50} for rats - 8.3 g/kg [5,55,70,86].

1,2,3-Trichlorobenzene; $C_6H_3Cl_3$ (181.46)

White powder; m.p. 53.5°C, b.p. 218°C, CLE 2.5-5.0%, ρ 1.46; soluble in benzene; insoluble in water.

Present in wastes from the production of some organic chemicals, pesticides, dichlorophenols, and lubricants.

Toxicity: has an overall toxic effect: does damage to the liver; also causes irritation of the eyes and respiratory tracts [55,70]; ASL - 0.008 mg/m^3 [18].

2,3,6-Trichlorobenzoic acid; $C_7H_3Cl_3O_2$ (225.46)

White powder; m.p. 125°C; soluble in organic solvents and water.

Present in wastes from the production of some pesticides.

Toxicity: has an overall toxic, irritating effect. $TC_{irritation}$ - 8.5 mg/m^3; LD_{50} for rats - 500-730 mg/kg [67]; MPSDC and MPADC - 0.01 and 0.006 mg/m^3 respectively [24].

1,1,1-Trichloroethane (methyl chloroform); $C_2H_3Cl_3$ (133.42)

Liquid; m.p. -32.8°C, b.p. 74.1°C, f.p. 570°C, ρ 1.389; soluble in organic solvents; slightly soluble in water (0.132%).

Present in wastes from the production of varnishes and paints, printing inks; also in wastes of metal-working works.

Toxicity: has an overall toxic, irritating, and narcotic effect [55,70,98].

Trichloroethylene (ethylene trichloride); $C_2H_3Cl_3$ (131.40)

Liquid with an odor of chloroform; m.p. -73°C, b.p. 87°C, f.p. 32°C, s.-i.p. 416°C, CLE 10.6-41%, ρ 1.46; soluble in organic solvents, very slightly soluble in water.

Present in wastes from the production of some fats, waxes, resins, oils, paints, and cellulose esters.

Toxicity: has an overall toxic and, at high concentrations, a narcotic effect [55, 70]; suspected of being carcinogenic [52]. MPSDC and MPADC - 4 and 1 mg/m^3 respectively; toxicity classification - III [20,21]; MPSDC and MPADC established in Bulgaria, GDR, and Yugoslavia - 4 and 1 mg/m^3 respectively; MPC established in FRG - 20 mg/m^3 [51].

Removal: by adsorption on activated carbon [34]; with the aid of filters made of polyethylene, polypropylene, nylon, poyesters, poly(vinylidene chloride), or teflon [34]; by combustion [402].

Determination: gas-liquid chromatography (sensitivity 0.12 mg/m^3) [63]; spectrometry [403].

Trichlorometaphos (O-methyl-O-(2,4,5-trichlorophenyl)-O-ethylthiophosphate); C_9H_{10}-Cl_3O_3PS (335.6)

Oily liquid; b.p. 127-133°C(≅20 Pa); soluble in water (40 mg/ℓ).

Present in wastes from the production of some insecticides.

Toxicity: has an overall toxic effect: causes hypometabolism; also causes irritation of the eyes, the respiratory tracts, and the skin. LD_{50} for laboratory animals - 150-500 mg/kg [67]; MPSDC and MPADC - 0.009 and 0.004 mg/m^3 respectively [24].

Determination: thin-layer chromatography [401].

Trichloromethylbenzene (phenyl chloroform); $C_7H_5Cl_3$ (195.48)

Oily liquid with a sharp odor; m.p. -4.8°C, b.p. 220.6°C, f.p. 91°C, s.-i.p. 433°C, ρ 1.372; soluble in organic solvents; insoluble in water.

Present in wastes from the production of some basic organic compounds and dyes.

Toxicity: has an overall toxic effect; causes irritation of the skin and mucous membranes [55,70,86]. ASL - 0.005 mg/m^3 [18].

2,4,5-Trichlorophenol; $C_6H_3Cl_3O$ (197.46)

White crystals; m.p. 68-70°C, b.p. 244-248°C; soluble in water (0.2 g/ℓ).

Present in wastes from the production of some pesticides and organic chemicals.

Toxicity: has an overall toxic and irritating effect; is adsorbed by undamaged skin [55,70]. LD_{50} for rats - 2.96 g/kg [55,70]; ASL - 0.003 mg/m^3 [18].

2,4,6-Trichlorophenol (omal); $C_6H_3Cl_3O$ (197.46)

White crystals; m.p. 69°C, b.p. 246°C; soluble in water (800 mg/ℓ).

Present in wastes from the production of some pesticides and organic chemicals.

Toxicity: has an overall toxic and irritating effect; is adsorbed by undamaged skin [55,70]. LD_{50} for rats - 820 mg/kg [55,70]. ASL - 0.003 mg/m^3 [18].

2,4,5-Trichlorophenoxyacetic acid; $C_8H_5Cl_3O_3$ (255.49)

White crystals; m.p. 158-159°C, soluble in water (0.02%).

Present in wastes from the production of some pesticides.

Toxicity: has an overall toxic effect; is mutagenic, carcinogenic, and teratogenic; is harmful to embryos; is adsorbed by healthy skin. LD_{50} for rats and dogs - 0.5 and 0.1 g/kg respectively [5,47,70,374].

1,2,3-Trichloropropane (glyceryl trichlorohydrin); $C_3H_5Cl_3$ (147.44)

Liquid; m.p. -14.7°C, b.p. 156.85°C, CLE 3.2-12.6%, f.p. 79°C, s.-i.p. 304°C, ρ 1.388; soluble in organic solvents; slightly soluble in water.

Present in wastes from the production of some organic compounds.

Toxicity: has an overall toxic and irritating effect; is adsorbed by undamaged skin; has a cumulative effect [55]. MPADC - 0.05 mg/m^3; toxicity classification - III [23].

Triethanolamine; $C_6H_{15}NO_3$ (149.19)

Liquid; m.p. 21.2°C, b.p. 360°C, f.p. 179°C; soluble in water and ethanol.

Present in wastes from the production of some surface-active agents, plastics, articles made of rubber, pharmaceuticals, cellulose and paper, textiles, dyes, pesticides, cement, solvents for casein and shellac, and petrochemicals; also in wastes of metal-working plants.

Toxicity: has an overall toxic and irritating effect; is adsorbed by undamaged skin. LD_{50} for rats and rabbits - 5.2-9.11 and 5.2 g/kg respectively [5,55,70].

Determination: thin-layer chromatography [404].

Triethylamine; $C_6H_{15}N$ (101.19)

Liquid with an odor of ammonia; b.p. 89.7°C, f.p. -12°C, s.-i.p. 510°C, CLE 1.2-8%, ρ 0.729; soluble in water and ethanol.

Present in wastes from the production of some organic chemicals, polymers, and ammonia.

Toxicity: has an overall toxic effect: causes incoordination of voluntary movements, respiratory disorders, pathological changes in the liver and kidneys; also causes irritation of the skin and chemical burns. MPSDC and MPADC - 0.14 mg/m^3; toxicity classification - III [20,21].

Determination: see [20,26,405].

Trifluoroacetic acid (trifluoroethanoic acid); $C_2HF_3O_2$ (112.03)

Liquid with a sharp odor; m.p. -15.4°C, b.p. 72.4°C, ρ 1.48; soluble in water.

Present in wastes from the production of fluorinated organic compounds.

Toxicity: has an overall toxic effect: causes dystrophic changes in the lungs, liver and brain; also causes irritation of mucous membranes, the respiratory tracts, and the skin. $TC_{irritation}$ and $TCDC_{acute}$ (for rats) - 0.25 and 1.5 mg/ℓ respectively [5].

Trifluoromethyl-3-nitrobenzene; $C_7H_4F_3NO_2$ (167.12)

Yellow liquid; m.p. -2.4°C, b.p. 198.0-201.5°C, ρ 1.431; soluble in ethanol and diethyl ether; insoluble in water.

Present in wastes from the production of some basic organic chemicals and pharmaceuticals.

Toxicity: is adsorbed by undamaged skin; has a toxic effect: causes cyanosis narcosis, dystrophic changes in the liver, heart, and kidneys; and disorders of the central nervous system. LC_{50} (after 2-hr exposure) for rats and mice - 0.87 and 0.88 mg/m^3 respectively; $TC_{irritation}$ for rabbits - 0.033 mg/ℓ [5].

Trifluralin (2,6-dinitro-N,N-diphenyl-4-trifluoromethylaniline); $C_{13}H_{16}F_3N_3O_4$ (335.3)

Yellow crystals; m.p. 48.5-49.0°C, b.p. 96-97°C(≅24 Pa); soluble in organic solvents; slightly soluble in water (1 mg/ℓ).

Present in wastes from the production of some pesticides.

Toxicity: has an overall toxic effect: causes respiratory disorders, supression of the functioning of the central nervous system. LD_{50} for rats and mice - 4.6 and 3.3 g/kg respectively [5,70].

Trifuraldiamine (hydrofuramide; furfural hydramide; furfuramide); $C_{15}H_{12}N_2O_3$ (268.27)

Crystalline powder; m.p. 117°C, b.p. 250°C(d.); soluble in organic solvents; insoluble in water.

Present in wastes from the production of articles made of rubber and some pesticides.

Toxicity: has an overall toxic effect: causes loss of appetite, pulmonary edema and emphysema; also has an irritating effect. LD_{50} for rats and mice - 0.4 and 0.95 g/kg respectively [5,55,70].

Trimethylamine; C_3H_9N (59.11)

Gas; m.p. -117.1°C, b.p. 2.9°C, f.p. 3.3°C, s.-i.p. 190°C, CLE 2.0-11.6%, ρ 0.632; soluble in water and organic solvents.

Present in wastes from the production of some organic compounds, pesticides, ammonia derivatives, aminazine, and dyes; also in wastes of beet-sugar refineries.

Toxicity: has an overall toxic effect: causes diarrhea, changes in the blood composition, bronchopneumonia, bleeding from the liver, kidneys, and spleen; also has an irritating effect; is precarcinogenic and mutagenic [79]. TC_{odor} and $TC_{chronic}$ - 0.002 [5] and 0.075 mg/ℓ [79] respectively; recommended MPSDC - 0.02 mg/m^3 [49].

Removal: see [400].

2,2,4-Trimethylpentane (isooctane); C_8H_{18} (114.2)

Liquid with an odor of kerosine; m.p. -107.38°C, b.p. 99.24°C, f.p. -9°C, s.-i.p. 430°C, CLE 0.95-6.0%, ρ 0.691; slightly soluble in ethanol and diethyl ether; insoluble in water.

Present in wastes from the production of some solvents, articles made of rubber; also in wastes of petroleum refineries.

Toxicity: has an irritating and, at high concentrations, a narcotic effect.

Removal: by thermal catalytic oxidation (efficiency 100%) [89].

Trinitromethane (nitroform); CHN_3O_6 (151.04)

White crystals; m.p. 15°C, b.p. 45-47°C(explodes), ρ 1.59; soluble in water.

Present in wastes from the production of some explosives.

Toxicity: has an overall toxic effect: causes methemoglobinemia, dystrophic changes in the liver and kidneys; also causes irritation of the respiratory tracts. $TC_{c.n.s.}$ - 0.04-0.05 mg/ℓ [5,70].

2,4,6-Trinitrotoluene (TNT); $C_7H_5N_3O_6$ (227.13)

Yellowish crystals; m.p. 80.85°C, f.p. 290°C, ρ 1.64; soluble in organic solvents; very slightly soluble in water (0.013%).

Present in wastes from the production of some explosives.

Toxicity: is adsorbed by undamaged skin; has an overall toxic effect: causes headache, dizziness, nausea, cataract; does damage to the liver; causes the skin and hair to turn yellow; also has an irritating effect; is allergenic. ASL - 0.007 mg/m^3 [18]; in USA recommended MPSDC - 0.01 mg/m^3 [46].

Urea (carbamide); CH_4N_2O (60.06)

White crystals; m.p. 132.7°C, ρ 1.335; soluble in water and ethanol; slightly soluble in diethyl ether.

Present in wastes from the production of some fertilizers, hydrazine, cyanates, cyanuric acid, articles made of rubber, paper, and some plastics and pharmaceuticals.

Toxicity: has a slight irritating effect [55]. MPADC - 0.20 mg/m^3; toxicity classification - IV [23].

n-Valeric acid (pentanoic acid); $C_5H_{10}O_2$ (102.14)

Oily liquid with a specific odor; m.p. -34.5°C, b.p. 185.4°C, f.p. 87°C, ρ 0.939; soluble in ethanol, diethyl ether, and in water (3.78 g/100 ml).

Present in wastes from the production of some pharmaceuticals and perfumes.

Toxicity: MPSDC and MPADC - 0.03 and 0.01 mg/m^3 respectively; toxicity classification - III [21,22,263]; MPSDC and MPADC established in Bulgaria, GDR, and Yugoslavia - 0.003 and 0.01 mg/m^3 respectively; ASL established in FRG - 20 mg/m^3 [44,51].

Determination: paper chromatography (sensitivity 3 μg) [27].

n-Valeric aldehyde (pentanal); $C_5H_{10}O$ (86.13)

Liquid; m.p. -91°C, b.p. 102-103°C, f.p. 12°C, ρ 0.809; soluble in water, ethanol, and diethyl ether.

Present in wastes from the production of articles made of rubber, some plastics, and aromatic substances.

Toxicity: has an overall toxic, irritating, and, at high concentrations, a narcotic effect. LD_{50} for mice - 6.4-12 mg/kg [47,55,70,72].

Vamidothion (O,O-dimethyl-S-[2-(N-methylcarbamoylethylthio)ethyl] thiophosphate) $C_8H_{18}NO_4PS$ (287.3)

White crystals; m.p. 46-48°C; soluble in ethanol, diethyl ether, and water.

Present in wastes from the production of some pesticides.

Toxicity: very toxic; is adsorbed by undamaged skin; causes ptyalism, running nose, tremors, cramps, and irritation of the respiratory tracts [67]. SC_{odor}, TC_{odor}, $SC_{irritation}$, and $TC_{irritation}$ - 0.016, 0.026, 0.009, and 0.016 mg/m^3 respectively [264]; MPSDC established in USA - 0.1 mg/m^3 [40].

Vapam (N-methyldithio sodium carbamate dihydrate); $C_2H_8NNaO_2S_2$ (165.2)

Pale cream-colored crystals with a sharp odor; soluble in water and ethanol; insoluble in diethyl ether.

Present in wastes from the production of some pesticides and latexes.

Toxicity: is adsorbed by undamaged skin; has an overall toxic effect: causes functional disturbances of the central nervous system, and pathological changes in the liver and kidneys; also causes irritation of mucous membranes and the skin; suspected of being carcinogenic. LD_{50} for rats, mice, and rabbits - 0.76, 0.143, and 0.32 g/kg respectively [5,52,67,70]; MPSDC and MPADC - 0.005 and 0.001 mg/m^3 respectively [20,21,34].

Vinyl acetate; $C_4H_6O_2$ (86.09)

Liquid with a sharp odor; m.p. -100.2°C, b.p. 73°C, f.p. 1.1°C, s.-i.p. 380°C, CLE 2.6-13.4%, ρ 0.934; soluble in organic solvents and in water (2.5%).

Present in wastes from the production of glues, varnishes and paints, vinyl acetate, plastics, petrochemicals, ethylene, and acetic acid.

Toxicity: has an overall toxic and irritating effect. SC_{odor} and TC_{odor} - 0.7 and 1 mg/m^3 respectively; $SC_{sensitivity\ to\ light}$ and $TC_{sensitivity\ to\ light}$ - 0.60 and 0.77 mg/m^3 respectively; $SC_{c.n.s.}$ and $TC_{c.n.s.}$ - 0.21 and 0.32 mg/m^3 respectively [265] MPSDC and MPADC - 0.15 mg/m^3; toxicity classification - III [21, 22]; MPSDC and MPADC established in Bulgaria, Czechoslovakia, and Yugoslavia - 0.2 mg/m^3, and in GDR - 0.4 and 0.15 mg/m^3 respectively [44,51]; MPC established in FRG - 150 mg/m^3 [51].

Determination: paper chromatography (sensitivity 3 μg) [27,29]; polarography (sensitivity 0.008 mg/m^3) [58,97].

N-Vinylcarbazole (N-vinyldiphenyleneimine; N-vinyldibenzopyrrole); $C_{14}H_{12}N$ (195.11)

White solid; m.p. 67°C, b.p. 150-155°C(133 Pa), ρ 1.094; slightly soluble in water.

Present in wastes from the production of some plastics.

Toxicity: very toxic; is allergenic; causes skin ulcers. LD_{50} for guinea pigs - 0.5 g/kg [5,55].

Vinyl chloride (chloroethylene) C_2H_3Cl (62.50)

Gas with a slight odor; m.p. -158.4°C, b.p. -13.8°C, f.p. -61°C, s.-i.p. 472°C, CLE 3.6-33.0%, ρ 0.911; soluble in organic solvents; slightly soluble in water.

Present in wastes from the production of some polymers and basic organic chemicals.

Toxicity: has an overall toxic effect: causes angiosarcoma, hepatocirrhosis, osteomalacia; has an irritating and, at high concentrations, a narcotic effect; is carcinogenic, mutagenic, teratogenic; is harmful to embryos [47,52,55,70,267-270]. SC_{odor} and TC_{odor} - 0.58-3.87 and 0.66-2.24 mg/m^3 respectively [271]; MPSDC has not been established either in the USSR or in other countries; MPC established in USA - 1 t/year [92], and in FRG - 150 mg/m^3 (with a flow rate at exhaust of 3 kg/hr) [51].

Removal: by steam stripping or combustion [272], by stripping off on a special column (efficiency 82.0-95.1%) [273].

Determination: see [272,274-278].

Vinylene carbonate; $C_3H_2O_3$ (86.05)

Crystals with an odor of ether; m.p. 20-21°C, b.p. 162°C, ρ 1.354; soluble in organic solvents; slightly soluble in water.

Present in wastes from the production of some plastics.

Toxicity: has an overall toxic effect: causes respiratory disorders, incoordination of voluntary movements, cramps, degeneration of the liver, kidneys, spleen, and cardiac muscle; also causes irritation of the skin and mucous membranes. LD_{50} for rats and mice - 1.86 and 0.5 g/kg respectively [5].

Vinylidene chloride (1,1-dichloroethylene); $C_2H_2Cl_2$ (96.95)

Liquid with an odor of chloroform; m.p. -122.1°C, b.p. 31.7°C, f.p. 13.9°C, s.-i.p. 642°C, CLE 5.6-11.4%, ρ 1.218; soluble in organic solvents; slightly soluble in water (0.04%).

Present in wastes from the production of some plastics, freons, and methylchloroform.

Toxicity: very toxic; causes pathological changes in the liver and kidneys; also causes irritation of the skin, mucous membranes, and the air passages; at high concentrations has a narcotic effect [47,55,70].

Determination: photometry (sensitivity 10 mg/m^3) [266].

Vinyl propionate; $C_5H_8O_2$ (100.12)

Liquid with a mild odor; freezing point -81.1°C, b.p. 95°C, f.p. 1°C, ρ 1.404;

soluble in organic solvents; slightly soluble in water (0.8%).

Present in wastes from the production of some plastics, solvents, leather, varnishes and paints.

Toxicity: has an overall toxic and irritating effect; causes pulmonary edema and burns of the cornea. $TC_{chronic}$ (for mice) and $TC_{sensitivity\ to\ light}$ (for rats) - 1 and 0.016 mg/ℓ respectively [5].

2-Vinylpyridine; C_7H_7N (105.15)

Liquid with an unpleasant odor; b.p. 158°C(d.), ρ 0.9746; soluble in organic solvents; slightly soluble in water.

Present in wastes from the production of synthetic rubber, ion-exchange resins, photographic chemicals, poly(vinylpyridine), and some pharmaceuticals.

Toxicity: has an overall toxic effect: causes incoordination of voluntary movements and cramps, and does damage to the liver; also has an irritating effect. LD_{50} for rats and mice 0.10.2 and 0.4-0.8 g/kg respectively [5,72].

m-Xylene (1,3-dimethylbenzene); C_8H_{10} (106.2)

Liquid with a characteristic odor; m.p. -47.9°C, b.p. 139.1°C, CLE 3.0-7.6%, ρ 0.864; soluble in organic solvents; insoluble in water.

Present in wastes from the production of some solvents, benzoic acid, phthalic anhydride, isophthalic and terephthalic acid, polyester fibers, dyes, and Canada balsam.

Toxicity: is adsorbed by undamaged skin; causes irritation of the air passages; at high concentrations has a narcotic effect [55,70]; suspected of being carcinogenic [45].

Xylenes (a mixture of dimethylbenzene isomers); C_8H_{10} (106.3)

Liquid; b.p. 138-144°C, CLE 3.0-7.6%, ρ 0.86; soluble in organic solvents; slightly soluble in water.

Present in wastes from the production of some basic organic chemicals.

Toxicity: SC_{odor}, TC_{odor}, and $SC_{irritation}$ - 0.41, 0.6, and 0.21 mg/m^3 respectively [326]; MPSDC and MPADC - 0.2 mg/m^3; toxicity classification - III [20, 21]; MPSDC and MPADC established in Bulgaria, Hungary, and Yugoslavia - 0.2 mg per m^3, and in GDR - 0.6 and 0.2 mg/m^3 respectively [44,51,327]; MPC established in FRG - 150 mg/m^3 [51].

Removal: by adsorption on activated carbon [46].

Determination: gas chromatographic analysis [29].

2,6-Xylenol (_vic_-m-xylenol; 2,6-dimethyl phenol); $C_8H_{10}O$ (122.17)

White needles; m.p. 49°C, b.p. 212°C, ρ 1.07; soluble in organic solvents and hot water.

Present in wastes from the production of some plastics, solvents, herbicides, pharmaceuticals, and perfumes.

Toxicity: has an overall toxic effect: causes vomiting, collapse, cramps, paralysis, and does damage to the cardiac muscle; also has an irritating effect. $SC_{chronic}$ and $TC_{chronic}$ - 0.01 and 0.055 mg/m^3 respectively [70,325]; recommended MPSDC - 0.02 mg/m^3; toxicity classification - II [325]; ASL - -.10 mg/m^3 [18].

Zectran (4-dimethyl-3,5-dimethylphenyl-N-methylcarbamate); $C_{12}H_{18}N_2O_2$ (222.29)

White powder; m.p. 85°C; soluble in organic solvents; slightly soluble in water.

Present in wastes from the production of some pesticides.

Toxicity: very toxic; does damage to the central nervous system and the lungs; also causes irritation of the respiratory tracts. LD_{50} for rats and mice - 75 and 39 mg/kg respectively [5].

Zineb (poly-N,N-ethylen-bis-zinc(dithiocarbamate)); $(C_4H_6N_2S_4Zn)_n$ (275.7)

White or yellow crystals with an an unpleasant odor; m.p. 140-160°C; insoluble in water.

Toxicity: has an overall toxic effect: causes headache, loss of appetite, chest pains, impairment of the blood formation process, and does damage to the liver and kidneys; also has an irritating effect: causes conjunctivitis, dermatitis; is harmful to embryos and the gonads; is allergenic and carcinogenic [5,67,73]. MPSDC and MPADC - 0.01 and 0.006 mg/m^3 respectively [24].

Determination: photocolorimetric and spectrophotometric analysis [434].

Ziram (zinc dimethyldithiocarbamate); $C_6H_{12}N_2S_4Zn$ (305.82)

White powder; m.p. 240°C, ρ 1.66; soluble in organic solvents; slightly soluble in water.

Present in wastes from the production of some fungicides.

Toxicity: very toxic; causes flaccidity, dyspnea, lacrimation, and does damage to the liver; also causes irritation of the eyes and air passages; is harmful to the gonads and embryos; is teratogenic, carcinogenic and mutagenic [374]. LD_{50} for mice, rabbits, and guinea pigs - 337, 100-200, and 100-200 mg/m^3 respectively [67, 73].

APPENDIX

Harmful organic substances present in the wastes of various industries

Industries producing:	Organic substances contained in the wastes	References
Acrylates	methyl acrylate, butyl acrylate, methyl methacrylate, methanol, acetone, butanol	87
Adipic acid	cyclohexane, acetic acid, acetyl aldehyde	19
Aluminum	hydrocarbons, resinous substances, 1,2-benzopyrene, phenanthrene, anthracene, alkylanthracenes, fluoranthene, pyrene, dihydrobenzofluoranthene, benzoanthracene, chrysene triphenylene, 11,12-benzofluoranthene, 1,2-dibenzoanthracene, 3,4-dibenzoanthracene, 1,12 -benzoperylene, 1,3,5-triphenylbenzene	17,61,100
Antibiotics	penicillin, streptomycin, butyl acetate, butanol, phenol, xeroform	168
Asphalt concrete	carbon black, hydrocarbons, 1,3-benzopyrene	17,47,90, 168
Basic organic chemicals	amines, nitro compounds, phenols, organic acids, pyridine, c-naphthol	19
Bitumen	formaldehyde, propanal, methylethyl ketone, nonane, heptane, butanol, dodecyl mercaptan, butyl mercaptan, phenol, propionic acid, benzaldehyde	194
ε-Caprolactam	benzene, toluene, phenol	18
Carbophos	carbophos	17
Coal-tar chemicals	benzene, toluene, xylene, methane, phenols	17,58,100
Construction materials	phenols	9,105
Dyes	aldehydes, ketones, organic acids including acetic acid, phenols, terpines, alkyl sulfates, butyl mercaptan, thiophenes, polychlorinated biphenyls	17,100,103
Fat substitutes	butanol, hexanol, methanol, pentanol, propanol, ethanol	102
Fats and oils	aldehydes, amines	34
Foodstuffs	volatile organic acids, hydrocarbons including polychlorinated ones	9,47,101
Hexamethylene diamine	adiponitrile	19
Metals	hydrocarbons, phenols, resins, organic polycyclic compounds, cyanides, benzene, xylene, naphthalene, pyridine	19,168
Methanol	methanol	17

Pesticides	hydrocarbons, phenols, various pestcides, ethyldichloro thiophosphate, methyldichloro thiophosphate, diethylchloro thiophosphate	19,47,64
Petrochemicals	aldehydes, 1,2-benzopyrene, pyrene, phenols, formaldehyde, acetone, toluene, benzene, styrene, 1,3-butadiene, propane, butane, butene, other hydrocarbons including polychlorinated ones	9,47,101
Plastics	phenols, sulfur-containing organic compounds, acrolein, formaldehyde, polyethylene, polyvinylchloride, poly(vinyl butyral), polystyrene, polyesters, urea, melamine, resins, polypropylene, polyurethane, styrene	34,193,195
Polyethylene	hydrocarbons	47
Polyisocyanates	2,4-diphenylmethyldiamine, methyldiamine, phosgene, chlorobenzene	177
Polymers	polyethylene, polypropylene, polyvinylchloride	19
Polystyrene	styrene, ethylbenzene, xylenes, cumene, benzene	191
Polyvinylchloride	vinyl chloride, polyvinylchloride, vinylidene chloride, vinyl acetate	47
Products of thermal cracking of solid fuels	phenols, resins, amines, volatile organic acids, pyridine, methanol	19
Refined oil products	oil, alkenes, organic acids, isopropylbenzene, acetone, alcohols, 1,2-benzopyrene, phenols, benzene	17,100,101, 168
Refrigerants	53 different organic compounds	58
Rubber, articles made of	carbon black, benzene, naphthylamine, chlorohydrocarbons	168
Streptomycin	pectin, formaldehyde, acetaldehyde, acetone, methylethyl ketone, ethanol, 2-propanol, methylbutyl ketone, methylamyl ketone, acetic acid, propionic acid, butyric acid, isovaleric acid, valeric acid, isocaproic acid, caproic acid, phenol, p-cresol	179
Sulfate pulp	mercaptans, phenols, alcohols, dimethyl sulfide, dimethyl disulfide, other hydrocarbons	17,100
Synthetic fibers	carbon disulfide	17
Synthetic fuels	methane, ethane, ethylene, benzene, toluene, ethylbenzene, xylene, indene, naphthalene, 2-methylnaphthalene, 1-methylnaphthalene, biphenyl, acenaphthene, phenanthrene, anthracene, fluoranthene, pyrene, dioctyl adipate	178

Synthetic materials	phenol, mixed esters, volatile fatty acids, formaldehyde, acetone, styrene, ε-caprolactam, hexamethylene diamine	19
Synthetic rubber	butadiene, butene, methylpropene, pentane, pentadiene, acetonitrile, acrylonitrile, acrolein, dichloroethane, methyl methacrylate, methanol, acetone cyanohydrin, methacrylic acid, phosgene, chlorobenzene, cyclohexene, ε-caprolactam	47,104,168
Textiles	formaldehyde, other hydrocarbons	34
Toluylene diisocyanate	chlorobenzene, phosgene, toluylene diamine	177
Vinyl chloride	vinyl chloride	101
Wood chemicals	phenols, volatile organic acids, pyridine	19

REFERENCES

1. Maximum permissible concentration of harmful substances in the air over populated areas. In: A Handbook of Environmental Protection. (Ed: L.P. Sharikov). Leningrad: Sudostroenie, 1978, pp. 425-445.
2. Sanitary-Protected Zones. Criteria for Selecting Construction Sites and Drawing up of General Plans. Moscow: Stroiizdat, 1972.
3. Environmental Protection: Collection of Party and Government Documents. Moscow: Politizdat, 1979.
4. L.P. Sharikov (Ed.). A Handbook of Environmental Protection. Leningrad: Sudostroenie, 1978, 558 pp.
5. N. V. Lazareva and E. N. Levina (Eds.). Harmful Substances in Industry. Organic Substances. A Handbook. 7th ed. Leningrad: Khimiya, 623 pp.
6. V. A. Ryazanov. Sanitary Measures for Air Protection. Moscow: Meditsina, 1954, 236 pp.
7. K. A. Bushtueva (Ed.). A Handbook on the Prevention of Air Pollution. Moscow: Meditsina, 1976.
8. Yu. A. Izrael and L. M. Filippova. In: Monitoring the State of the Environment. Transactions of the Soviet-British Symposium, Cardington, England, November-December, 1976. Leningrad: Gidrometeoizdat, 1977, pp.34-40.
9. Fixing of Standards and Control of Industrial Wastes Discharged into the Atmosphere. (Ed. M. E. Berlyand). All-Union Seminar, Moscow, October 1975. Leningrad: Gidrometeoizdat, 1977, 123 pp.
10. Nature Protection. Air. Classification of Wastes According to their Composition. GOST 17.2.1.01.76. Moscow: USSR State Committee of Standards, 1976, 5 pp.
11. Nature Protection. Air. Meteorological Aspects of Air Pollution and Industrial Wastes. Main Terms and Definitions. GOST 17.2.1.04.77 Moscow: USSR State Committee of Standards, 1977.
12. Nature Protection. Air. Air Quality Control Regulations for Populated Areas. GOST 17.2.3.01.77. Moscow: USSR State Committee of Standards, 1977, 4 pp.
13. M. I. Birger, A. Yu. Valdberg, and B. I. Myagkov. A Handbook of Dust and Ash Trapping. Moscow: Energiya, 1975, 296 pp.
14. R. S. Breiman. In: Chromatographic Analysis of the Environment. (Translated from the English). Moscow: Khimiya, 1979, pp. 92-131.
15. S. N. Ganz (Ed.). Purification of Industrial Gases. A Handbook. Dnepropetrovsk: Promin, 1977, 115 pp.
16. G. M. Gordon and I. L. Peisakhov. Dust Trapping and Purification of Gases in Nonferrous Metallurgy. 3rd ed. Moscow: Metallurgiya, 1977, 455 pp.

17. I. E. Kuznetsov and T. M. Troitskaya. Prevention of Air Pollution by Harmful Chemicals. Moscow: Khimiya, 1979, 349 pp.
18. Approximate Safe Levels of Pollutants in the Air over Populated Areas. No.1430-76. Moscow: Medgiz, 1976.
19. A. V. Pavlov (Ed.). Laboratory Investigations of the Environment. Kiev: Zdorovya, 1978, 288 pp.
20. E. A. Pregud. Sanitary-Chemical Air Pollution Control. A Handbook. Leningrad: Khimiya, 1978, 336 pp.
21. Maximum Permissible Concentration of Harmful Substances in the Air over Populated Areas. No. 1342-75. Moscow: Medgiz, 1975.
22. Maximum Permissible Concentration of Harmful Substances in Bodies of Water. A Handbook. 2nd ed. Leningrad: Khimiya, 1975, 455 pp.
23. Maximum Permissible Concentration of Harmful Substances in the Air Over Populated Areas. No. 1441-76. Moscow: Medgiz, 1976.
24. Estimated Maximum Permissible Concentration of Pesticides in the Air Over Populated Areas. No. 1233-75. Moscow: Medgiz, 1975.
25. A. A. Rusanov (Ed.). A Handbook of Dust and Ash Trapping. Moscow: Energiya, 1975, 296 pp.
26. T. A. Semenov and I. L. Leites (Eds.). Purification of Technological Gases. Moscow: Khimiya, 1977, 388 pp.
27. T. V. Soboleva and V. A. Khrustaleva. Methods for Determining Harmful Substances in the Air. A Handbook. Moscow: Meditsina, 1974, 300 pp.
28. Instructions on Estimating the Dispersion of Industrial Pollutants. SN-369-74. Moscow: Stroiizdat, 1975, 41 pp.
29. G. I. Sidorenko and M. T. Dmitriev (Eds.). Unified Methods of Determining Air Pollutants. Moscow: CMEA, 1976, 198 pp.
30. The ABC of Environmental Protection and Pollution Engineering. Leipzig: Quarg, 1976, 207 pp.
31. A. Bjorseth, G. Lunde, and A. Lindskog. Atmospheric Environment, 1979, vol. 13., No. 1, pp.45-53.
32. R. G. Bond and C. P. Straub (Eds.). Handbook of Environmental Control, vol. XII. Cleveland: CRC Press, 1972, 576 pp.
33. A. J. Buonicore and L. Theodore. Industrial Control Equipment for Gaseous Pollutants, vol. 1, p. 204, and vol 2, p.128. Cleveland: CRC Press Inc., 1972.
34. P. N. Cheremisinoff and R. I Young. Pollution Engineering: Practice Handbook. Michigan: Ann Arbor Science, 1975, 1073 pp.
35. R. Coleman. Proceedings of International Symposium. Fort Collins, Colorado,

1973. Washington, 1976, pp. 162-169.

36. R. W. Coughlin, R. D. Siegel, and Ch. Rai (Eds.). Recent Advances in Air Pollution Control. ALChE Symposium Series No. 137, vol. 70, 1974. New York: American Institute of Chemical Engineering, 1974, 528 pp.
37. L. Fishbein. Chromatography of Environmental Hazards. Carcinogens, Mutagens and Teratogens, vol. 1, 1972, 499 pp.
38. R. B. Frieberg and R. Cederlof. Environmental Health Perspectives. 1978, vol. 22, pp. 45-66.
39. R. B. King, A. C. Altolne, and J. S. Fordyce. J. Air Pollution Control Association, 1977, vol. 27, No. 9, pp.867-871.
40. R. E. Lee (Ed.). Air Pollution from Pesticides and Agricultural Processes. Cleveland: CRC Press, 1976, 264 pp.
41. B. J. Liptak (Ed .). Environmental Engineers' Handbook, vol. 2. Air Pollution. Radnor, Pa.: Chilton Book Co., 1974, 1340 pp.
42. G. Lunde and A. Bjorseth. Nature, 1977, vol. 268, No. 562, pp. 518-519.
43. J. M. Marchello. Control of Air Pollution Sources. New York: Dekker Co., 1976, 630 pp.
44. W. Martin and A. C Stern. The World's Air Quality Management Standards, vol. 1. The Air Quality Management Standards of the World Including United States Federal Standards, vol. XXIV, 1974, 382 pp.
45. D. F. Natush. Environmental Health Perspectives, 1978, vol. 22, pp.79-90.
46. H. W. Parker. Air Pollution. Englewood Cliffs: Prentice Hall, 1977, 287 pp.
47. N. I. Sax (Ed.). Industrial Pollution. New York: Van Nostrand Co., 1974, 702 pp.
48. M. Sittig. How to Remove Pollutants and Toxic Materials from Air and Water. A Practical Guide. New York: Data Corp., 1977, XII, 621 pp.
49. A. C. Stern (Ed). Air Pollution. 2nd ed., vol. 3.. Sources of Air Pollution and Their Control. New York: Academic Press, 1968, 866 pp.
50. A. C. Stern (Ed.). Air Pollution. 3rd ed., vol. 2. Effects of Air Pollution, Environmental Sciences. New York: Academic Press, 1977, 684 pp.
51. A. C. Stern (Ed.). Air Pollution. 3rd ed., vol. 5. Air Quality Management. Environmental Sciences. New York: Academic Press, 1977, 700 pp.
52. J. Tarr and C. Damme. Chem. Eng., 1978, No. 10, pp. 86-89.
53. H. U. Wanner, A. Deuber, and J Statish. In: Proceedings of the 12th International Colloquium. Paris, May 5-7, 1976. (Ed. M. Benarie). Amsterdam: Elsevier Scientific Publishing Co., 1976, pp. 99-108.
54. The ABC of Environmental Protection and Pollution Engineering. Leipzig: Quarg, 1976.

55. N. I. Sax. Dangerous Properties of Industrial Materials, 4th ed. New York: Van Nostrand Publishing Co., 1975, 1258 pp.
56. C. J. Hilado and H. J. Cumming. J. of Fire Flammability, 1979, vol. 10, pp. 252-260.
57. K. Sexton and H. Westberg. J. Air Pollution Control Association, 1980, vol. 30, No. 8, pp. 911-914.
58. A. J. Buonicore. Environmental Science Technology, 1979, vol. 13, No. 11. pp. 1340-1342.
59. H. F. Lund (Ed.). Industrial Pollution Control Handbook. New York: McGraw-Hill Book Co., 1971, pp. 1-26.
60. I. Ya. Sigal. Protecting Air from Wastes from the Combustion of Fuels. Leningrad: Nedra, 1977, 294 pp.
61. I. M. Andron'ev and O. V. Filip'ev. Dust and Gaseous Discharges from Metallurgical Plants. Moscow: Metallurizdat, 1977, 328 pp.
62. S. B. Stark. Dust Trapping and the Purification of Gases in Metallurgy. Moscow: Metallurgizdat, 1977, 328 pp.
63. M. D. Manita, R. M. Salikhdzhaniva, and S. F. Yavorovskaya. Modern Methods for Determining Air Pollutants in Populated Areas. Moscow: Meditsina, 1980, 254 pp.
64. V. Leite. Determination of Pollutants in Air and at the Work Area. (Translated from the 2nd German edition). Leningrad: Khimiya, 1980, 342 pp.
65. I. V. Ryabov (Ed.). Fire Hazard of Substances and Materials Used in the Chemical Industry. A Handbook. Moscow: Khimiya, 1970, 335 pp.
66. B. M. Fedyushin, S. A. Anurov, and S. A. Keltsev. Khim. Prom., 1977, No. 8, pp. 597-598.
67. L. I. Medved (Ed.). A Handbook of Pesticides. Hygiene and the Use of Pesticides and Their Toxicology. Kiev: Urozhai, 1974, 448 pp.
68. A. A. Zuikov and O. I. Yakovleva. In: Protection of the Environment from Industrial Pollution. Leningrad: Khimiya, 1975, pp.21-25.
69. J. A. Danielson (Ed.). Air Pollution Engineering Manual. 2nd ed. Washington, 1973, 987 pp.
70. The Merck Index. An Encyclopedia of Chemicals and Drugs. 8th ed. New York: Merck Co., Inc., 1968, 1713 pp.
71. A. Yu. Korolchenko (Ed.). Inflammable and Dangerously Explosive Substances and Materials. Moscow: Protivopozhar. Oborona, 1979.
72. F. A. Patty. Industrial Hygiene and Toxicology. Vol. 2. New York: Interscience Publishers, 1967.

73. O. P. Chepinoga. In: Hygiene and the Use of Pesticides, and the Clinical Toxicology of Pesticides. Toxicology of Pesticides. Clinical Features of Poisoning (Ed. L. I. Medved). Kiev: VNIIGNTOKS, 1970, No. 8, pp. 30-40.
74. A. P. Dronin and I. A. Pugach. The Technology of Separation of Gaseous Hydrocarbons. Moscow: Khimiya, 1976, 176 pp.
75. E. N. Serpinova. Industrial Adsorption of Gases and Vapors. 2nd ed. Moscow: Vyshaya Shkola, 1969, 389 pp.
76. Inflammable and Dangerously Explosive Substances and Materials. Collection of Papers Prepared by the All-Union Research Institute of Fire-Prevention. Moscow, 1979, 174 pp.
77. BDR. Technische Einleitung zur Reinhalt und der Luft, 1974.
78. Inflammable and Dangerously Explosive Substances and Materials. Collection of Papers Prepared by the All-Union Research Institute of Fire-Prevention. Issue 1, Moscow, 1978, 183 pp.
79. E. Sawicki. In: Environmental Pollutants. Detection and Measurement (Eds.: T. Y. Toribara, J. R. Coleman, B. E. Dahneke, and Y. Feldman). New York: Plenum Press, 1977, pp.69-98.
80. A. C. Stern (Ed.). Air Poluution, 3rd ed., vol. 1. Air Pollutants, Their Transformation and Transport. Environmental Sciences. New York: Academic Press, 1976, 715 pp.
81. L. Van Valck and K. Van Cauwenberghe. Atmospheric Environment. Science, 1978, vol. 201, No. 4362, pp. 1200-1205.
83. V. A. Isidorov, I. G. Zenkevich, and B. V. Ioffe. Gig. i. San., 1981, No. 1, pp. 19-21.
84. J. A. Cooper. J. Air Pollution Control Association, 1980, vol. 30, No. 8, pp. 855-867.
85. E. Ahland and H. Merteus. VDI-Bericht, 1980, p. 358.
86. K. Verschueren. Handbook of Environmental Data on Organic Chemicals. New York: Van Nostrand Co., 1977, 659 pp.
87. L. I. Ishukova, S. N. Zueva, and L. S. Kuritsyna. Khim. Prom., 1980, No. 7, p. 402.
88. R. W. Serth and T. W. Hughes. Environmental Sci. Technol., 1980, vol. 12, No. 3, pp. 298-301.
89. T. G. Aanova. Zhur. Vsesoyuz. Khim. Obshch. im. D. I. Mendeleev, 1969, vol. 14, No. 4, pp. 381-387.
90. Approximate Safe Level of Pollutants in the Air over Populated Areas. No.1890-78, 1978.

91. Approximate Safe Level of Pollutants in the Air over Populated Areas. No.2062-79, 1976.

92. H. M. Ellis and A. R. G. Greenway. J. Air Pollution Control Association, 1981, vol. 31, No. 2, pp. 136-138.

93. L. Fishbein. Potential Carcinogens and Mutagens in Environmental Sciences. Amsterdam: Elsevier Scientific Publ. Co., 1979, 537 pp.

94. M. T. Dmitriev and G. M. Kolesnikova. Gig. i San., 1981, No. 8, pp. 56-58.

95. L. V. Mel'nikova. Gig. i San., 1981, No. 1, p.94.

96. Maximum Permissible Concentration of Pollutants in the Air over Populated Areas. No. 1892-78, 1981.

97. G. I. Aranovich (Ed.). Handbook of Physico-Chemical Methods of Investigating the Environment. Leningrad: Sudostroenie, 1979, 147 pp.

98. Industrial Chlororganic Products. A Handbook. Moscow: Khimiya, 1978, 654 pp.

99. M. M. Milyutin. Khim Prom., 1977, No. 3, pp. 163-166.

100. D. B. Meeker. Air Quality Control. Natural Issues, Standards, and Goals. Washington: Natl. Assoc. of Manufact. Research and Technol. Dept., 1975, 93 pp.

101. M. D. High. J. Air Pollution Control Association, 1976, vol. 26, No. 5, pp. 471-479.

102. F. I. Dubrovskaya and Ya. K. Yushko. Gig. i San., 1964, No. 7, pp. 6-11.

103. I. M. Nazarov. In: Monitoring the State of the Environment. Transactions of the Sov.-British Symposium, Cardington, 1976. Leningrad, 1977, pp. 81-95.

104. M. Sittig. Environmental Sources and Emissions Handbook. London, 1975, 523 pp.

105. A. I. Pirumov. Trapping Dust from Air. Moscow: Stroiizdat, 1974, 207 pp.

106. Yu. A. Izrael, I. M. Nazarov, and L. M. Filippov. In: A Comprehensive Analysis of the Environment. Proceedings of the Soviet-US Symposium, Tashkent, 1977. Leningrad, 1978, pp. 101-130.

107. L. M. Shabad. Circulation of Carcinogens in the Environment. Moscow: Meditsina, 1973, 364 pp.

108. R. Cederloef, R. Doll, and B. Fowler. Environmental Health Perspectives, 1978, vol. 22, pp. 1-12.

109. B. Commoner. Chem. Technol., 1977, vol. 7, No. 2, pp. 76-82.

110. N. P. Dubinin and Yu. V. Pashin. Mutagenesis and the Environment. Moscow: Nauka, 1978, 127 pp.

111. H. Rosenkranz. Environmental Health Perspectives, 1977, vol. 21, pp. 79-84.

112. H. M. Bartsch, M. Molaveille, and A Barfin. Environmental Health Perspectives, 1976, No. 12, pp. 193-198.
113. M. M. Muratov and S. I. Gus'kova. Gig. i San., 1978, No. 7, pp. 111-112.
114. R. I. Khil'chevskaya. Zhur. Vsesoyuz. Khim. Obshch. im. D. I. ______ 1979, vol. 24, No. 1, pp. 18-25.
115. R. E. Munn. J. Air Pollution Control Assn., 1977, vol. 27, No. 9, pp. 842-843.
116. E. I. Goncharuk. Sanitary Protection of the Soil from Chemical Pollutants. Kiev: Zdorov'ya, 1977, 160 pp.
117. J. A. Naegele. In: Industrial Pollution (Ed: N. I. Sax). New York: Van Nostrand Co., 1974, pp. 82-100.
118. J. Riederer. In: New Concepts in Air Pollution Research (Ed. J. O. Willums). Base: Birkhouser Verlag, 1974, pp. 74-85.
119. J. H. Seinfild. Air Pollution. Physical and Chemical Fundamentals. New York: McGraw-Hill Book Co., 1975, XVI, 523 pp.
120. M. E. Berlyand (Ed.). Meteorological Aspects of Air Pollution. Leningrad: Gidrometeoizdat, 1971, pp. 337-344.
121. A. Ya. Khesina, G. A. Smirnov, and L. M. Shabad. Gig. i San., 1979, No. 6, pp. 39-43.
122. R. L. Byers, B. B. Crocker, and D. W. Cooper (Eds.). Dispersion and Control of Atmospheric Emissions: New Energy-Source Pollution Potential. Vol. 73. New York: Amer. Inst. of Chem. Eng., 1977, 365 pp.
123. M. H. Mareau. In: Atmospheric Pollution. Proceedings of the 12th International Colloquium,Paris, May 5-7, 1976. (Ed.: M. M. Benarie). Amsterdam: Elsevier Scientific Publ. Co., 1976, pp. 405-418.
124. H. S. Stoker and S. L. Seager. Environmental Chemistry. Air and Water Pollution. 2nd ed. Glenview: Foresman Co., 1976, 233 pp.
125. K. A. Bushtueva. In: Maximum Permissible Concentration of Air Pollutants. Issue 5. Moscow: Khimiya, 1961, pp. 118-125.
126. U. G. Pogosyan. In: Biological Effect and Hygienic Aspects of Air Pollutants. Moscow: Khimiya, 1967, pp. 135-154.
127. I. G. Gracheva, Ya. S. Kochan, and R. I. Onikul. Trudy Glav. Geofiz. Obser. im. A. I. Volikov, No. 238, pp. 14-26.
128. V. A. Zaitsev and A. P. Tsygankov. Zhur. Vsesoyuz. Obshch. im. D. I. Mendeleev, 1979, vol. 24, No. 1, pp. 3-12.
129. F. G. Banit and A. D. Mal'gin. Dust Trapping and Purification of Gases in the Construction Materials Industry. Moscow: Stroiizdat, 1979, 345 pp.
130. A. S. Stern (Ed.). Air Pollution. Vol. 4. Engineering Control of Air Pollu-

tion, 3rd ed. New York: Academic Press, 1977, 946 pp.

131. H. E. Hesketh. Understanding and Controlling Air Pollution, 2nd ed. Michigan, Ann Arbor, 1974, XVI, 413 pp.

132. C. J. Stairmand. In: International Clean Air Conference. Melbourne, May 15-18, 1972, pp. 199-204.

133. R. J. Joyce, J. R. Luthko, and R. K. Sinha. New Development in Ambient Odor Control by Adsorption. New York: Acad. Sci., 1974, 237 pp.

134. R. W. Williams and K. White. Chemical Process, 1978, vol. 41, No. 5, p. 18.

135. B. R. Perez. GYP, 1975, vol. 4, No. 34, pp.77-82.

136. V. A. Speisher. Moscow: Energiya, 1977, 263 pp.

137. Yu. D. Znamensky and V. P. Balabanov. Khim. Prom. , 1976, No. 7, pp. 503-505.

138. L. P. Shariko (Ed.). In: Environmental Protection. Leningrad: Dudostroenie, 1978, pp. 471-475.

139. L'exportation de pollution atmospherique des effort pour la reduire. Observ. OCDE, 1977, No. 8, pp. 6-8.

140. A. M. Stepanov. In: Geography and Practice. Irkutsk, 1976, pp. 108-111.

141. W. Strauss. Industrial Gas Cleaning. The Principles and Practice of the Control of Gaseous and Particulate Emissions, 2nd ed. Oxford: Pergamon Press, 1975, 621 pp.

142. A. V. Lysak, I. M. Nazarov, and A. G. Ryaboshapko. Zhur. Vsesoyuz. Khim. Obshch. im. D. I. Mendeleev, 1979, vol. 24, No. 1, pp. 25-29.

143. I. I. Datsenko, V. Z. Martynyuk, and B. M. Shtabsky. Gig. i San, 1976, No. 9, pp. 91-93.

144. A. Roffman. J. Environmental Science, 1977, vol. 20, No. 4, pp. 9-12.

145. C. W. Smith. Hydrocarbon Process, 1977, vol. 56, No. 1, pp. 213-226.

146. A. N. Sherban, A. V. Primak, and V. I. Kopeikin. Automated Systems of Controlling Air Pollution. Kiev: Tekhnika, 1979, 158 pp.

147. M. E. Berlyand and O. I. Kurenbin. Trudy Glav. Observ. im. A. I. Voeikov, No. 238, pp. 3-13.

148. R. W. Dunlap and M. K. Deland. Hydrocarbon Processing, 1978, vol. 57, No. 10, pp. 91-97.

149. N. L. Vasil'eva and N. A. Gozhenko. Khim. Tekhnol., 1977, No. 6, pp.10-12.

150. T. Schneider (Ed.). Automatic Air Quality Monitoring System.. Proceedings of the Conference held at Bilthower, the Netherlands, June 5-8, 1973. Amsterdam: Elsevier Scientific Publ., Co., 1974, 265 pp.

151. E. J. Lillis and J. J. Schueneman. J. Air Pollution Control Association, 1975,

vol. 25, No. 8, pp. 804-809.

152. P. O. Warker. Analysis of Air Pollutants. New York: Wiley Interscience, 1976, 329 pp.

153. W. Muckli. In: Atmospheric Pollution. Proceedings of the 12th Intern. Colloquium, Paris, May 5-7, 1976. (Ed: M. M. Benarie). Paris, 1976, pp. 173-178.

154. Automatic Air Pollution Control in the North Rhein Region (in German). Techn. Mitteil., 1978, vol. 71, No. 7, p. 380.

155. Monitoring Air Pollution in the Nenedig Region (in German). Industry Anzeiger, 1978, vol. 100, No. 29, pp. 24-26.

156. G. Defrance and J. M. Dejardin. Bull. bimestr. Associat. amic. anciens elevess Ecole nat. techn. indust. et miner. Donais. 1978, No. 83, pp. 1079-1088.

157. M. Terable. In: Atmospheric Pollution Proceedings of the 12th Intern. Colloquim, Paris, May 5-7, 1976 (Ed.: M. M. Benarie). Amsterdam: Elsevier Scientific Publishers, vol. 1, 1976, pp. 487-510.

158. J. W. Scales (Ed.). Air Quality Instrumentation. Pittsburg, vol. 1, 1972, pp. 89-137.

159. A. E. O'Keefe. Water, Air and Soil Pollution, 1974, vol. 3, No. 4, pp. 447-450.

160. A. C. Gibby, W. B. Telfair, and D. T. Tomason. Chem. Technol., 1977, vol. 6, No. 1, pp. 27-31.

161. T. M. Royal, C.E. Decker, and J. B. Tommerdahl. In: Proceedings of IECE Nat. Aerospace and Electron. Conf., Washington, 1975, No. 4, pp. 697-703.

162. Chemical Proces, 1976, vol. 39, No. 1, p.23.

163. L. Moegele and M. Birkle. Chemie-Ingen. Technol., 1979, vol. 51, pp. 415-419.

164. H. J. Heinz. Rev. frank. corpor. gras, 1976, vol. 23, No. 12, pp. 657-663, 675.

165. M. Lorincz. Munkavelem, 1979, vol. 25, Nos. 4-6, pp. 31-38.

166. G. A. Foksha, A. P. Chernyaev, and V. K. Leonenko. In: Industrial and Sanitary Purification of Gases. 1977, pp. 23-24.

167. A. M. Rozenshtein, A. M. Pol'chitsky, and A. I. Zhuikova. In: Hygiene in Populated Areas, 1979, pp. 45-47.

168. A. D. Lebedev (Ed.). The Environment and Human Health. Problems of Constructive Geography. Moscow: Nauka, 1979, 212 pp.

169. V. I. Nikolaevsky and A. T. Miroshnikova. Gig. I San., 1974, No. 4, pp. 16-18.

170. Ya. M. Grushko. Harmful Organic Compounds in Industrial Affluents. 2nd ed. Leningrad: Khimiya, 1982.

171. I. B. Kotlyar and A. I. Seregina. Plastmassy, 1978, No. 10, pp. 58-61.
172. H. J. Rafson. Food Technol., 1977, vol. 31, No. 6, pp. 32-35.
173. H. Juentgen. Tech. Mitteilungen, 1977, vol. 70, No. 1, pp. 55-64.
174. A. V. Sakharov and I. P. Zege. Purification of Affluents and Gaseous Wastes in the Paint and Varnish Industry. 2nd ed. Moscow: Khimiya, 1979, 179 pp.
175. R. G. Leites and B. Ya. Gorovoi. Gig. i San., 1938, Nos. 7,8, pp. 34-41.
176. V. V. Elisina. Khim. Prom., 1973, No. 4, p. 74.
177. T. G. Lipina, A. Ya. Tubina, and L. V. Kinilova. Gig. i Prof. Zabol., 1979, No. 8, pp. 55-56.
178. R. S. Dzierlenga, F. G. Mesich, and R. A. Magee. Environmental Science Technology, 1979, vol. 13, No. 3, pp. 288-293.
179. L. V. Kondakova, V. N. Krasnov, and Yu. K. Shapashnikov. Gig. i San., 1973, No. 8, p.89.
180. Y. Kuwata. Taiki osen gakkaishi, 1979, vol. 14, No. 5, pp. 197-202.
181. Y. Hoshika. Taiki osen gakkaishi, 1979, vol. 14, No. 15, pp. 210-219.
182. A. Colmsjoe and U. Sternberg. Polynuclear Aromatic Hydrocarbons. 3rd ed. 1979, pp. 121-139.
183. P. I. Bresler, N. Sh. Volberg, and T. A. Goranina. Gig. i San., 1979, No. 5, pp. 61-63.
184. B. N. Ames. Science, 1979, vol. 204, No. 4393, pp. 587-593.
185. T. A. Maugh. Science, 1978, vol. 201, p. 1200.
186. J. Namiesnik and E. Kozlovski. Chem. Anal. (Poland), 1980, vol. 25, No. 2, pp. 301-305.
187. L. Van Vaeck and K. Van Cauwenberghe. Atmosph. Environment, 1978, vol. 12, No. 11, pp. 2229-2239.
188. R. Raffenbeul. U. das technische Umweltmagazin, 1976, vol. 4, pp. 18-19.
189. N. N. Mel'nikov, A.I. Volkov, and O. L. Korotkova. Pesticides and the Environment. Moscow: Khimiya, 1985, 351 pp.
190. Toxic Materials News, 1979, vol. 6, No. 12, pp. 89-96.
191. M. H. Richmond and D. C. Wooden. Soc. Plast. Eng. Techn. Pap., 1979, No. 25, pp. 845-848.
192. D. D. Berlinrat, P. N. Cheremisinoff, and R. B. Trattner. Adv. Instrum. 1979, vol. 34, No. 2, pp. 357-364.
193. P. Pfaeffly, H. Vainio, and A. Hesso. Scand. J. Work, Environ. Health, 1979, vol. 5, No. 2, pp. 158-161.
194. Yu. B. Nadzharov and B. E. Smelova. Khim. i Tekhnol. Topliva i Masel, 1976, No. 11, pp. 38-39.

195. Harmful Effect of New Pollutants on Human Health. Geneva: World Health Organization, 1978.

196. K. N. Grigor'eva and N. N. Bainazarov. Khim. i Tekhnol. Topliv i Masel, 1976, No. 11, pp.7-8.

197. V. V. Fryazinov and I. B. Grudinkov. Khim. i Tekhnol Topliv i Masel, 1976, No. 11, pp. 39-41.

198. D. Staats. Great Lakes Communications, 1979, vol. 10, No. 3, pp. 3-4.

199. Coal Age, 1979, vol. 84, No. 11, p. 13.

200. Handbook of Air Quality in Cities. Copenhagen: World Health Organization, 1980, pp. 36.

201. J. N. Pitts, K. A. Cauwenberghe,, D. Grosien, and P. J. Schmidt. Science, 1978, vol. 202, No. 43, pp. 515-518.

202. H. H. E. Schroeder. J. Mine Vent. Soc. Africa, 1980, vol. 33, No. 7, pp. 105-117.

203. M. T. Dmitriev and V. A. Mishchikhina. Gig. i San., 1982, No. 4, pp. 65-68.

204. A. K. Berzin. In: Urgent Problems in Labor Hygiene and Occupational Diseases. Proceedings of a Conference. Riga, 1968, pp. 47-48.

205. Gig. i San., 1981, No. 7, pp. 51-52.

206. M. M. Plotnikov. Gig. i. San., 1957, No. 6, pp. 10-15.

207. M. I. Gusev, I. S. Dronov, and A. I. Svechnikov. In: Hygienic Protection from the Biological Effect of Air Pollution (Ed.: V. A. Rayzanov). Issue 10. Moscow: Meditsina, 1967, pp. 122-135.

208. D. S. Sinkuvene. Gig. i San., 1970, No. 3, pp. 6-10.

209. S. A. Ipatova and E. V. Deyanova. Gig. i San., 1973, No. 10, pp. 65-67.

210. J. N. Pitts, D. Grosjean, and Van Cauwenberghw. Eviron. Sci. Technol., 1978, vol. 12, No. 8, pp. 946-953.

211. F. I. Dubrovskaya, A. I. Pinigina, and V. M. Styazhkin. Gig. i San., 1966, No. 1, pp. 97-98.

212. Chem. Process, 1976, vol. 39, No. 1, p.29.

213. B. K. Baikov. M. Kh. Khachaturyan, and E. V. Borodina. Gig. i San., 1973, No. 9, pp. 10-14.

214. L. Kh. Tsyganovskaya, L. I. Volokhova, and V. M. Voronin. In: Hygienic Aspects of Environmental Protection. Issue 6. Moscow: Meditsina, 1978, pp. 88-92.

215. A. A. Makarenko. In: Proceedings of a Scientific Conference Attended by Graduate Students and Interns of the 1st Moscow Med. Inst., 1964, pp. 107-109.

216. S. Naishtein and G. Mironyuk. Protection of the Environment from Pesticides. Kishinev, 1971.
217. Yu. G. Ozerskii, Ya. A. Akimov, and I. N. Kabrelyan. Gig. i San., 1976, No. 1, pp. 69-71.
218. P. G. Tkachev. Gig. i San., 1963, No. 4, pp. 3-11.
219. P. G. Demidov and V. S. Saushev. Combustion and the Properties of Fuels. Moscow: Nedra:, 1975, 275 pp.
220. L. V. Mel'nikova and A. A. Belyakov. Gig. i San., 1971, No. 9, pp. 67-69.
221. J. Driscol. Flue Gas Monitoring Techniques. Manual Determination of Gaseous Pollutants. Michigan: Ann Arbor Sci. Publ., IX, 1974, 366 pp.
222. R. Villalobos and R. L. Chapman. In: Air Quality Instrumentation (Ed.: S. W. Scales). Pittsburg: I. S. Instr. Soc. Amer., 1972, pp. 114-128.
223. R. V. Lindal and I. V. Ermakov. Gig. i San., 1963, No. 9, pp. 51-54.
224. Yu. G. Fel'dman. In: Maximum Permissible Concentrations of Air Pollutants (Ed.: V. A. Ryazanov). Moscow: Medgiz, 1962, pp. 109-127.
225. T. G. Alanova and L. Ya. Margolin. In: Abstracts of Reports of a Conference on Methods for Removing Harmful Substances from Gaseous Wastes and Affluents. Dzerzhinsk, 1967, pp. 228-234.
226. L. A. Mikheeva, A. F. Terekhina, and S. G. Sigaev. Nefteprerab. i Neftekhim., 1978, No. 11, p. 45.
227. N. B. Imasheva. In: Hygienic Protection From the Bilogical Effect of Air Pollution (Eds.: V. A. Ryazanov and M. S. Gol'dberg). Moscow: Meditsina, 1966, pp. 101-118.
228. Schadstoffe un der Atmosphere aus onkologischer und toxicologisher Sicht. Berlin: Akad. Verlag, 1978, 128 pp.
229. N. F. Izmerov. In: Maximum Permissible Concentrations of Air Pollutants. (Ed.: V. A. Ryazanov). Moscow: Medgiz, 1961, pp. 72-93.
230. I. S. Gusev. In: Hygienic Protection from the Biological Effect of Air Pollutants (Ed.: V. A. Ryazanov). Moscow: Meditsina, 1967, pp. 96-108.
231. Yu. V. Novikov. In: Maximum Permissible Concentrations of Air Pollutants. (Ed.: V. A. Ryazanov). Moscow: Medgiz, 1957, pp. 85-107.
232. D. Henschler. Angew. Chemie, 1973, vol. 85, p. 317.
233. V. B. Dobrokhotov and M. I. Enikeeva. Gig. i San., 1977, No. 1, pp. 32-34.
234. A. V. Kuzevanova, A. L. Shtein, and S. P. Ryazanova. Koks i Khim., 1977, No. 2, pp. 55-58.
235. N. V. Kel'tsev and V. M. Mukhin. In: Industrial and Sanitary Purification of Gases. Scientific Abstracts, 1978, pp. 18-19.
236. Techn Mitteilungen, 1978, vol. 71, No. 7, p.380.

237. H. H. Gruhn. Ind. Anzeiger, 1978, vol. 100, No. 29, pp. 24-26.
238. E. Sh. Gronsberg and I. V. Gubanova. Gig. i San., 1973, No. 5, pp. 81-83.
239. N. Ya. Yanysheva. Gig. i San., 1972, No. 7, pp. 87-91.
240. G. Kh. Ripp. In: Hygienic Protection from the Biological Effect of Air Pollution. (Ed.: V. A. Ryazanov). Issue 10. Moscow: Meditsina, 1967, pp. 33-54.
241. M. L. Krasovitskaya and L. K. Malyarova. In: Hygienic Protection from the Biological Effect of Air Pollution. Issue 11. Moscow: Meditsina, 1968, pp. 43-50.
242. I. R. Major and R. Perry. In: International Symposium on the Chemical Aspects of Air Pollution. Cortina d'Amprezzo, Italy, 1970. Pure and Appl. Chem., 1970, vol. 24, No. 4, pp 685-693.
243. T. I. Kravchenko and G. A. Chemer. Gig. i San., 1975, No. 3, pp. 78-80.
244. E. A. Komrakova and L. V. Kuznetsova, Gig. i San., 1981, No. 1, pp. 43-45.
245. P. A. Chebotarev. Gig. i San., No. 9, pp. 81-82.
246. V. A. Gofmekler. In: Maximum Permissible Concentrations of Air Pollutants (Ed.: V. A. Ryazanov). Moscow: Medgiz, 1961, pp. 142-168.
247. A. P. Rumyantsev, N. A. Ostroumova, and S. A. Astapova. Gig. i San. 1976, vol. 39, No. 1, p.29.
248. G. A. Ivanova, L. M. Doronina, and L. F. Eshenbakh. In: Industrial and Sanitary Purification of Gases, 1976, pp. 26-27.
249. S. Yu. Mukhamedov. Gig. i San., 1968, No. 12, pp.8-13.
250. N. G. Popova, Yu. U. Khasanov, and T. B. Muzykantova. Gig. i San., 1975, No. 6, p. 50.
251. Z. A. Leika. Gig. i San., 1983, No. 1, pp. 52-53.
252. F. I. Dubrovskaya and M. Kh. Khachaturyan. In: Air Quality Control in Town Planning. Moscow: NII Gigieny, 1968, pp. 10-15.
253. R. U. Ubaidullaev and U. A. Madzhidov. Gig. i San., 1978, No. 3, pp. 11-14.
254. E. Sh. Gronsberg. Gig. i San., 1975, No. 7, pp. 77-79.
255. C. H. B. Bins. J. Soc. Occup. Med., 1979, vol. 29, No. 4, pp.134-141.
256. M. M. Muratov and M. T. Takhirov. Gig. i San., 1979, No. 11, pp. 74-76.
257. Japan Plastics, 1976, vol. 10, No. 5, pp. 9-12.
258. E. Burghardt, R. Eltes, and H. J. Van der Wiel. Atmosph. Environ., 1979, vol. 13, No. 7, pp. 1057-1060.
259. J. S. Nadler. In: Atmospheric Pollution. Proceedings of the 13th Intern. Colloquium, Paris, May 5-7, 1976 (Ed.: M. M. Benarie). Amsterdam, 1976,

pp. 257-269.

260. M. T. Dmitriev and V. A. Mishukhin. Gig. i San., 1981, No. 4, pp.46-49.
261. G. V. Il'icheva and L. V. Kuznetsova. Gig. i San., 1981, No. 7, pp. 52-54.
262. L. A. Bazarova. A Toxicological Description of the General and Specific Effect of Piperidene and Hexamethylenimine. Abstract of Dissertation (Cand. of Sci.). Moscow, 1970.
263. A. E. Kulakov. In: Hygienic Protection from the Biological Effect of Air Pollution. Issue 9. (Eds.: V. A. Ryazanov and M. S. Gol'dberg). Moscow: Meditsina, 1968, pp. 28-41; ibid., Issue 10, 1967, pp.15-32.
264. M. N. Bernadiner. Abstract of Dissertation (Cand. of Sci.). Moscow, 1970.
265. K. V. Grigor'eva, V. E. Prisyazhnyuk, and G. A. Dmitrienko. In: Hygiene of Populated Areas. Kiev: Zdorovya, 1981, pp. 65-69.
266. C. Johnson, T. C. Yu, and M. L. Montogomery. Bull. Environ. Contam. and Toxicology, 1977, vol. 17, No. 3, pp. 369-372.
267. L. Kh. Tsyganovskaya, L. I Volokhova, and V. M. Voronin. Gig. i San., 1978, No. 5, pp. 3-9.
268. R. A. Kazymov. Gig. i San., 1977, No. 1, pp. 6-10.
269. L. G. Aleksandrova. Gig. i San., 1979, No. 12, pp. 40-42.
270. Yu. G. Fel'dman. Gig. i San., 1967, No. 1, pp. 11-14.
271. L. A. Gasyuk. Khim. Volokna, 1976, No. 6, pp. 58-59.
272. E. G. Kachmar. Gig. i San., 1966, No. 5, pp. 57-58.
273. N. I. Fomicheva and P. A. Mel'nikova. Gig. i San., 1952, No. 5, pp. 49-52.
274. T. S. Burenko, M. I. Dement'eva, and S. F. Yavorovskaya. Gig. i San, 1975, No. 6, pp. 46-47.
275. F. I. Dubrovskaya, M. Kh. Khachaturyan, and V. P. Levkin. Gig. i San., 1977, No. 4, pp. 7-11.
276. D. G. Odoshashvili. Gig. i San., 1962, No. 4, pp.3-7.
277. R. U. Ubaidullov and U. A. Madzhidov. Gig. i San., 1976, No. 9, pp.7-9.
278. G. I. Solomin. In: Maximum Permissible Concentrations of Air Pollutants (Ed.: V. A. Ryazanov). Moscow: Medgiz, 1962, pp. 146-164.
279. A. L. Kuchinsky and V. G. Kovalenko. Gig. i San., 1975, No. 1, pp.69-70.
280. T. S. Lipina, A. Ya. Turbig, and L. V. Kunilova. Gig. Truda i Prof. Zabol., 1979, No. 8, pp. 55-56.
281. N. G. Andreeshcheva. Gig. i San., 1968, No. 4, pp.12-16.
282. M. K. Borisova. Gig. i San., 1957, No. 3, 13-19.
283. V. V. Elisina. Khim. Prom., 1977, No. 4, p.74.
284. I. M. Starshov and G. Ya. Ivanova. Gig. i San., 1969, No. 7, pp.54-55.

285. L. F. Shishkina. Gig. Truda i Prof. Zabol., 1965, No. 12, pp.13-18.
286. L. I. Rapoport, I. Sh. Kofman, and L. V. Gortseva. Gig. I San., 1976, No. 1, pp. 59-61.
287. N. P. Kosiborod. Gig. I San., 1968, No. 1, pp. 28-31.
288. I. I. Pochina. Gig. i San., 1975, No. 4, pp. 58-60.
289. K. A. Bushtueva, L. E. Bezpal'ko, and M. M. Gasilina. Gig. i San., 1982, No. 5, pp. 8-13.
290. G. V. Selyuzhitsky. Gig. Truda i Prof. Zabol., 1972, No. 6, pp.46-47.
291. V. A. Gofmekler. Gig. i San., 1960, No. 4, pp. 9-15.
292. V. P. Tsulaya, N. V. Morenkova, and L. I. Volokhova. Gig. i San., 1978, No. 5, pp. 6-9.
293. V. R. Tsulueva, L. I. Zefirova, and E. L. Pereverzeva. Gig. i San., 1972, No. 8, pp. 16-19.
294. B. K. Baikov, O.E. Gorlova, and M. I. Gusev. Gig. i San., 1974, No. 4, pp. 6-13.
295. N. M. Krichevskaya. Gig. i San., 1968, No. 1, pp. 22-28.
296. F. I. Dubrovskaya. Gig. i San., 1969, No. 6, pp. 14-18.
297. D. H. Martin, R. Levis, and F. A. Tibbitts. Bull. Environ. Contamination and Toxicology, 1978, vol. 20, No. 2, pp. 155-158.
298. B. Szucki and T. Nazimek. Bromatol. i Chem. Toxsykola, 1976, vol. 9, No.4, pp. 409-417.
299. Yu. A. Krotov, A. S. Lykova, and M. A. Skachkov. Gig. i San., 1981, No. 2, pp. 14-17.
300. V. N. Os'kina. Gig. i San., 1981, No. 2, pp. 50-51.
301. A. I. Larionov, Yu. M. Konstantinov, and G. L. Cherntsova. Gig. i San., 1983, No. 1, pp. 78-79.
302. Yun-Tai Chen. Gig. i San., 1963, No. 2, pp. 93-95.
303. G. I. Solomin. Gig. i San., 1964, No. 2, pp. 3-8; ibid., 1966, No. 5, pp. 101-103.
304. K. V. Grigor'eva. In: Hygienic Protection from the Biological Effect of Air Pollution (Eds.: V. A. Ryazanov and M. S. Gol'dberg). Moscow: Meditsina, 1966, pp. 42-57.
305. A. S. Agaev, Yu. B. Nadzhafov, and T. N. Kobylkina. Khim. i Tekh. Topli i Masel, 1976, No. 11, pp. 44-45.
306. L. A. Tepkina. Gig. i San., 1968, No. 6, pp. 3-9.
307. N. G. Polezhaev and L. V. Kuznetsova. Gig. i San., 1967, No. 4, pp. 63-64.
308. A. P. Rumyantsev, I. Ya. Lobanova, and S. A. Astapova. Gig. i San., 1981, No. 5, p.10.

309. A. E. O'Keeffe, G. S. Orman, and R. K. Stevens. In: Air Quality Instrumentation. Vol. 1. (Ed.: J. W. Scales). Pittsburg, 1972, pp.26-38.

310. L. E. Bezpal'ko. Gig. i San., 1967, No. 10, pp. 3-7.

311. E. A. Komrakova and L. V. Kuznetsova. Gig. i San., 1981, No. 1, pp. 43-45.

312. T. I. Kravchenko and G. A. Chemer. In: Hygiene and the Use of Polymeric Materials. Kiev, 1976, pp. 254-256.

313. E. V. Eremyan and K. A. Davtyan. Prom. Armeniya, 1976, No. 10, pp. 58-60.

314. I. N. Mukhlenov, G. P. Buzanova, N. A. Saukhin, and E. A. Kulish. Bum. Prom., 1978, No. 7, pp. 27-28.

315. Technocrat, 1976, vol. 9, p. 84.

316. V. P. Yakimova. Gig. i San., 1977, No. 5, pp.55-60.

317. V. I. Filatova. In: Maximum Permissible Concentrations of Air Pollutants (Eds.: V. A. Ryazanov and M. S. Gol'dberg). Moscow: Meditsina, 1964, pp. 59-76.

318. V. N. Dmitrieva, L.A. Kotok, and N. S. Stepanova. Gig. i San., 1976, No. 12, pp. 73-74.

319. Chzen-Tsi Chzhao. Gig. i San., 1959, No. 10, pp. 7-12.

320. R. Ubaidullaev. In: Hygienic Protection from the Biological Effect of Air Pollutants (Eds.: V. A. Ryazanov and M. S. Gol'dberg). Moscow: Meditsina, 19-67, pp. 65-74.

321. T. S. Burenko, E. G. Zhravlev, and T. A. Miklashevich. Gig. Truda i Prof. Zabol., 1977, No. 3, pp. 55-65.

322. R. V. Gorskaya. Gig. i San., 1965, No. 11, pp. 64-65.

323. Z. A. Krotova. Gig. i San., 1975, No. 4, pp.61-63.

324. M. T. Dmitriev and V. A. Mishchikhin. Gig. i San., 1983, No. 3, pp. 53-55.

325. N. G. Andreeshcheva. Gig. i San., 1964, No. 8, pp. 5-10.

326. N. G. Andreeshcheva. Gig. i San., 1968, No. 12, p. 93.

327. B. K. Akhmedov. Gig. i San., 1968, No. 10, pp. 10-15.

328. I. A. Zibireva. Gig. i San., 1967, No. 7, pp. 3-9.

329. A. Dahms and W. Metzner. Holz als Roh und Werkstoff, 1979, vol. 37, No. 9, pp. 341-344.

330. M. I. Gusev and K. N. Chelikanov. Gig. i San., 1963, No. 5, pp. 3-8.

331. V. R. Tsulaya, T. I. Bonashevskaya, and V. V. Zykova. Gig. i San., 1977, No. 8, pp. 50-53.

332. A. L. Kuchinsky. Gig. i San., 1977, No. 4, pp. 57-58.

333. J. Huber. U. tech. Umwelt, 1979, No. 2, pp. 36-37.
334. V. A. Gofmekler and L. A. Safonova. Gig. i San., 1945, No. 9, pp. 17-19.
335. E. S. Koen. Gig. Truda i Prof. Zabol., 1977, No. 3, pp. 56-57.
336. N. N. Sirotkina and L. I. Molyavko. Gig. i San., 1981, No. 7, pp. 46-47.
337. M. L. Krasovitskaya and L. K. Malyarova. In: Hygienic Protection from the Biological Effect of Air Pollution (Eds.: V. A. Ryazanov and M. S. Gol'dberg). Issue 10. Moscow: Meditsina, 1966, pp. 74-100.
338. N. I. Kaznina and N. P. Zinov'eva. Gig. i San., 1976, No. 9, pp. 68-69.
339. Sh. E. Tokanova. Gig. i San., 1982, No. 4, pp. 10-13.
340. I. A. Guseinov. Gig. i San., 1968, No. 11, pp. 46-48.
341. L. V. Mel'nikova and A. A. Belyakov. Gig. i San., 1971, No. 9, pp. 67-69.
342. Li-Shen. Gig. i San., 1961, No. 8, pp. 11-17.
343. L. D. Pribytkov. Neftepererab. i Neftekhim., 1979, No. 1, pp. 47-48.
344. S. K. Osokina and N. A. Markina. Gig. i San., 1977, No. 10, pp. 57-59.
345. M. D. Babina. Gig. i San., 1962, No. 3, pp. 50-51.
346. Kh. Z. Avrutova, K. P. Privalova, and N. Ya Khlopin. Gig. i San., 1954, No. 3, pp. 50-51.
347. V. B. Dorogova and A. A. Kachaeva. In: Modern Methods for Determining Toxic Substances in the Air. Moscow, 1981, pp. 57-59.
348. Sh. S. Khikmatulleva. Gig. i San., 1967, No. 6, pp. 3-6.
349. E. P. Aigina, G. S. Lapukhova, and Sh. S. Khimatulleva. Gig. i San., 1968, No. 6, pp. 98-100.
350. V. A. Chizhikov. In: Maximum Permissible Concentrations of Air Pollutants (Eds.: Ryazanov and M. S. Gol'dberg). Moscow: Meditsina, 1964, pp. 21-41.
351. Yu. G. Ozersky, Yu. A. Akimov, and S. N. Kadrel'yan. Gig. i San., 1976, No. 1, pp. 69-71.
352. V. B. Dobrokhotov and M. I. Enikeeva. Gig. i San., 1977, No. 1, pp. 32-34.
353. Zh. Detri. Air Must be Clean. Moscow: Progress, 1973, 878 pp.
354. G. G. Luk'yanova, O. P. Baburova, and T. A. Kruzhkova. In: Scientific Papers Prepared by VTsPS Inst of Labor Protection. Issue 104(6), Moscow, 1976, pp. 52-59.
355. T. A. Lebedeva. Proceedings of the Second All-Union Conference on Pesticide Residues and the Prevention of Contamination of Foodstuffs, Animal Feed, and the Environment. Analytical Methods. Moscow, 1971, pp. 179-181.
356. N. V. Dmitrieva. Gig. i San., 1962, No. 8, pp. 47-50.
357. V. A. Tsendrovakaya. Gig. I San., 1979, N0. 6, pp. 85-86.
358. V. A. Ivanov and P. S. Khal'zov. In: A Toxic-Hygienic Description of Harmful Substances Present in Wastes from the Production of Synthetic Rubber.

Voronezh: Med. Inst., 1966, pp. 56-62.

359. V. A. Gofmekler. Gig. i San., 1968, No. 4, pp. 96-97.
360. M. T. Takhirov. Gig. i San., 1969, No. 4, pp. 103-106.
361. N. A. Krylova. Gig. i San., 1968, No. 4, pp. 46-48.
362. B. M. Mukhitov. Gig. San., 1962, No. 6, pp. 16-24.
363. H. Quillmann. Chem. Technol., 1978, vol. 7, No. 5, pp. 203-204.
364. E. A. Druyan. Gig. i San., 1975, No. 10, pp. 62-65.
365. E. Auernan, M. Kneuer, and R. Meyer. Z. Gesamte Hygiene und Grenzgebiete, 1976, vol. 22, No. 1, pp. 9-12.
366. I. I. Demin and V. I. Eremenchuk. Tsement, 1979, No. 5, p. 12.
367. R. Faidullaev and Kh. A. Kamil'dzhanov. Gig. i San., 1976, No. 7, pp. 10-14.
368. V. P. Melekhina. Gig. i San., 1958, No. 8, pp. 10-14.
369. V. S. Nikolaevsky and A. T. Miroshnikova. Gig. i San., 1974, No. 4, pp. 16-18.
370. E. N. Gantwell, E. S. Jacobs, and J. M. Pierrard. Alternative Automative Emission Control Systems. A.C.S. Symposium Series, 1974, pp. 99-158.
371. M. A. Klosenko, M. V. Pis'mennya, and L. P. Novitskaya. Gig. San., 1967, No. 6, pp. 67-70.
372. L. R. Slavogorodskaya. In: Problems of Hygiene in Populated Areas. Kiev, 1964, pp. 175-180.
373. N. F. Beloborodova and V. G. Likho. Gig. i San., 1965, No. 3, pp. 61-62.
374. A. V. Mnatsakan'yan. Gig. i San., 1964, No. 9, pp.13-18.
375. A. G. Sukasyan, K. T. Geodakyan, G. S. Khzanyan, and O. S. Berudzhnyan. Armenian Khim. Zhur., 1976, vol. 29, No. 8, pp. 728-730.
376. A. A. Dobrinsky. In: Hygienic Protection from the Biological Effect of Air Pollutants. (Eds.: V. A. Ryazanov and M. S. Gol'dberg). Issue 9. Moscow: Meditsina, 1966, pp. 119-132.
377. N. A. Krylova. Gig. i San., 1969, No. 1, pp. 50-52.
378. A. N. Fomin. In: Air Quality Control in Town Planning. Moscow: NII Gigieny, 1968, pp. 50-55.
379. R. S. Kamalov. Gig. i San., 1979, No. 2, pp. 50-53.
380. Mod. Plant Oper. and Maint., 1976, vol. 17, No. 3, pp. 28-30.
381. V. I. Andreev. Proceedings of the XII Sci. Conference of Young Hygienists and Sanitary Workers. Moscow, 1969, pp. 109-111.
382. P. G. Tkachev. Gig. i San., 1969, No. 8, pp. 7-10.
383. T. Yuldashev. Gig. i San., 1965, No. 10, pp. 3--7.
384. N. A. Krylova. Gig. i San., 1961, No. 10, pp. 48-49.

INDEX

Part I

Inorganic Substances

Part II

Organic Substances